Agrifood Waste as Biofertilizer

Agrifood Waste and the Environment Series

A huge amount of solid agricultural waste is produced globally by the agrifood industry and this can have severe implications for the environment if not treated promptly and efficiently. Despite this potential for pollution, however, these agrifood wastes are rich potential sources of carbohydrates, sugars, phenols, flavonoids and other bioactive compounds that can be effectively transformed and valorized into value-added products for different environmental and industrial applications.

The 'Agrifood Waste and the Environment' series looks at how these waste resources can be transformed into useful products, such as feedstock, sustainable biocontrol agents, bioenzymes bioenergy substrates and effective fertilizers, which can help reduce the environmental burden of the agriculture industry. The series provides a valuable and informative resource for those working and researching in the areas of agrifood waste valorization, agriculture and improving the circular economy.

Series Editor

Neha Srivastava,
Department of Chemical Engineering and Technology, Indian Institute of Technology (BHU), Varanasi, India
Email: sri.neha10may@gmail.com

Agrifood Waste as Biofertilizer

Manju M. Gupta

Abha Kumari

Anirudh Sharma

CABI

CABI is a trading name of CAB International

CABI
Nosworthy Way
Wallingford
Oxfordshire OX10 8DE
UK

Tel: +44 (0)1491 832111
E-mail: info@cabi.org
Website: www.cabi.org

CABI
200 Portland Street
Boston
MA 02114
USA

Tel: +1 (617)682-9015
E-mail: cabi-nao@cabi.org

A catalogue record for this book is available from the British Library, London, UK.

ISBN-13: 9781836991007 (hardback)
9781836991014 (ePDF)
9781836991021 (ePub)

DOI: 10.1079/9781836991021.0000

Commissioning Editor: Jamie Lee
Editorial Assistant: Theresa Regueira
Production Editor: Rosie Hayden

Typeset by Straive, Pondicherry, India
Printed in the USA

Contents

About the Authors

Dr Manju M. Gupta is a Professor of Botany at Sri Aurobindo College, University of Delhi, where she has been affiliated for over three decades. She holds an MSc and PhD from the University of Delhi and completed her postdoctoral research under Professor L.K. Abbott at the University of Western Australia. She is currently a Visiting Professor at Northern Arizona University, USA, since October 2024. Dr Gupta has over 37 years of research experience in Arbuscular Mycorrhizal (AM) fungi, Bioinformatics, Soil Microbiology, and Microbiome studies with a focus on sustainable agriculture and environmental change. She has completed four research projects on AM fungal databases and fuel from waste, published 35 research papers in high-impact journals, and contributed extensively to scientific publishing through books and editorial work. She is on the editorial team of the Springer journal *Symbiosis*. Her recent work includes the development of a globally accessible AM fungal barcoding tutorial in collaboration with the Pathogen and Microbiome Institute, Northern Arizona University, USA. She is Vice President of the International Symbiosis Society (USA), an advisor at SPUN (an international initiative to map global mycorrhizal networks), and has received numerous awards, including the BOYSCAST Fellowship from the Government of India. Dr Gupta has delivered 38 invited lectures across the world and has convened several national and international symposia on microbial ecology and plant–soil interactions.

Dr Abha Kumari is currently working as an Associate Professor at the Centre for Biotechnology and Biochemical Engineering, Amity Institute of Biotechnology, Amity University, Noida, India. She has earlier served as an Associate Professor in the Department of Biotechnology at Delhi Technological University, Delhi, India, a visiting researcher at the Tohoku Institute of Technology, Sendai, Japan, a DBT Postdoctoral Fellow at the National Chemical Laboratory, Pune, and a Postdoctoral Fellow at Lakehead University, Thunder Bay, Ontario, Canada. She was awarded a Pool Officer fellowship from the Council of Scientific Industrial Research, India. Dr Abha earned her PhD and MSc (Engg) from the Indian Institute of Science (IISc), Bangalore, India. She received a fellowship during her master's and doctoral programs from IISc, Bangalore. Her research interests are in biochemical engineering and biotechnology with a specialization in waste-to-value, municipal solid waste-to-biogas, and carotenoid production, their applications and biorefinery, bioenzyme from waste, waste management, bioremediation, industrial effluent treatment using photochemical techniques, and metals recovery from waste. She has more than 20 years of teaching experience in bioprocess technology, biochemical engineering, downstream processing, enzyme technology, and principles of chemical engineering. She has published research articles in peer-reviewed international journals and authored numerous book chapters. She has edited a book. She has worked as a Principal Investigator on two government-funded projects and as a

Co-investigator on one government-funded project. She is a life member of the Environment and Social Development Association (ESDA) India, the Biotechnology Research Society of India (BRSI) India, the Indian Institute of Metals (IIM) India, and the Indian Science Congress Association.

Dr Anirudh Sharma has been working as Assistant Professor at the Department of Biotechnology, Jaypee Institute of Information Technology, since July 2023. Prior to this, he served as an Assistant Professor at the University Institute of Biotechnology, Chandigarh University, Mohali, Punjab, from January 2019 to June 2023. He earned his PhD from the Department of Biotechnology, Thapar Institute of Engineering & Technology, in 2018, during which he qualified for the Council of Scientific & Industrial Research Senior Research Fellowship (CSIR-SRF) in 2014 and received a CSIR fellowship. Dr Sharma completed his Master's degree in Biotechnology from Chaudhary Charan Singh University, Meerut, in 2010. Earlier in his career, Dr Sharma worked as a Junior Research Fellow on a Department of Atomic Energy Board of Research in Nuclear Sciences (DAE-BRNS) sponsored project, where he spent time at the Bhabha Atomic Research Centre (BARC) in Mumbai as part of his assignments. He also worked as a project fellow at the Institute of Nuclear Medicine & Allied Sciences Defence Research and Development Organisation (INMAS-DRDO) from July 2010 to July 2011. His primary research focus is on developing processes for biofuels production from waste feedstocks through biocatalysis.

Preface

Twenty-first-century agriculture is at a moment of profound transition. Driven by a growing global population and the pressing need to combat climate change, resource scarcity, and soil erosion, the search for sustainable, environmentally friendly, and economically viable agricultural solutions has become critical. This book offers a thorough, yet modest, contribution to this discussion by exploring the potential of microbial inoculants and agrifood waste in enhancing soil health and fertility and promoting sustainable agricultural practices.

With a rapidly expanding global population and a quickly rising demand to address climate change, shortages of resources, and soil erosion, a quest for sustainable, environmental, and commercially viable agrarian answers is as urgent as ever. This volume is a humble yet comprehensive work in the discourse on microbial inoculants and agrifood waste, examining their interactive capabilities in enhancing soil health, fertility, and sustainable agrarian cultures.

The chapters are widely scoped, from fundamental soil–microbe interactions to applied biofertilizer production technologies in liquid and solid forms. We have included field-level approaches of application, profitability, and higher-level policy and environmental implications of biofertilizer in India. The particular emphasis on agrifood waste valorization is notable, a theme growing in importance environmentally, and through input minimization and agrisystems' nutrient circuit closure.

Biofertilizers' properties, impacts, findings, and market assessments have been thoroughly examined. The book also features chapters on both economics and the scope of biofertilizer. It is a vital resource for anyone interested in sustainable fertilizers, soil chemistry, agronomy, agribusiness, and agroecology.

The book integrates the wisdom and contributions of practitioners who have been personally involved in a hands-on capacity in the lab, in the field, and in the classroom. It is written for students, researchers, extensioners, and practitioners of sustainable agriculture who are attempting to learn or adopt integrated, nature-based ways of managing nutrients.

As we look toward a planet in which higher accountability in how we care for our planet's natural resources is necessary, our desire is that this work serve as a technical guide as well as a source of inspiration toward action. May it serve as a foundation upon which productive, regenerative, and equitable farm and food systems are built.

Acknowledgments

We gratefully acknowledge the support of all institutions, colleagues, and collaborators who made this book possible.

Professor Manju M. Gupta is deeply grateful to the Principal of Sri Aurobindo College, University of Delhi, for granting her a sabbatical that enabled work in a new academic setting. She extends special thanks to Professor Nancy Collins Johnson, School of Earth & Sustainability and Department of Biology, Northern Arizona University, USA, for providing official institutional support and a vibrant research environment whose excellent infrastructure supported the writing and completion of this book. I am especially thankful to my husband, Mukesh Gupta, and my son, Akshat Gupta, for their unwavering support throughout my stay in the United States during my sabbatical. I am deeply grateful to Dr Vasilis Kokkoris, Amsterdam Institute for Life and Environment, Vrije Universiteit Amsterdam, for his generous support and for sharing photographs that enliven several concepts in this volume. I also thank Ms Sedona Spun, Northern Arizona University, for contributing photographs that highlight the connections between nature and scientific inquiry.

Dr Abha Kumari expresses deep gratitude to the Indian Institute of Science, Bangalore, and eminent professors with whom I worked for imparting extensive knowledge in advanced science and engineering and instilling in-depth understanding in me. I would also like to acknowledge the Department of Biotechnology, Government of India, for funding projects on waste. I take this opportunity to acknowledge and express my sincere appreciation to the Founder President, Amity Group of Institutions, Chancellor, Amity Group of Institutions, and Vice Chancellor, Amity Group of Institutions, for their continuous encouragement and support. Special thanks to my team, Surbhi, Reet, Apeksha, Smita, Mannat, and Nishika, for their assistance in editing, their continuous support, and their sincere endeavour. My heartfelt gratitude to my late parents, brothers, and sisters for their love and affection and for being an unwavering source of strength, inspiration, support, and constant encouragement.

Special thanks are also due to the farmers and field practitioners whose practical insights and observations kept our work grounded in reality. We also thank several students and research scholars who assisted in experiments, data collection, and literature reviews at various stages.

Our sincere gratitude goes to the reviewers and editorial staff for their guidance and suggestions during the preparation of the manuscript. Finally, we thank our families for their patience and encouragement, which made it possible for us to bring this volume to fruition.

This book is a product of collective effort, and we dedicate it to all those working toward a more sustainable and resilient agricultural future.

1 Introduction

Manju M. Gupta*
Sri Aurobindo College, University of Delhi, Delhi, India

Abstract
Agrifood waste consists of both organic and inorganic residues generated across the agricultural supply chain—from production to consumption. This results in huge economic losses and significantly contributes to environmental problems such as greenhouse gas emissions, soil degradation, and eutrophication. At the same time, hunger and food insecurity remain widespread. This chapter introduces the concept of agrifood waste and explores its conversion into sustainable biofertilizers within a circular economy framework. It is structured into six key sections, which cover the sources and impacts of agrifood waste, provide definitions of biofertilizers, assess commercial microbial inoculants, and highlight the necessity of transitioning to alternative soil inputs. The chapter also examines waste valorization processes and outlines the overall objectives of the book. Biofertilizers, including microbial agents like *Rhizobium* species, phosphate-solubilizing bacteria, and arbuscular mycorrhizal fungi, are clearly distinguished from conventional organic and chemical fertilizers. Unlike traditional inputs, these microorganisms enhance nutrient cycling, improve soil health, and increase crop productivity, making them a sustainable alternative. The chapter further introduces methods for converting agrifood waste into valuable products through anaerobic digestion, composting, and pyrolysis. These processes produce digestate and biochar—by-products that enrich soil fertility, enhance microbial diversity, and reduce reliance on synthetic chemical fertilizers. It emphasizes their agronomic, environmental, and economic advantages, positioning biofertilizers as essential tools for sustainable agriculture. Ultimately, this "waste-to-wealth" approach supports resilient, low-impact, and resource-efficient food systems for a more sustainable future.

1.1 What is Agrifood Waste?

Agrifood waste includes both organic and inorganic by-products created during the entire process of getting food from the farm to your plate. This includes everything from leftover crops in the fields to food scraps discarded after you eat. Unlike issues like nutrient runoff from fertilizers, agrifood waste refers to wasted food itself. This wasted food pollutes our soil, water, and air. In fact, the United Nations (2024) reported that 13.2% of food is lost before it even reaches stores, and another 19% is wasted after purchase. This isn't just an environmental problem; it also poses health risks and safety concerns, and it is not sustainable. Current food waste disposal methods are inadequate. This emphasizes the need for innovative recycling and reuse technologies.

*Corresponding author: mmgupta@aurobindo.du.ac.in

DOI: 10.1079/9781836991021.0001

On a global scale, agrifood waste has reached the level of a crisis. In 2022 alone, the global estimate of meals discarded by households was over 1 billion daily. This is despite the fact that 783 million people faced hunger and one-third of the global population experienced food insecurity (UN, 2024). Per capita food wasted varies according to country. For example, Europe and North America generate 95–115 kg annually, compared to just 6–11 kg in Africa and South/South-east Asia (Roka, 2019). We need to have a proper method for agrifood disposal because the food waste in landfills releases methane, contributing significantly to global greenhouse gas emissions, accounting for 8–10% of the total. Similarly, the runoff of nutrients from waste to aquatic bodies leads to eutrophication—a process that harms aquatic life and disrupts microbial balance (UNEP, 2024). Global food waste is estimated to be US$1 trillion yearly due to the value of the 1.3 billion tonnes of food wasted as well as waste management expenses (UNICEF, 2024). Managing agrifood waste could help to redirect agro resources to combat human hunger and reduce food insecurity (UN, 2024). These impacts show the pressing need to improve existing waste management practices.

As illustrated in Fig. 1.1, which maps the supply chain from farm to dining table, agrifood waste production occurs at every stage. During production and harvesting, crops like wheat or soybeans may be lost because of pests, adverse weather, disease, or unharvested produce resulting from low market prices or labor shortages. In processing and packaging, by-products such as wheat bran, whey, or irregularly shaped produce contribute to waste. Further losses occur during transportation and due to retail rejection and household food waste. Understanding these sources is essential for designing targeted waste reduction strategies.

Agrifood waste is rich in solid components (such as cellulose and lignin) and dissolved nutrients (like sugars, amino acids, or minerals). This makes it a valuable resource for improving soil fertility (Kushwaha *et al.*, 2025). Rather than relying on synthetic fertilizers—which can degrade soil and cause nutrient runoff—agrifood waste can be transformed into sustainable biofertilizers such as digestate and biochar (Gupta, 2024).

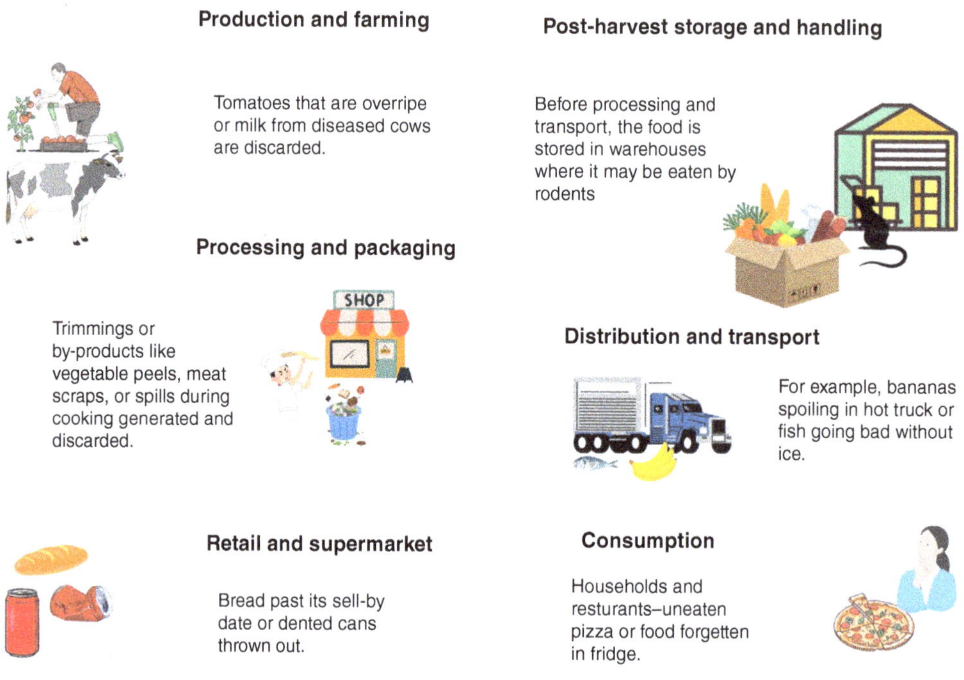

Production and farming
Tomatoes that are overripe or milk from diseased cows are discarded.
Post-harvest storage and handling
Before processing and transport, the food is stored in warehouses where it may be eaten by rodents
Processing and packaging
SHOP
Trimmings or by-products like vegetable peels, meat scraps, or spills during cooking generated and discarded.
Distribution and transport
For example, bananas spoiling in hot truck or fish going bad without ice.
Retail and supermarket
Bread past its sell-by date or dented cans thrown out.
Consumption
Households and resturants–uneaten pizza or food forgetten in fridge.

Fig. 1.1. This infographic illustrates the six key stages of the food supply chain—production, post-harvest handling, processing, distribution, retail, and consumption—highlighting critical points of food loss and waste at each step. The design was created and assembled by the author using Canva. Figure author's own.

Anaerobic digestion, biological process where microbes decompose the agrifood waste in oxygen-free conditions, produces digestate, a nutrient-rich material containing nitrogen, phosphorus, and potassium that enhances soil quality and supports beneficial microbes (Sharma and Tong, 2025). Pyrolysis is a thermal process that breaks down organic materials into simpler compounds, producing gases, liquids, and a solid residue called biochar. Biochar produced through pyrolysis of crop residues like rice husks improves soil structure, water retention, and microbial habitats, boosting long-term fertility (Pradhan *et al.*, 2024). Biochar enriched with microbial inoculants such as rhizobia or arbuscular mycorrhizal fungi (AMF) enhances nutrient uptake and soil stability. This not only adds to soil fertility but also promotes circular agriculture, a farming approach that aims to minimize waste and maximize resource utilization by creating a closed-loop system where outputs from one process become inputs for another (Kushwaha *et al.*, 2025; Pandey *et al.*, 2025). These innovations show how agrifood waste can be repurposed to support sustainable agriculture and soil health.

As populations grow and urban areas expand, the demand for food—and the volume of waste—continues to rise, placing additional pressure on waste management systems. Converting agrifood waste into biofertilizers not only mitigates environmental harm but also enhances soil health and contributes to food security. To scale these solutions, policy reforms, farmer education, and improved waste processing infrastructure are essential. This "waste to wealth" model (Kushwaha *et al.*, 2025) offers a promising path forward.

This chapter introduces the concept of agrifood waste and its transformation into organic fertilizers. It contrasts biofertilizers with chemical and organic alternatives, emphasizing their critical role in sustainable agriculture through improved soil fertility and plant growth. Biofertilizers represent a key strategy for building a more resilient and sustainable food system.

1.2 Defining Biofertilizers

Sustainable agriculture needs fertilizers to enhance soil fertility and crop productivity; these fertilizers fall into three categories: organic, biofertilizers, and chemical (inorganic). Organic fertilizers are derived from natural sources like agrifood waste (e.g. compost, livestock manure), plant residues (e.g. rice husks), or non-edible de-oiled seed cakes. They release nutrients gradually through microbial decomposition, thereby improving the soil's physical, chemical, and biological properties. Chemical fertilizers such as urea, potash, and superphosphate, on the other hand, are synthetically manufactured with high, readily available NPK (nitrogen, phosphorus, potassium) ratios (20–60%) (Gupta, 2024). Biofertilizers, in contrast, are formulated products containing live or dormant beneficial microorganisms, such as nitrogen-fixing rhizobia, phosphate-solubilizing *Pseudomonas*, or cellulolytic *Bacillus*, which enhance nutrient availability through several natural processes, like nitrogen fixation, phosphate solubilization, or organic matter decomposition (Vessey, 2003). Fig. 1.2 visually highlights the differences between three fertilizer types, examining their origin (natural, microbial, or synthetic), form (such as organic matter, microbial preparations, or chemical compounds), and function (including slow nutrient release, enhanced microbial activity for nutrient availability, or immediate nutrient supply).This clear distinction highlights biofertilizers' unique role in sustainable agriculture. Microbial inoculants are synonymous with biofertilizers.

Unlike organic fertilizers, which supply nutrients through organic matter breakdown, biofertilizers rely on microbial activity to improve plant nutrient uptake, and they are distinct from biostimulants (which enhance plant growth via non-nutrient mechanisms) and biopesticides (which control pests). Many misuse the term "biofertilizer" to describe organic fertilizers like compost or manure; however, scientifically and legally, it refers to preparations containing live or dormant cells of efficient microbial strains such as *Azotobacter* for nitrogen fixation, *Pseudomonas* for phosphate solubilization, or AMF for nutrient uptake and soil aggregation via glomalin production (Gupta and Abbott, 2021; Tiong *et al.*, 2024; Sharma and Tong, 2025). Agrifood waste, such as vegetable scraps or rice husks, serves as a cost-effective growth medium for cultivating these microorganisms, linking biofertilizers to the "waste to wealth" model (Kushwaha *et al.*, 2025). Biofertilizers may also include microbial

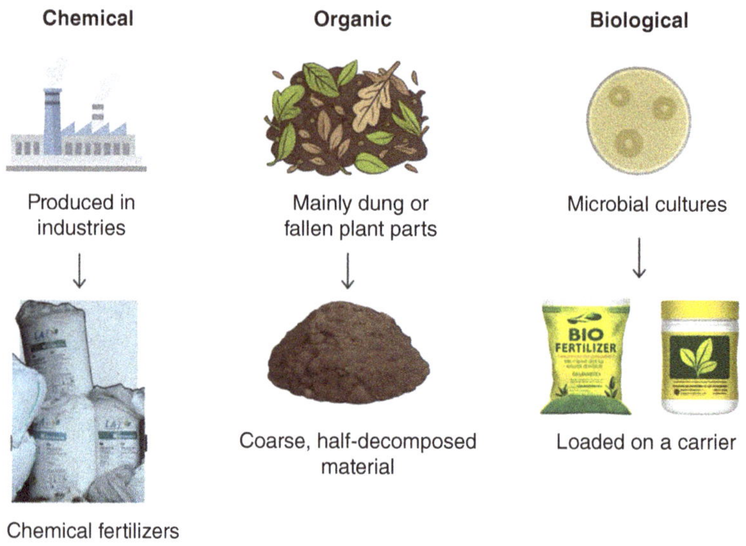

Fig. 1.2. Three types of fertilizers, their origin and form. Figure author's own.

consortia, where multiple strains (e.g. rhizobia and AMF) work synergistically to enhance nutrient cycling and plant growth, as seen in digestate applications that boost soil microbial diversity (Hernández-Álvarez *et al.*, 2023). This microbial focus makes biofertilizers a sustainable alternative to chemical fertilizers, which degrade soil and reduce microbial diversity.

Biofertilizers do not directly add nutrients like traditional fertilizers but enhance nutrient availability, complementing rather than replacing organic or chemical fertilizers to reduce overall chemical inputs (Vessey, 2003). Historically, definitions of biofertilizers have evolved: Okon and Labandera-Gonzalez (1994) argued that rhizosphere organisms that assist in nutrient utilization—but do not directly replace nutrients—such as mycorrhizal fungi or plant growth-promoting rhizobacteria should not be classified as biofertilizers, while Vessey (2003) proposed a broader definition encompassing any microbial preparation that colonizes the rhizosphere and increases primary nutrient availability. Fuentes-Ramirez and Caballero-Mellado (2005) further expanded this to include microbes with direct or indirect growth-promoting effects. These evolving definitions reflect the growing recognition of biofertilizers' diverse mechanisms in sustainable agriculture.

Given their rising importance, biofertilizers require clear legal definitions and quality control measures to ensure product reliability and encourage farmer adoption (Malusá & Vassilev, 2014). Formulations must ensure stability (e.g. shelf life from production to application), effective delivery to soil or plants, and inclusion of carriers or additives to maintain microbial viability, as seen in commercial products using agrifood waste-derived digestate or biochar (Pradhan *et al.*, 2024). Regulatory frameworks should standardize registration and marketing requirements, guaranteeing the quality of biofertilizers as marketable products containing beneficial strains, such as *Bacillus* or AMF, in a carrier matrix (Malusá and Vassilev, 2014). By leveraging agrifood waste to produce biofertilizers, farmers can enhance soil fertility, reduce environmental impact, and address global challenges like food security and soil degradation, making biofertilizers an indispensable tool for sustainable agriculture.

1.3 Commercial Biofertilizers and Soil Microbial Inoculants

Many different biofertilizer products are sold and used by farmers to improve soil health and crop yields sustainably. Commercially available biofertilizers are formulated products with live or dormant beneficial microorganisms that enhance nutrient availability and thrive on different growth mediums (Kushwaha *et al.*, 2025). Factors such as availability, cost, the microorganism, and desired biofertilizer form (liquid or solid) frequently determine the choice of growth medium. It is crucial that the chosen medium provides the necessary nutrients, is non-toxic, and supports the viability and activity of the beneficial microbes. Biofertilizer production uses diverse organic and inorganic materials as growth mediums. Agricultural waste (rice husks, sawdust, compost, etc.) forms organic growing media, differing from inorganic options of peat, clay, and perlite. Biofertilizers, categorized as liquid or carrier-based, boast advantages in liquid form: 12–24-month shelf life, storage at up to 45°C, no contamination, distinctive fermented odor, improved seed and soil survival, and high commercial/export potential (Malusá and Vassilev, 2014). These formulations make biofertilizers a practical, sustainable tool for farmers leveraging agrifood waste.

We categorize biofertilizers by microbial type—bacterial, fungal, or algal—as shown in Fig. 1.3. Bacterial inoculants, such as *Rhizobium*

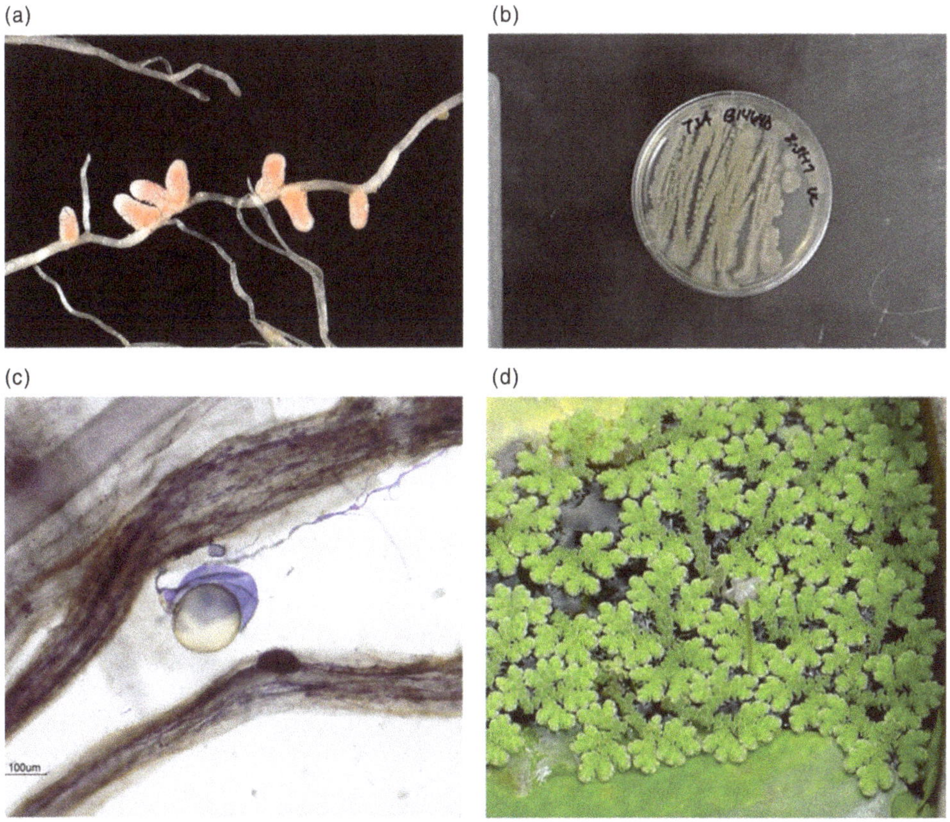

Fig. 1.3. Representative biofertilizer-associated components and organisms. (a) The root nodules of a 4-week-old *Medicago italica* inoculated with *Sinorhizobium meliloti*. From https://commons.wikimedia.org/w/index.php?search=root+nodule&title=Special%3AMediaSearch&type=image; (b) Bacterial colonies of *Azotobacter beijerinckii* NRRL B-14640 (Type Strain) on agar plate. From https://commons.wikimedia.org/wiki/File:Azotobacter_beijerinckii_NRRL_B-14640_(Type_Strain).jpg#Licensing; (c) Arbuscular mycorrhizal fungal (AMF) *Funneliformis* sp. spore entering the root cortex, also showing a characteristic vesicle (blue-stained) and hyphae, involved in phosphorus uptake and soil aggregation. Photo credit Sedona Spun @NAU; (d) Azolla caroliniana. From https://upload.wikimedia.org/wikipedia/commons/8/80/Azolla_caroliniana0.jpg (These images have been identified as being free of known restrictions under copyright law, including all related and neighboring rights.)

(for legumes), *Azotobacter* (for non-legumes), and *Azospirillum* (for cereals and other crops), provide nitrogen and enhance plant growth. Fungal inoculants, such as AMF, form symbiotic associations with plant roots, taking carbohydrates while improving phosphorus uptake and resilience to water stress, pathogens, salts, and heavy metals. Algal inoculants, like the *Azolla-Anabaena* combination, provide nitrogen to paddy fields. Soil microbial inoculants, which include a broader range of plant growth–promoting microbes, enhance plant development through direct and indirect mechanisms beyond nutrient supply. These microbial types highlight the diverse applications of biofertilizers in crop production.

The effectiveness and popularity of these microbial inoculants is supported by several examples in literature, showcasing their wider use compared to chemical fertilizers. In Brazilian soybeans, *Bradyrhizobium*, a nitrogen-fixing bacteria, replaces expensive chemical fertilizers, whereas *Azospirillum* increases cereal yields by 5–30% and lessens fertilizer requirements via hormone production (Tiong *et al.*, 2024). Phosphorus-solubilizing microbes, such as *Pseudomonas* and *Bacillus* in products like JumpStart (*Penicillium bilaii*), unlock bound soil phosphorus (Malusá and Vassilev, 2014). Inoculants like *Trichoderma* and *Pseudomonas chlororaphis* produce siderophores and pathogen-suppressing compounds, protecting cereals against diseases like *Fusarium* and enhancing biomass under drought. AM fungi vary in their effectiveness in increasing the crop productivity (Gupta and Abbott, 2021). Entomopathogenic fungi (*Metarhizium anisopliae*) and bacteria (*Serratia entomophila*) control root-zone pests like wireworms and grass grubs. These multifaceted benefits make microbial inoculants essential for sustainable crop production.

Microbial inoculants, frequently grown on agrifood waste products such as digestate or biochar, enhance stress resistance, bioremediation, and soil quality (Pradhan *et al.*, 2024). For example AM fungi and exopolysaccharide-producing *Pseudomonas putida* enhance stress resistance through soil aggregation and water retention by improving texture in stress-prone areas (Gupta *et al.*, 2018; Gupta, 2020). *Azospirillum brasilense* and *Trichoderma harzianum* have been proposed to enhance salinity and drought tolerance by adjusting nutrient uptake and root morphology, which awaits field validation. Microbes like *Burkholderia* and *Pseudomonas* aid bioremediation by degrading pollutants or reducing metal toxicity, with consortia showing promise but inconsistent field results. Inoculants like SoilBuilder (*Bacillus* species) and *Bacillus amyloliquefaciens* reduce nitrous oxide (N_2O) emissions in controlled settings, particularly in acidic soils, but further field trials are needed (Pradhan *et al.*, 2024). Emerging bio-cementation using *Sporosarcina pasteurii* stabilizes soils, though large-scale applications require development (Malusá and Vassilev, 2014). Ongoing research is needed to validate these applications for broader farmer adoption.

1.4 Biofertilizers: Why Their Need is Inevitable

Chemical fertilizers fueled India's Green Revolution in the 1960s, leading to food self-sufficiency. Declining soil health due to over-reliance on chemical fertilizers is driving a shift towards traditional farming methods, including the use of manure and biofertilizers (Pingali, 2012). Overuse of chemical fertilizers, such as urea, has diminished soil fertility. For example, a farmer's yield decreased from 30 to 25 bags of grain per acre when fertilizer application increased from six to seven bags (FAO, 2021). This reflects a fertilizer response relationship where excess nutrients beyond a critical concentration become toxic, reducing yields (FAO, 2021). The soil has gradually lost productive capacity because of overreliance on chemical inputs (FAO, 2017). Biofertilizers, made from microbes and agrifood waste, offer a sustainable way to improve soil health (Kushwaha *et al.*, 2025). This shift underscores the urgent need for eco-friendly inputs to sustain agricultural output.

Chemical fertilizers harm soil health, deplete trace elements, and reduce crop nutritional quality, necessitating biofertilizers (Das *et al.*, 2023; Bhardwaj *et al.*, 2024). Prolonged usage of chemical fertilizers acidifies topsoil (pH below 5.5). This further disrupts beneficial microbes that provide disease resistance, reduces earthworm activity, and compacts soil by dissolving soil crumbs, limiting water infiltration (Gupta,

2024). Excess nitrogen and insufficient trace elements both negatively impact plant health. For example, high nitrogen levels increase disease susceptibility, while trace element deficiencies lower nutritional value and flavor (e.g. reduced vitamin C in citrus; Mozafar, 1993). Nutrient leaching pollutes groundwater and causes eutrophication, where runoff fuels algal blooms, depleting oxygen and harming aquatic ecosystems and fishing industries (Akinnawo, 2023; Saha *et al.*, 2024). These environmental and agronomic impacts highlight biofertilizers as a critical solution to restore soil and crop quality.

Non-renewable resources, including oil and natural gas, are depleted in chemical fertilizer production. This, coupled with their widespread use, has negatively impacted microbial habitats, pollinators, and human health (Gupta, 2024). With 783 million people facing hunger (UN, 2024), chemical fertilizers alone cannot sustainably feed a growing population without further environmental degradation. Their overuse contributes to soil loss and ecosystem decline, despite the recognized importance of soil's natural capital for societal wellbeing (Humphries and Brazier, 2018). Biofertilizers, cultivated using agrifood waste like rice husks or manure, minimize ecological disturbance and offer a safe alternative to chemical inputs (Pretty and Bharucha, 2015). By substituting a portion of chemical nitrogen and phosphorus, biofertilizers support sustainable agricultural productivity.

Biofertilizers are cost-effective and eco-friendly, elevating crop yields by 10–40% by increasing proteins, amino acids, and vitamins through microbial activity (Bhardwaj *et al.*, 2014). Beneficial microbes such as *Rhizobium* (for legumes), *Azotobacter* (for non-legumes like wheat and maize), and AMF (for phosphorus uptake) boost soil fertility. These microbes deliver nutrients, secrete growth hormones, improve soil structure, and increase microbial diversity, while also reducing the negative effects of chemical fertilizers (Ahemad and Kibret, 2014). They provide micronutrients, organic matter, and resilience against salinity and chemical runoff, unlike chemical fertilizers, which produce less nutritious crops (Gaur, 2010; Gorain *et al.*, 2022; Tiong *et al.*, 2024). These benefits make biofertilizers essential for sustainable crop production and improved food quality.

The environmental issues of overuse of chemical inputs, their impact on water quality, and pesticide residues in food chains are receiving global attention (O'Callaghan *et al.*, 2022). Biofertilizers, supported by microbes like *Pseudomonas*, offer technical feasibility, economic viability, and social acceptability, but challenges like field validation and farmer education remain (Malusá and Vassilev, 2014). We need policy support and standardized testing to scale adoption and ensure biofertilizers meet quality standards. By leveraging agrifood waste, biofertilizers enhance soil fertility and address global challenges like food security and environmental degradation, ensuring a healthy future for agriculture (UN, 2024). Thus the need for microbial inoculants is inevitable for sustainable agriculture. They provide an eco-friendly alternative to chemical fertilizers and pesticides, boosting soil health, nutrient absorption, and crop resilience.

1.5 The Concept of Agrifood Waste Conversion and Circular Economy

Agrifood waste serves as a critical asset in sustainable agriculture, providing nutrients and supporting plant growth and pathogen protection (Kushwaha *et al.*, 2025). This section highlights how agrifood waste can be used to make valuable products. It also defines key terms for its circular economy transformation, which focuses on eliminating waste and pollution, extending product life cycles, and regenerating sustainable agriculture.

Agrifood waste, such as fruit rinds, vegetable scraps, and crop residues, is rich in nutrients like nitrogen, phosphorus, potassium, and trace elements (like iron and zinc) (Khanyile *et al.*, 2024; Voss *et al.*, 2024). Composting or vermicomposting these wastes yields a slow-release nutrient source. Several examples could be cited, for instance, adding rice straw or coffee grounds increases soil organic matter by 1–2% per season, enhancing aeration, water-holding capacity, and root growth in cereals like wheat and rice (Pradhan *et al.*, 2024). Fermented citrus residues release growth hormones (e.g. auxins, gibberellins), improving root elongation and flowering in peppers, with 10–15% yield increases (Shahrajabian *et al.*, 2021). Compost from onion skins or potato peels contains microbes like *Bacillus*, *Trichoderma*, *Rhizobium*, and AMF, suppressing pathogens (*Fusarium*, *Rhizoctonia*). This induces systemic resistance in plants, resulting, for example, in reduced fungal infections in grapevines via grape pomace compost (Iqbal *et al.*, 2024; Toma *et al.*, 2024). These benefits make agrifood waste a valuable resource for soil health and crop productivity.

Agrifood recycling exemplifies the circular economy's waste minimization and resource maximization through reuse, recycling, and regeneration (Colla *et al.*, 2025). Unlike the linear "take-make-dispose" model, it promotes sustainability by converting agrifood waste into biofertilizers, bioenergy, or soil amendments. Fig. 1.4 shows this cycle, illustrating the transformation of waste into resources like compost, digestate, and biochar that support agriculture. However, scaling these processes faces several challenges (Malusá and Vassilev, 2014): high costs deter investment in circular technologies; waste collection and processing face logistical hurdles; infrastructure, like recycling facilities and transportation, is often limited; farmers lack knowledge and awareness; and regulatory inconsistencies exist across regions.

Below, key terms are defined for agrifood waste conversion and the circular economy to clarify their use in the book's remaining chapters.

- **Anaerobic digestion**: a biological process where microbes decompose organic matter in oxygen-free conditions, producing biogas and nutrient-rich digestate for fertilizer use.
- **Biochar**: carbon-rich material from pyrolysis of biomass, improving soil fertility, water retention, and carbon sequestration when used as a soil amendment.
- **Bioenergy**: energy from agricultural waste, such as biofuels (biogas, bioethanol).
- **Biorefinery**: a system converting biomass into biofuels, biochemicals, and biomaterials, maximizing resource efficiency via processes like pretreatment and fermentation.
- **Composting**: an aerobic process where microbes break down organic materials into nutrient-rich compost.

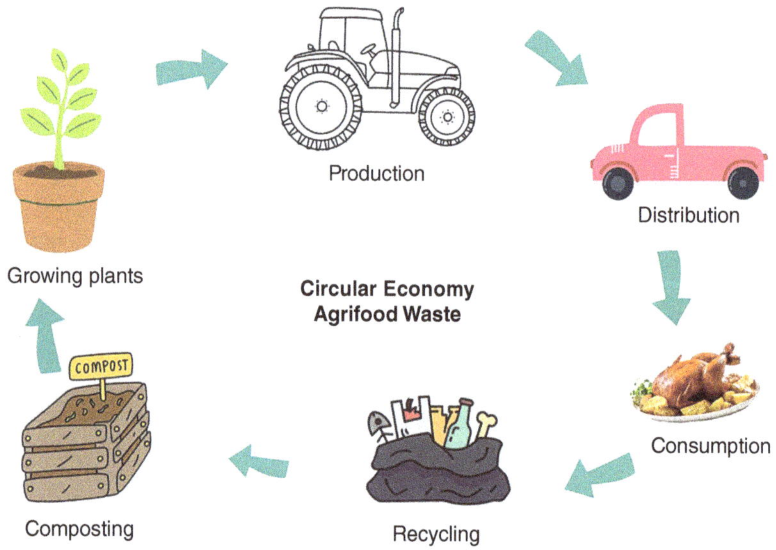

Fig. 1.4. Circular economy model for agrifood waste management. The diagram illustrates the flow of resources in a circular economy framework applied to the agrifood system. Starting with production, crops are grown using inputs such as seeds, water, and biofertilizers. These are transported during the distribution phase. In the consumption stage, food is used by households, markets, and restaurants. Leftovers and scraps from this phase move into the recycling stage, where agrifood waste is collected and processed. Through composting, anaerobic digestion, or fermentation, the waste is transformed into bioproducts such as compost, biochar, or biofertilizers. These bioproducts return to the production stage, completing the cycle and supporting sustainable agriculture. Figure author's own.

- **Fermentation**: an anaerobic process where microbes break down organic compounds, producing alcohols, acids, or gases to support plant growth.
- **Food recovery hierarchy**: a framework prioritizing waste management: source reduction, feeding people, animal feed, industrial uses, composting, and landfill as a last resort.
- **Pretreatment**: techniques to break down complex waste structures (e.g. lignocellulose) for efficient conversion in biorefineries or digestion processes.
- **Pyrolysis**: a thermochemical process for waste conversion and resource recovery.
- **Source reduction**: preventing food waste through improved practices in production, storage, and marketing.
- **Valorization**: transforming agricultural waste into high-value products (e.g. biofertilizers, bioenergy) to reduce environmental harm and create economic opportunities.

These concepts are interlinked within the circular economy. Source reduction minimizes waste, pretreatment enables efficient valorization, and processes like anaerobic digestion, composting, and biochar creation recycle nutrients and carbon into soil, supporting sustainable agriculture (Tiong *et al.*, 2024). The food recovery hierarchy prioritizes these processes for environmental and economic benefits. Agrifood waste conversion thus drives a sustainable, resilient food system.

1.6 Aims and Scope of the Book

This book deeply explores agrifood waste, microbial inoculants, and organic fertilizers. It examines their microbial diversity, how they are produced, their uses, and how they can be integrated into sustainable agriculture. This resource is for students, researchers, and practitioners in agriculture and microbiology. It discusses agrifood waste utilization and looks at innovative methods of creating valuable products from waste within a circular economy. This method addresses global challenges, including food scarcity and environmental damage. The book also explores the science behind biofertilizers and examines their economic benefits and positive environmental impact. Biofertilizers are key to sustainable development by turning waste into useful resources. This information helps us to understand how biofertilizers can transform agriculture.

The book guides readers through the varied role of biofertilizers in sustainable agriculture. Chapter 1 introduces agrifood waste and biofertilizers, differentiating biofertilizers from chemical and organic options. The chapter highlights microbial inoculants' contribution to nutrient availability and soil health in present-day agriculture and their link to agrifood waste. The soil complex, its organic components, and the microbes' role in biofertilizer success are the focus of Chapter 2. Chapter 3, which focuses on key microbial groups as inoculants, examines microbes such as *Rhizobium* for nitrogen fixation in legumes, *Azotobacter* for non-leguminous crops, plant growth-promoting rhizobacteria for growth enhancement, and AMF for phosphorus uptake. Chapter 4 discusses mass multiplication of these inoculants. Chapters 5 and 6 provide detailed techniques for making solid and liquid microbial inoculants, which includes using agrifood waste. It also addresses the challenges of scaling up production. Chapter 5, on solid biofertilizers from agrifood waste, focuses on producing solid biofertilizers like compost from banana peels and crop residues, evaluating their nutrient release and field performance. Liquid biofertilizers from agrifood waste are evaluated in Chapter 6 for effectiveness, storage, and usability. The practical uses and constraints of biofertilizers across different soils and climates are covered in Chapter 7, which also addresses the issue of inconsistent field validation. Chapter 8 discusses green manuring, biocomposting, and vermicomposting with agrifood waste, as well as national sustainability initiatives, within the context of organic farming. Chapter 9 details the environmental advantages of reduced chemical use, lower greenhouse gas emissions, and support for climate-smart agriculture. Chapter 10, on biofertilizer economics, covers cost–benefit analyses, market potential, and obstacles, including biorefinery costs. Chapter 11 covers adoption, policy frameworks, farmer education challenges, and opportunities for regional innovation with reference to India.

In conclusion, this book emphasizes the innovative and transformative potential of biofertilizers in promoting sustainable agricultural practices. Their application would ensure that productivity is maintained while actively managing and protecting the environment. Biofertilizers, using microbial inoculants and agrifood waste, alongside organic farming, lessen the dependence on chemical fertilizers and improve soil health parameters. The challenges include different soil types, expensive biorefineries, and farmers needing more training. However, there are also good opportunities. Support from the government is beneficial. Innovations also create new possibilities. India shows great promise in this area. Overcoming these problems can lead to better biofertilizers and farming.

References

Ahemad, M. and Kibret, M. (2014) Mechanisms and applications of plant growth promoting rhizobacteria: current perspective. *Journal of King Saud University - Science* 26(1), 1–20. DOI: 10.1016/j.jksus.2013.05.001

Akinnawo, S.O. (2023) Eutrophication: causes, consequences, physical, chemical and biological techniques for mitigation strategies. *Environmental Challenges* 12, 100733. DOI: 10.1016/j.envc.2023.100733

Bhardwaj, D., Ansari, M.W., Sahoo, R.K., and Tuteja, N. (2014) Biofertilizers function as key player in sustainable agriculture by improving soil fertility, plant tolerance and crop productivity. *Microbial Cell Factories* 13, 66. DOI: 10.1186/1475-2859-13-66

Bhardwaj, R.L., Parashar, A., Parewa, H.P., and Vyas, L. (2024) An alarming decline in the nutritional quality of foods: the biggest challenge for future generations' health. *Foods (Basel, Switzerland)* 13(6), 877. DOI: 10.3390/foods13060877

Colla, L.M., Rempel, A., Simon, V., Berwian, G., Braun, J. *et al.* (2025) Circular economy achievements in agroindustrial waste managements: a current opinion. In: Rai, S., Bhardwaj, A.K., and Colla, L.M. (eds) *Sustainable Management of Agro-Food Waste*. Academic Press, London, pp. 1–11.

Das, H., Devi, N., Venu, N., and Borah, A. (2023) Chemical fertilizer and its effects on the soil environment. *Research and Review in Agriculture Sciences*, Vol. 7. pp. 31–51.

Food and Agriculture Organization of the United Nations (2017) *The Future of Food and Agriculture: Trends and Challenges*. FAO, Rome.

Food and Agriculture Organization of the United Nations (2021) *Soil Health for Paddy Rice: A Manual for Farmer Field School Facilitators*. FAO, Rome.

Fuentes-Ramirez, L.E. and Caballero-Mellado, J. (2005) Bacterial biofertilizers. In: Siddiqui, Z.A. (ed.) *PGPR: Biocontrol and Biofertilization*. Springer, New York, pp. 143–172.

Gaur, A.C. (2010) *Biofertilizers in Sustainable Agriculture*. Indian Council of Agricultural Research, New Delhi.

Gorain, B., Paul, S., and Parihar, M. (2022) Role of soil microbes in micronutrient solubilization. In: Singh, H. and Vashnav, A. (eds) *New and Future Developments in Microbial Biotechnology and Bioengineering*. Elsevier, Amsterdam, pp. 131–150.

Gupta, M.M. (2020) Arbuscular mycorrhizal fungi: the potential soil health indicators. In: Giri, B. and Varma, A. (eds) *Soil Health*. Springer, Cham, Switzerland, pp. 183–195.

Gupta, M.M. (2024) *Biofertilizers NEP Curricula*. Swaraj Prakashan, Delhi.

Gupta, M.M. and Abbott, L.K. (2021) Exploring economic assessment of the arbuscular mycorrhizal symbiosis. *Symbiosis* 83(2), 143–152. DOI: 10.1007/s13199-020-00738-0

Gupta, M.M., Aggarwal, A., and Asha, A. (2018) From mycorrhizosphere to rhizosphere microbiome: the paradigm shift. In: Giri, B., Prasad, R., and Varma, A. (eds) *Root Biology*. Springer, Cham, Switzerland, pp. 487–500.

Hernández-Álvarez, C., Peimbert, M., Rodríguez-Martin, P., Trejo-Aguilar, D., and Alcaraz, L.D. (2023) A study of microbial diversity in a biofertilizer consortium. *PLOS One* 18(8), e0286285. DOI: 10.1371/journal.pone.0286285

Humphries, R.N. and Brazier, R.E. (2018) Exploring the case for a national-scale soil conservation and soil condition framework for evaluating and reporting on environmental and land use policies. *Soil Use and Management* 34(1), 134–146. DOI: 10.1111/sum.12400

Iqbal, S., Ashfaq, M., Rao, M.J., Khan, K.S., Malik, A.H. *et al.* (2024) Trichoderma viride: an eco-friendly biocontrol solution against soil-borne pathogens in vegetables under different soil conditions. *Horticulturae* 10(12), 1277. DOI: 10.3390/horticulturae10121277

Khanyile, N., Dlamini, N., Masenya, A., Madlala, N.C., and Shezi, S. (2024) Preparation of biofertilizers from banana peels: their impact on soil and crop enhancement. *Agriculture* 14(11), 1894. DOI: 10.3390/agriculture14111894

Kushwaha, D., Katiyar, P., Singh, R., Verma, Y., Singh, D. *et al.* (2025) Value-added bioproduct: a sustainable way for food waste valorization and circular bioeconomy. In: Rai, S., Bhardwaj, A.K., and Colla, L.M. (eds) *Sustainable Management of Agro-Food Waste*. Academic Press, London, pp. 127–139.

Malusá, E. and Vassilev, N. (2014) *Biofertilizers: A Handbook on Recent Trends*. Springer, Cham, Switzerland.

Mozafar, A. (1993) Nitrogen Fertilizers and the Amount of Vitamins in Plants: A Review. *Journal of Plant Nutrition* 16(12), 2479–2506. DOI: 10.1080/01904169309364698

O'Callaghan, M., Ballard, R.A., and Wright, D. (2022) Soil microbial inoculants for sustainable agriculture: limitations and opportunities. *Soil Use and Management* 38(3), 1340–1369. DOI: 10.1111/sum.12811

Okon, Y. and Labandera-Gonzalez, C.A. (1994) Agronomic applications of *Azospirillum*: an evaluation of 20 years worldwide field inoculation. *Soil Biology and Biochemistry* 26(12), 1591–1601. DOI: 10.1016/0038-0717(94)90311-5

Pandey, A.K., Thakur, S., Mehra, R., Kaler, R.S.S., Paul, M. *et al.* (2025) Transforming agri-food waste: innovative pathways toward a zero-waste circular economy. *Food Chemistry* 28, 102604. DOI: 10.1016/j.fochx.2025.102604

Pingali, P.L. (2012) Green revolution: impacts, limits, and the path ahead. *Proceedings of the National Academy of Sciences* 109(31), 12302–12308. DOI: 10.1073/pnas.0912953109

Pradhan, S., Parthasarathy, P., Mackey, H.R., Al-Ansari, T., and McKay, G. (2024) Food waste biochar: a sustainable solution for agriculture application and soil–water remediation. *Carbon Research* 3(1), Article 41. DOI: 10.1007/s44246-024-00123-2

Pretty, J. and Bharucha, Z.P. (2015) Integrated pest management for sustainable intensification of agriculture in Asia and Africa. *Insects* 6(1), 152–182. DOI: 10.3390/insects6010152

Roka, K. (2019) Environmental and social impacts of food waste. In: Filho, W.L., Azul, A.M., Brandli, L., Özuyar, P., and Wall, T. (eds) *Responsible Consumption and Production*. Springer, Cham, Switzerland, pp. 1–12.

Saha, B., Fatima, A., Saha, S., Sahoo, S. K., and Poddar, P. (2024). Environmental pollution due to improper use of chemical fertilizers and their remediation. In: Ganguly, P., Mandal, J., Paramsivam, and M., Patra, S. (eds) *Environmental Contaminants*. Apple Academic Press, Palm Bay, Florida, pp. 203–219.

Shahrajabian, M.H., Chaski, C., Polyzos, N., and Petropoulos, S.A. (2021) Biostimulants application: a low input cropping management tool for sustainable farming of vegetables. *Biomolecules* 11(5), Article 698. DOI: 10.3390/biom11050698

Sharma, P. and Tong, Y.W. (2025) Management and problems of food waste in Asian countries. In: In: Rai, S., Bhardwaj, A.K., and Colla, L.M. (eds) *Sustainable Management of Agro-Food Waste*. Academic Press, London, pp. 141–155.

Tiong, Y.W., Sharma, P., Xu, S., Bu, J., An, S. *et al.* (2024) Enhancing sustainable crop cultivation: the impact of renewable soil amendments and digestate fertilizer on crop growth and nutrient composition. *Environmental Pollution* 342, 123132. DOI: 10.1016/j.envpol.2023.123132

Toma (Sărdărescu), D.-I., Manaila-Maximean, D., Fierascu, I., Baroi, A.M., Matei (Brazdis), R.I. *et al.* (2024) Applications of natural polymers in the grapevine industry: plant protection and value-added utilization of waste. *Polymers* 17(1), 18. DOI: 10.3390/polym17010018

UNICEF (2024) *The State of Food Security and Nutrition in the World 2024*. FAO, Rome.

United Nations (2024) International Day of Awareness on Food Loss and Waste Reduction. Available at: https://www.un.org/en/observances/end-food-waste-day (accessed September 4, 2025).

United Nations Environment Programme (2024) Reducing food loss and waste for a healthier planet. Available at: https://www.genevaenvironmentnetwork.org/resources/updates/reducing-food-loss-and-waste-for-a-healthier-planet/ (accessed September 4, 2025).

Vessey, J.K. (2003) Plant growth promoting rhizobacteria as biofertilizers. *Plant and Soil* 255(2), 571–586. DOI: 10.1023/A:1026037216893

Voss, M., Valle, C., Calcio Gaudino, E., Tabasso, S., Forte, C. *et al.* (2024) Unlocking the potential of agri-food waste for sustainable innovation in agriculture. *Recycling* 9(2), 25. DOI: 10.3390/recycling9020025

2 Soil Fertility and Functions: Soil Complex, Agrifood Waste, and Microbes

Manju M. Gupta*
Sri Aurobindo College, University of Delhi, Delhi, India

Abstract
Soil is a dynamic living system central to terrestrial ecosystem functioning, supporting plant productivity, regulating water and nutrient cycles, and sustaining biodiversity. This chapter explores the interlinked roles of the soil structure complex, microbial communities, and agrifood waste in maintaining soil fertility and health. Soil structure, comprising mineral particles, organic matter, water, and air, provides the physical and chemical framework for nutrient retention and microbial activity. Microbiota—including bacteria, fungi, and actinomycetes—drive critical processes such as nutrient cycling, organic matter decomposition, aggregation, and plant–microbe symbioses. Among these, arbuscular mycorrhizal fungi contribute significantly through glomalin production and root associations that enhance aggregation, nutrient uptake, and ecosystem resilience. Agrifood waste, when converted into digestate or biochar, offers sustainable alternatives to chemical fertilizers. Digestate enriches organic matter, supplies nutrients, and stimulates beneficial microbial activity, while biochar provides long-term improvements in soil structure, water retention, and microbial habitat. Together, these amendments foster soil fertility, mitigate degradation, and contribute to climate change resilience by enhancing carbon sequestration and reducing dependence on synthetic inputs. The integration of soil structure management, microbial ecology, and waste recycling thus represents a holistic approach to sustainable agriculture and food security.

2.1 Introduction

Soil, a dynamic living system, is the keystone of terrestrial ecosystems, distinct from the inert regolith of planetary bodies like Mars and the Moon (Young and Crawford, 2004). Soil is the foundation of life on Earth by supporting plant growth, providing habitat for a diverse array of organisms, cycling essential nutrients, filtering water, and mitigating climate change. A soil's fertility—its capacity to supply plants with essential nutrients, water, and oxygen—is vital to its ecological role (Hodges, 2010). Good soil quality is essential for plant, animal, and environmental health and is evaluated by its physical, chemical, and biological properties. Soil health views soil as a living ecosystem that functions optimally and sustainably (Gupta, 2020; Lehmann *et al.*, 2020). Healthy, high-quality soils promote robust crop yields, enhance resilience to erosion and drought, and address global challenges like food security, biodiversity loss, and climate change.

Comprising mineral particles, organic matter, water, air, and diverse microbiota, soil

*Corresponding author: mbansalsac@outlook.com

DOI: 10.1079/9781836991021.0002

operates as a complex, dynamic system that governs its health and functionality. However, chemical fertilizers have often degraded soil health, reducing fertility and disrupting ecosystem processes. Sustainable practices, such as applying agrifood waste and microbial inoculants, offer solutions to restore soil quality and support long-term agricultural productivity. These practices enhance soil structure, boost microbial activity, and promote nutrient cycling, contributing to resilient ecosystems capable of addressing global challenges like food security and climate change.

Maintenance of good soil health requires a clear understanding of nutrient availability, soil structure, and the biological interactions that support nutrient uptake by plants and soil organisms. This chapter explores three key components of soil health. The first is the soil structure complex, which forms the physical and biological foundation for nutrient retention and plant growth. The second is soil microbiota, which boost nutrient availability through processes such as nitrogen fixation and phosphorus solubilization. The third is agrifood waste, which enhances soil fertility and structure, promoting a resilient soil ecosystem for sustainable agriculture.

2.2 Soil Structure Complex

Soil structure describes how soil particles, aggregates, and pores are arranged. This arrangement impacts fertility, water, air, and root growth (Young and Crawford, 2004; Bronick and Lal, 2005). Soil, a complex of minerals, organic matter, water, and air, supports vital processes. These include nutrient cycling, gas exchange, and microbial activity, boosted by soil microbiota and organic matter (Fig. 2.1).

2.2.1 Mineral matrix

Constituting 45–50% of soil volume, the mineral matrix—formed from weathered parent rock—consists of sand, silt, and clay particles that define soil texture (Al-Kaisi *et al.*, 2017). Micro-aggregates (<250 μm) and macro-aggregates (0.25–2 mm) are formed from the soil particles. These aggregates create pore spaces that regulate water and air movement, and provide microhabitats for diverse microbial communities like Actinobacteria and Glomeromycota. These microbes further stabilize the aggregates through biofilm formation and glomalin production (Fig. 2.2; Bach *et al.*, 2018; Custódio *et al.*, 2022). Sand promotes drainage and aeration but has low nutrient-holding capacity. Clay and silt, with higher

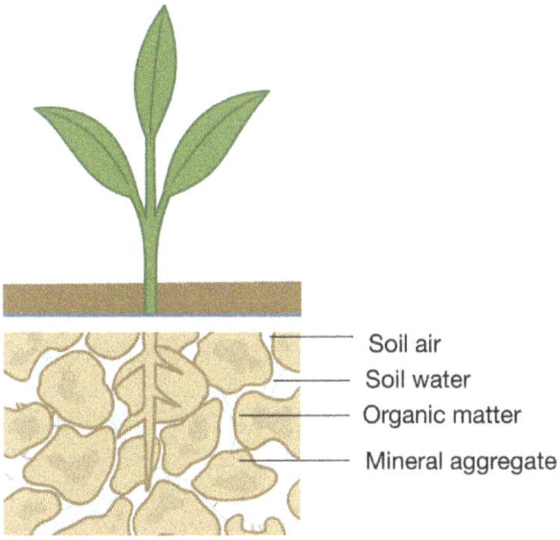

Fig. 2.1. The soil structure complex. Figure author's own.

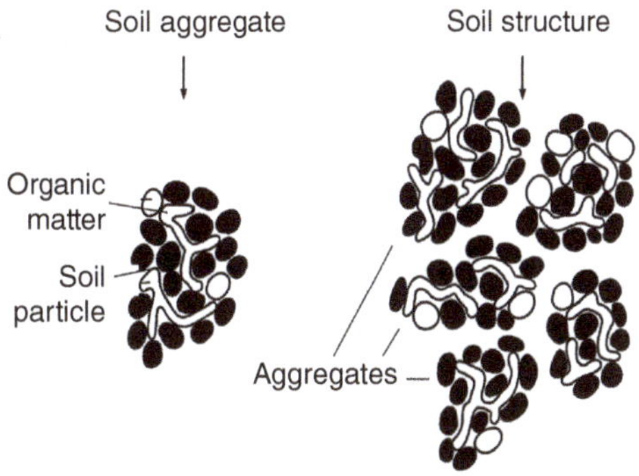

Fig. 2.2. Formation of soil aggregates and their contribution to soil structure. The illustration shows the hierarchical organization of soil components. On the left, a soil aggregate is formed by the binding of soil particles with organic matter, creating a cohesive unit. On the right, multiple soil aggregates combine to form the soil structure, which determines the arrangement, porosity, and overall health of the soil. This structural organization is essential for water retention, root penetration, aeration, and microbial activity in the soil ecosystem. Figure author's own.

surface areas, enhance cation exchange capacity (CEC), ranging from 10 to 150 meq/100g in clay-rich soils (Hodges, 2010). CEC, the soil's ability to retain positively charged ions like calcium (Ca^{2+}), magnesium (Mg^{2+}), sodium (Na^{+}), and potassium (K^{+}), supports nutrient availability and structure stability by adsorbing cations via negatively charged clay and organic matter surfaces (Fig. 2.3; Gupta, 2024).

2.2.2 Organic matter

Making up 1–5% of soil volume, soil organic matter is a complex mix. It includes decomposed plants and animals, plus living microbes. Humus is a final, stable product of decomposition of organic matter, rich in carbon. It can stay in soil for decades to centuries. Humus plays a key role in long-term carbon storage (Pettit, 2004). Humus supplies 20–80% of soil nitrogen, phosphorus, and sulfur, while its colloidal nature enhances soil aggregation, reduces erosion, and supports microbial activity (Gupta, 2024).

Food waste-derived digestate significantly enriches soil organic matter content, which in turn boosts nutrient availability for plants and promotes the flourishing of beneficial microbes, including *Proteobacteria* and *Azotobacter*. Increased organic matter stimulates microbial activity, leading to exopolysaccharide (EPS) production. This improves soil aggregation and structure (Cheong *et al.*, 2020; Sharma *et al.*, 2023). Fertile soils typically contain 2–8% organic matter. Adding agrifood waste increases this organic matter content. This boosts water retention and carbon sequestration (Pettit, 2004; Tiong *et al.*, 2024).

2.2.3 Soil water and air

Occupying 20–30% of soil volume each, water and air fill pore spaces, balancing nutrient transport, root respiration, and microbial activity (Gavrilescu, 2021; Gupta, 2024). Soil water exists in three forms: gravitational, capillary, and hygroscopic. These forms dissolve nutrients (like nitrates and phosphates) which plants absorb. Retention varies with soil texture; sandy soils hold around 5% water, whereas clay soils can hold up to 40% (Fig. 2.3b). Soil air supplies oxygen (optimal at 10–15% by volume) for aerobic processes, but compacted soils may reduce aeration to below 5%, limiting microbial activity. Agrifood waste-derived biochar enhances pore structure, increasing water retention and aeration, while providing habitats for microbes like Firmicutes that support nutrient cycling (Pradhan *et al.*, 2024). Well-structured soils, like loams, have a good water–air balance. Microbial biofilms and digestate improve moisture retention and gas exchange, boosting plant growth and soil health (Al-Kaisi *et al.*, 2017; Cai *et al.*, 2019).

The soil structure complex integrates mineral particles, organic matter, and pore spaces to create physicochemical niches that foster microbial communities and support agrifood waste applications. Soil microbes improve soil aggregation via glomalin and biofilms. Agrifood waste (e.g. digestate and biochar) boosts organic matter, microbial activity, and soil stability, fostering fertility, nutrient cycling, and ecological resilience (Fox *et al.*, 2018; Sharma and Tong, 2025).

2.3 Soil Microbiota and Their Role in Soil Structure and Function

Soil microbiota—the living component of the soil—comprises microorganisms (bacteria, fungi, actinomycetes), macrofauna (earthworms, insects), and plant roots. It is estimated that 80–90% of soil functions are microbially driven (Benbi and Nieder, 2003), with 1 g of soil containing approximately 10^{8}–10^{10} microbial cells (Zhang *et al.*, 2017). The diversity and composition of these communities vary across soil types, shaped by soil properties, which in turn are influenced by microbial activity. These microbes contribute to soil processes through nutrient cycling, organic matter decomposition, soil aggregation, and interactions with plants, thereby shaping soil's physical, chemical, and biological properties.

The composition and activity of soil microbiota are regulated by complex biotic interactions, including microbe–microbe and plant–microbe relationships, as well as abiotic factors such as soil properties and environmental conditions.

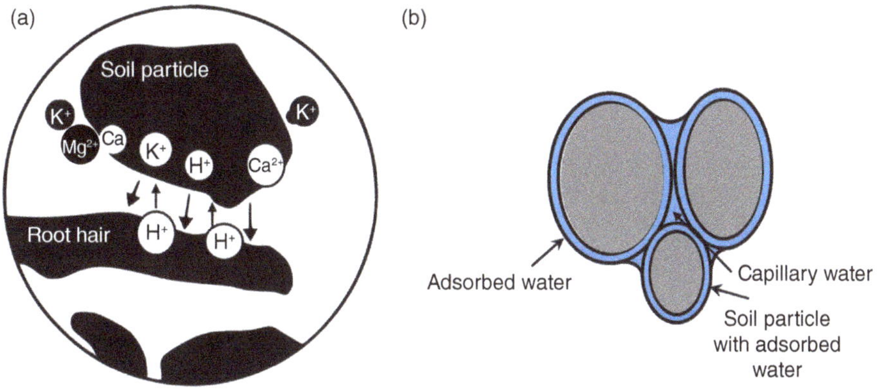

Fig. 2.3. Mechanisms of nutrient and water interaction in the rhizosphere. (a) Ion exchange at the root–soil interface. Plant roots obtain positively charged nutrients like K^+, Ca^{2+}, and Mg^{2+} from soil particles. This happens through an exchange with H^+ ions released by root hairs. This cation exchange process facilitates nutrient uptake by plants. (b) Water retention around soil particles. The diagram illustrates adsorbed water, clinging to soil particles, and capillary water, filling pores and usable by plants. Figures author's own.

Soil structure, an important abiotic factor, creates microhabitats that shape microbial communities (Fig. 2.2; Custódio *et al.*, 2022). Micro-aggregates tend to harbor bacterial phyla like Gemmatimonadetes and Actinobacteria (Rubrobacteridae), fungal orders Onygenales and Chaetosphaeriales, and families such as Trichosporonaceae. In contrast, macro-aggregates support filamentous fungi, especially arbuscular mycorrhizal fungi (AMF) of the Glomeromycota phylum (Bach *et al.*, 2018). Soil moisture strongly affects the types of microbes present. Fungi tend to survive droughts better than bacteria, perhaps because they store osmoregulators and have extensive networks of hyphae (Amend *et al.*, 2016; Nguyen *et al.*, 2018). Other critical abiotic drivers include nutrient availability, pH, and temperature, all of which alter microbial structures and the biogeochemical processes in soils (Buzzard *et al.*, 2019).

Biotic interactions, including bacteria–bacteria, fungi–bacteria, and plant–microbe relationships, also shape microbial dynamics (Custódio *et al.*, 2022). Advances in next-generation sequencing have enabled detailed profiling of microbiomes under various conditions, while microbial network analyses reveal how interactions within communities influence ecology and functionality (Agler *et al.*, 2016).

Fungal and bacterial interactions are often synergistic or antagonistic, playing vital roles in ecosystem functions and plant health (Deveau *et al.*, 2018; Durán *et al.*, 2018). AMF form mutualistic relationships with plants, which boosts nutrient cycling and creates habitats for other microbes (Emmett *et al.*, 2021). Plant root exudates and decomposition enrich the rhizosphere, supporting microbial life and soil health (Gupta *et al.*, 2018). Symbiotic relationships, like those between rhizobia and legumes or arbuscular mycorrhizal fungi and plant roots, improve nutrient uptake and stress tolerance. For instance, AMF in *Medicago truncatula* promote the enrichment of rhizobia and non-symbiotic endophytes like *Pseudomonas* in the rhizosphere (Wang *et al.*, 2021).

Soil microbiota, including bacteria, fungi, actinomycetes, and other microorganisms, are critical drivers of soil structure and function, influencing soil health, fertility, and ecosystem sustainability.

2.3.1 Soil structure enhancement

Soil microbiota significantly influence soil structure by promoting aggregation and improving soil stability.

Mycorrhizal fungi and glomalin production: AMF improve soil structure. They produce glomalin and related proteins that bind soil particles, enhancing porosity, water infiltration, aeration, and root penetration (Rillig *et al.*, 2002; Rillig and Mummey, 2006). These proteins, along with AMF hyphae, necromass, and exudates, contribute to mineral-associated organic matter formation and carbon sequestration. The glomalin and related proteins persists in soil for decades to support long-term aggregate stability and ecosystem resilience (Liu *et al.*, 2022). AMF hyphal networks are larger than root systems. They store and move soil carbon, creating a better soil structure that retains water and helps plants grow (Camenzind and Rillig, 2013). Additionally, AMF affect plant growth, root architecture, and rhizodeposition, indirectly promoting aggregation (Jones *et al.*, 2004). They may also shape microbial communities that further influence aggregation.

Exopolysaccharide (EPS)-producing bacteria: Bacteria such as *Pseudomonas putida* and *Bacillus polymyxa* secrete EPS, which acts as a glue to bind soil particles, increasing aggregate stability and water retention. This is particularly beneficial in arid or saline soils, where EPS maintains a hydrated environment around roots, improving plant water and nutrient uptake (Sandhya *et al.*, 2009). Many soil bacteria form biofilms—structured communities embedded in extracellular polymers—that enhance adhesion, resilience, and soil structure by promoting aggregation and stability (Cai *et al.*, 2019).

Bio-cementation: Urease-producing bacteria like *Sporosarcina pasteurii* facilitate microbially induced carbonate precipitation, strengthening soil by forming calcium carbonate bridges between particles. This emerging technique shows promise for soil stabilization but requires further field validation (Carter *et al.*, 2023).

2.3.2 Nutrient cycling and availability

Microbial communities drive nutrient cycling, making essential elements available to plants.

Nitrogen fixation: Symbiotic rhizobia and free-living bacteria, such as *Azospirillum* and *Azotobacter*, convert atmospheric nitrogen into forms usable by plants, thus lessening the need for synthetic fertilizers. For example, *Bradyrhizobium* inoculation in soybeans significantly boosts nitrogen availability.

Phosphorus solubilization: Bacteria (e.g. *Pseudomonas*, *Bacillus*) and fungi (e.g. *Penicillium bilaii*) solubilize insoluble phosphorus through organic acid secretion and enzyme production (e.g. phytases), enhancing plant uptake (Richardson and Simpson, 2011).

Organic matter decomposition: Soil microbes, particularly fungi and actinomycetes, break down organic matter, releasing nutrients like carbon, nitrogen, and sulfur into the soil. This supports plant growth and adds to soil fertility (Gupta and Germida, 1988).

Litter degradation: Soil microbes, including fungi (like *Aspergillus* and *Penicillium*) and bacteria (such as *Bacillus* and *Streptomyces*), are essential for litter decomposition. They break down complex plant matter (cellulose, hemicellulose, and lignin) into simpler substances. This process releases essential nutrients such as nitrogen, phosphorus, and potassium into the soil, enhancing nutrient availability for plants and supporting soil microbial communities. Fungal hyphae penetrate litter, secreting enzymes like cellulases and ligninases, while bacteria contribute through extracellular enzyme production, accelerating decomposition and carbon cycling (Berg and McClaugherty, 2014). AMF lack saprotrophic capabilities and don't directly degrade organic matter (Gui *et al.* 2017). However, they do influence decomposition. Hodge *et al.* (2001) showed how AMF affect decomposition. They found that AMF increased host plant nitrogen uptake and carbon loss from the litter. However, the mechanisms by which AMF are able to influence litter decomposition remain unclear.

2.3.3 Bioremediation and soil health

Microbes mitigate soil contamination and improve soil quality.

Pollutant degradation: Microbes like *Burkholderia* and *Pseudomonas* degrade organic pollutants (e.g. dioxins, hydrocarbons) and reduce heavy metal toxicity by binding or transforming metals like cadmium (Ayilara and Babalola, 2023).

Greenhouse gas mitigation: Certain microbial inoculants, such as *Bacillus amyloliquefaciens* and commercial products like SoilBuilder (containing *Bacillus* spp.), reduce nitrous oxide (N_2O) emissions by inhibiting nitrification or immobilizing nitrogen. However, their effectiveness in the field requires further investigation (Wu *et al.*, 2018).

2.3.4 Plant–microbe interactions and soil function

Soil microbiota enhance plant growth and resilience, indirectly supporting soil function.

Root microbiome assembly: The root microbiota, influenced by plant genetics (e.g. NRT1.1B in rice or PHR1-PSR in legumes), enhances nutrient uptake and stress tolerance. For instance, indica rice varieties recruit nitrogen-metabolizing bacteria, improving nitrogen-use efficiency (Zhang *et al.*, 2019).

Stress tolerance: Microbes like *Azospirillum brasilense* and *Trichoderma harzianum* improve plant tolerance to abiotic stresses. These microbes alter root morphology, produce protective compounds, and maintain soil–root interactions during drought and salinity (Fukami *et al.*, 2018).

Pathogen suppression: Microbes such as *Trichoderma* and *Pseudomonas chlororaphis* produce antibiotics and siderophores, suppressing soil-borne pathogens and maintaining a healthy soil environment (Schuster and Schmoll, 2010).

2.4 Enhancing Soil Structure and Nutrient Balance through Agrifood Waste Applications

As a continuation of Chapter 1, where the nature and potential of agrifood waste were introduced, this section focuses on its role in improving soil structure and contributing additional nutrients. Agrifood waste, a by-product of food production and consumption, presents a serious environmental challenge due to its volume and improper disposal. However, when biologically converted, this waste becomes a valuable resource for enhancing soil health and promoting sustainable agriculture. Sharma and Tong (2025) outlines the diverse benefits of transforming food waste into biofertilizers, especially its impact on soil structure and function.

Soil structure is vital for agricultural productivity and is shaped by its physical composition, porosity, water retention, and microbial activity. Application of food waste-derived digestate improves these attributes by enriching organic matter and stimulating beneficial microbes. Digestate, generated through anaerobic digestion, enhances aggregation, water infiltration, and aeration—key traits for soil resilience (Sharma *et al.*, 2023). It is rich in nutrients such as nitrogen, phosphorus, potassium, calcium, and magnesium, which aid in forming and stabilizing soil aggregates (Häfner *et al.*, 2022; Tiong *et al.*, 2024). The organic matter acts as a binder, forming stable aggregates that improve tilth, reduce erosion, and support root growth and nutrient uptake.

Food waste anaerobic digestate is nutrient dense, though composition varies by source and treatment (Ries *et al.*, 2023). High ammonium nitrogen (NH_4^+) content can require dilution or nitrification to prevent toxicity (Pelayo *et al.*, 2021). Excess sodium and chloride may hinder water and nutrient absorption, especially in hydroponic systems (Bergstrand *et al.*, 2020). Phosphorus and sulfur availability can be limited by the formation of insoluble compounds during digestion (Mickan *et al.*, 2022). Effective digestate application involves managing pH and supplementing micronutrients such as magnesium, boron, and molybdenum (Ries *et al.*, 2023).

These structural improvements offer direct agronomic benefits, including improved porosity, enhanced gas exchange, and increased water-holding capacity, all contributing to higher crop yields and input efficiency. For instance, tomato cultivation using food waste digestate has shown yield increases of up to 17.3%, credited to improved soil conditions (Tiong *et al.*, 2024).

Food waste digestate fosters beneficial microbial communities like Proteobacteria, Firmicutes, and nitrogen-fixing bacteria such as *Azotobacter* (Cheong *et al.*, 2020; Sharma *et al.*, 2023). These microbes decompose organic residues and release exopolysaccharides that bind soil particles, further enhancing moisture retention and compaction resistance. Digestate-derived inoculants also support nutrient cycling and long-term fertility through humus formation. Compared to inorganic fertilizers, digestate results in greater microbial biodiversity—crucial for a resilient soil ecosystem (Coelho *et al.*, 2020). Increased microbial biomass and enzymatic activity improve nitrogen and carbon cycling (Ren *et al.*, 2020). Additionally, it reduces the harmful effects of synthetic fertilizers, such as soil acidification and organic matter loss (Kibblewhite *et al.*, 2008).

Unlike digestate, which delivers readily available nutrients, biochar contributes in a more sustained way. Produced via pyrolysis, biochar is carbon-rich and enhances soil structure, water retention, and microbial habitat. Though lower in immediate nutrients, it contains small amounts of phosphorus, potassium, and micronutrients, depending on the feedstock. Biochar enhances cation exchange capacity, thus lessening nutrient leaching and promoting microbial activity. It acts as a soil enhancer rather than a fertilizer, offering slow but steady benefits to soil health.

Biochar made from food waste offers further benefits. As a stable carbon source, it improves fertility, retains water, stores carbon, and promotes microbial life (Pradhan *et al.*, 2024). It serves as a carrier for rhizobia, enhancing nutrient cycles and microbial interactions, especially in poor tropical soils. Even small applications can significantly enhance water retention, nutrient availability, plant growth, and microbial functions. This is achieved by reducing nutrient leaching and improving aeration (Zhang *et al.*, 2013; Agegnehu *et al.*, 2016; Pradhan *et al.*, 2022). Food waste biochar increases the abundance of beneficial microbes and enzymatic activities such as urease, phosphatase, and dehydrogenase (Ali *et al.*, 2021; Zheng *et al.*, 2021; Liu *et al.*, 2023; Zhou *et al.*, 2023).

In conclusion, converting agrifood waste into digestate offers an ecologically sound, sustainable approach to soil enhancement. Its dual benefit of physical improvement and microbial stimulation makes it essential for resilient agriculture. By integrating food waste-derived organic fertilizers into soil strategies, both farmers and policy makers can promote productivity and sustainability.

2.5 Challenges and Future Directions

Soil microbiota and agrifood waste applications vary in efficacy due to soil type, pH, and management practices. Tailored formulations and field trials are needed to optimize microbial inoculants and waste-derived products. Emerging techniques like bio-cementation and microbial consortia for bioremediation show promise but require scalability validation. Integrating agrifood waste with microbial strategies offers a sustainable path to enhance soil health, supporting global food security and environmental sustainability.

References

Agegnehu, G., Bass, A.M., Nelson, P.N., and Bird, M.I. (2016) Benefits of biochar, compost and biochar-compost for soil quality, maize yield and greenhouse gas emissions in a tropical agricultural soil. *The Science of the Total Environment* 543, 295–306. DOI: 101016/j.scitotenv.2015.11.054

Agler, M.T., Ruhe, J., Kroll, S., Morhenn, C., Kim, S.-T. *et al.* (2016) Microbial hub taxa link host and abiotic factors to plant microbiome variation. *PLoS Biology* 14(1), e1002352. DOI: 101371/journal.pbio.1002352

Al-Kaisi, M.M., Lal, R., Olson, K.R., and Lowery, B. (2017) Fundamentals and functions of soil environment. In: Al-Kaisi, M.M. and Lowery, B. (eds) *Soil Health and Intensification of Agroecosystems*. Academic Press, London, pp. 1–23.

Ali, L., Manzoor, N., Li, X., Naveed, M., Nadeem, S.M. *et al.* (2021) Impact of corn cob-derived biochar in altering soil quality, biochemical status and improving maize growth under drought stress. *Agronomy* 11(11), 2300. DOI: 103390/agronomy11112300

Amend, A.S., Martiny, A.C., Allison, S.D., Berlemont, R., Goulden, M.L. *et al.* (2016) Microbial response to simulated global change is phylogenetically conserved and linked with functional potential. *The ISME Journal* 10(1), 109–118. DOI: 101038/ismej.2015.96

Ayilara, M.S. and Babalola, O.O. (2023) Bioremediation of environmental wastes: the role of microorganisms. *Frontiers in Agronomy* 5, 1183691. DOI: 103389/fagro.2023.1183691

Bach, E.M., Williams, R.J., Hargreaves, S.K., Yang, F., and Hofmockel, K.S. (2018) Greatest soil microbial diversity found in micro-habitats. *Soil Biology and Biochemistry* 118, 217–226. DOI: 101016/j.soilbio.2017.12.018

Benbi, D. and Nieder, R. (2003) *Handbook of Processes and Modeling in the Soil-Plant System*. CRC Press, Boca Raton, Florida.

Berg, B. and McClaugherty, C. (2014) *Plant Litter: Decomposition, Humus Formation, Carbon Sequestration*, 3rd edn. Springer, Cham, Switzerland.

Bergstrand, K.-J., Asp, H., and Hultberg, M. (2020) Utilizing anaerobic digestates as nutrient solutions in hydroponic production systems. *Sustainability* 12(23), 10076. DOI: 103390/su122310076

Bronick, C.J. and Lal, R. (2005) Soil structure and management: a review. *Geoderma* 124(1–2), 3–22. DOI: 101016/j.geoderma.2004.03.005

Buzzard, V., Michaletz, S.T., Deng, Y., He, Z., Ning, D. *et al.* (2019) Continental scale structuring of forest and soil diversity via functional traits. *Nature Ecology & Evolution* 3(9), 1298–1308. DOI: 101038/s41559-019-0954-7

Cai, P., Sun, X., Wu, Y., Gao, C., Mortimer, M. *et al.* (2019) Soil biofilms: microbial interactions, challenges, and advanced techniques for ex-situ characterization. *Soil Ecology Letters* 1(3–4), 85–93. DOI: 101007/s42832-019-0017-8

Camenzind, T. and Rillig, M.C. (2013) Extraradical arbuscular mycorrhizal fungal hyphae in an organic tropical montane forest soil. *Soil Biology and Biochemistry* 64, 96–102. DOI: 101016/j.soilbio.2013.04.011

Carter, M.S., Tuttle, M.J., Mancini, J.A., Martineau, R., Hung, C.S. *et al.* (2023) Microbially induced calcium carbonate precipitation by *Sporosarcina pasteurii*: a case study in optimizing biological $CaCO_3$ precipitation. *Applied and Environmental Microbiology* 89(8), e0179422. DOI: 101128/aem.01794-22

Cheong, J.C., Lee, J.T., Lim, J.W., Song, S., Tan, J.K. *et al.* (2020) Closing the food waste loop: food waste anaerobic digestate as fertilizer for the cultivation of the leafy vegetable, xiao bai cai (Brassica rapa). *Science of the Total Environment* 715, 136789. DOI: 10.1016/j.scitotenv.2020.136789

Coelho, J.J., Hennessy, A., Casey, I., Bragança, C.R., Woodcock, T., and Kennedy, N. (2020) Biofertilisation with anaerobic digestates: a field study of effects on soil microbial abundance and diversity. *Applied Soil Ecology* 147, 103403. DOI: 10.1016/j.apsoil.2019.103403

Custódio, V., Gonin, M., Stabl, G., Bakhoum, N., Oliveira, M.M. *et al.* (2022) Sculpting the soil microbiota. *The Plant Journal: For Cell and Molecular Biology* 109(3), 508–522. DOI: 101111/tpj.15591

Deveau, A., Bonito, G., Uehling, J., Paoletti, M., Becker, M. *et al.* (2018) Bacterial-fungal interactions: ecology, mechanisms and challenges. *FEMS Microbiology Reviews* 42(3), 335–352. DOI: 101093/femsre/fuy008

Durán, P., Thiergart, T., Garrido-Oter, R., Agler, M., Kemen, E. *et al.* (2018) Microbial interkingdom interactions in roots promote Arabidopsis survival. *Cell* 175(4), 973–983. DOI: 101016/j.cell.2018.10.020

Emmett, B.D., Lévesque-Tremblay, V., and Harrison, M.J. (2021) Conserved and reproducible bacterial communities associate with extraradical hyphae of arbuscular mycorrhizal fungi. *The ISME Journal* 15(8), 2276–2288. DOI: 101038/s41396-021-00938-7

Fox, A., Ikoyi, I., Torres-Sallan, G., Lanigan, G., Schmalenberger, A. *et al.* (2018) The influence of aggregate size fraction and horizon position on microbial community composition. *Applied Soil Ecology* 127, 19–29. DOI: 101016/j.apsoil.2018.02.023

Gavrilescu, M. (2021) Water, soil, and plants interactions in a threatened environment. *Water* 13(19), 2746. DOI: 103390/w13192746

Gui, H., Hyde, K., Xu, J., and Mortimer, P. (2017) Arbuscular mycorrhiza enhance the rate of litter decomposition while inhibiting soil microbial community development. *Scientific Reports* 7, 42184. DOI: 101038/srep42184

Gupta, M.M. (2020) Arbuscular mycorrhizal fungi: the potential soil health indicators. In: Giri, B. and Varma, A. (eds) *Soil Health*. Springer, Cham, Switzerland, pp. 183–195.

Gupta, M.M. (2024) *Biofertilizers NEP Curricula*. Swaraj Prakashan, Delhi.

Gupta, M.M., Aggarwal, A., and Asha, A. (2018) From mycorrhizosphere to rhizosphere microbiome: the paradigm shift. In Giri, B., Prasad, R., and Varma, A. (eds) *Root Biology*. Springer International Publishing, Cham, Switzerland, pp. 487–500.

Gupta, V.V.S.R. and Germida, J.J. (1988) Distribution of microbial biomass and its activity in different soil aggregate size classes as affected by cultivation. *Soil Biology and Biochemistry* 20(6), 777–786. DOI: 101016/0038-0717(88)90082-X

Häfner, F., Hartung, J., and Möller, K. (2022) Digestate composition affecting N fertiliser value and C mineralisation. *Waste and Biomass Valorization* 13(8), 3445–3462. DOI: 101007/s12649-022-01754-8

Hodge, A., Campbell, C.D., and Fitter, A.H. (2001) An arbuscular mycorrhizal fungus accelerates decomposition and acquires nitrogen directly from organic material. *Nature* 413(6853), 297–299. DOI: 101038/35095041

Hodges, S.C. (2010) *Soil Fertility Basics*. North Carolina State University, Raleigh, North Carolina. Available at: https://www.online-pdh.com/file.php/562/FER_SG.pdf (accessed August 22, 2025).

Fukami, J., Cerezini, P., and Hungria, M. (2018) *Azospirillum*: benefits that go far beyond biological nitrogen fixation. *AMB Express* 8(1), 73. DOI: 10.1186/s13568-018-0608-1

Jones, D.L., Hodge, A., and Kuzyakov, Y. (2004) Plant and mycorrhizal regulation of rhizodeposition. *The New Phytologist* 163(3), 459–480. DOI: 101111/j.1469-8137.2004.01130.x

Lehmann, J., Bossio, D.A., Kögel-Knabner, I., and Rillig, M.C. (2020) The concept and future prospects of soil health. *Nature Reviews: Earth & Environment* 1(10), 544–553. DOI: 101038/s43017-020-0080-8

Liu, S., Wang, Q., Qian, L., Zhang, B., Chen, X. *et al.* (2022) Mapping the scientific knowledge of glomalin-related soil protein with implications for carbon sequestration. *Ecosystem Health and Sustainability* 8(1), 2085185. DOI: 10.1080/20964129.2022.2085185

Liu, X., Jiang, C., Qin, Y., Wang, C., Wang, J. *et al.* (2023) Production of biochar from squeezed liquid of fruit and vegetable waste: impacts on soil N2O emission and microbial community. *Environmental Research* 239, 117245. DOI: 101016/j.envres.2023.117245

Kibblewhite, M.G., Ritz, K., and Swift, M.J. (2008) Soil health in agricultural systems. *Philosophical Transactions of the Royal Society B* 363(1492), 685–701. DOI: 101098/rstb.2007.2178

Mickan, B.S., Ren, A.-T., Buhlmann, C.H., Ghadouani, A., Solaiman, Z.M. *et al.* (2022) Closing the circle for urban food waste anaerobic digestion: the use of digestate and biochar on plant growth in potting soil. *Journal of Cleaner Production* 347, 131071. DOI: 101016/j.jclepro.2022.131071

Nguyen, L.T.T., Osanai, Y., Lai, K., Anderson, I.C., Bange, M.P. *et al.* (2018) Responses of the soil microbial community to nitrogen fertilizer regimes and historical exposure to extreme weather events: flooding or prolonged-drought. *Soil Biology and Biochemistry* 118, 227–236. DOI: 101016/j.soilbio.2017.12.016

Pelayo Lind, O., Hultberg, M., Bergstrand, K.-J., Larsson-Jönsson, H., Caspersen, S. *et al.* (2021) Biogas digestate in vegetable hydroponic production: pH dynamics and pH management by controlled nitrification. *Waste and Biomass Valorization* 12(1), 123–133. DOI: 101007/s12649-020-00965-y

Pettit, R.E. (2004) Organic matter, humus, humate, humic acid, fulvic acid and humin: their importance in soil fertility and plant health. *CTI Research* 10, 1–7.

Pradhan, S., Mackey, H.R., Al-Ansari, T.A., and McKay, G. (2022) Biochar from food waste: a sustainable amendment to reduce water stress and improve the growth of chickpea plants. *Biomass Conversion and Biorefinery* 12(10), 4549–4562. DOI: 101007/s13399-022-02575-1

Pradhan, S., Parthasarathy, P., Mackey, H.R., Al-Ansari, T., and McKay, G. (2024) Food waste biochar: a sustainable solution for agriculture application and soil–water remediation. *Carbon Research* 3(1), 41. DOI: 10.1007/s44246-024-00123-2

Ren, A.T., Abbott, L.K., Chen, Y., Xiong, Y.C., and Mickan, B.S. (2020) Nutrient recovery from anaerobic digestion of food waste: impacts of digestate on plant growth and rhizosphere bacterial community composition and potential function in ryegrass. *Biology and Fertility of Soils* 56(7), 973–989. DOI: 10.1007/s00374-020-01477-6

Richardson, A.E. and Simpson, R.J. (2011) Soil microorganisms mediating phosphorus availability. *Plant Physiology* 156(3), 989–996. DOI: 101104/pp.111.175448

Ries, J., Chen, Z., and Park, Y. (2023) Potential applications of food-waste-based anaerobic digestate for sustainable crop production practice. *Sustainability* 15(11), 8520. DOI: 103390/su15118520

Rillig, M.C. and Mummey, D.L. (2006) Mycorrhizas and soil structure. *The New Phytologist* 171(1), 41–53. DOI: 101111/j.1469-8137.2006.01750.x

Rillig, M.C., Wright, S.F., and Eviner, V.T. (2002) The role of arbuscular mycorrhizal fungi and glomalin in soil aggregation: comparing effects of five plant species. *Plant and Soil* 238(2), 325–333. DOI: 101023/A:1014483303813

Sandhya, V.D., Shaik, Z.A., Grover, M., Reddy, G., and Venkateswarlu, B. (2009) Alleviation of drought stress effects in sunflower seedlings by the exopolysaccharides producing Pseudomonas putida strain GAP-P45. *Biology and Fertility of Soils* 46(1), 17–26. DOI: 10.1007/s00374-009-0401-z

Schuster, A. and Schmoll, M. (2010) Biology and biotechnology of Trichoderma. *Applied Microbiology and Biotechnology* 87(3), 787–799. DOI: 10.1007/s00253-010-2632-1

Sharma, P. and Tong, Y.W. (2025) Food waste-derived biofertilizers for agriculture sustainability. In: Ren, J. (ed.) *Waste-to-Energy*. Elsevier, Amsterdam, pp. 195–212.

Sharma, P., Tiong, Y.W., Yan, M., Tian, H., Lam, H.T. *et al.* (2023) Assessing *Stachytarpheta jamaicensis* (L.) Vahl growth response and rhizosphere microbial community structure after application of food waste anaerobic digestate as biofertilizer with renewable soil amendments. *Biomass and Bioenergy* 178, 106968. DOI: 101016/j.biombioe.2023.106968

Tiong, Y.W., Sharma, P., Xu, S., Bu, J., An, S. *et al.* (2024) Enhancing sustainable crop cultivation: the impact of renewable soil amendments and digestate fertilizer on crop growth and nutrient composition. *Environmental Pollution* 342, 123132. DOI: 10.1016/j.envpol.2023.123132

Wang, X., Feng, H., Wang, Y., Wang, M., Xie, X. *et al.* (2021) Mycorrhizal symbiosis modulates the rhizosphere microbiota to promote rhizobia-legume symbiosis. *Molecular Plant* 14(3), 503–516. DOI: 101016/j.molp.2020.12.002

Wu, S., Zhuang, G., Bai, Z., Cen, Y., Xu, S. *et al.* (2018) Mitigation of nitrous oxide emissions from acidic soils by *Bacillus amyloliquefaciens*, a plant growth-promoting bacterium. *Global Change Biology* 24(6), 2352–2365. DOI: 101111/gcb.14025

Young, I.M. and Crawford, J.W. (2004) Interactions and self-organization in the soil-microbe complex. *Science* 304(5677), 1634–1637. DOI: 101126/science.1097394

Zhang, M., Shan, S., Chen, Y., Wang, F., Yang, D. *et al.* (2019) Biochar reduces cadmium accumulation in rice grains in a tungsten mining area-field experiment: effects of biochar type and dosage, rice variety, and pollution level. *Environmental Geochemistry and Health* 41(1), 43–52. DOI: 101007/s10653-018-0138-8

Zhang, X.K., Li, Q., Liang, W.J., Zhang, M., Bao, X.L. *et al.* (2013) Soil nematode response to biochar addition in a Chinese wheat field. *Pedosphere* 23(1), 98–103. DOI: 101016/S1002-0160(12)60082-8

Zhang, Z., Qu, Y., Li, S., Feng, K., Wang, S. *et al.* (2017) Soil bacterial quantification approaches coupling with relative abundances reflecting the changes of taxa. *Scientific Reports* 7, 4837. DOI: 101038/s41598-017-05261-2

Zheng, Z., Ali, A., Su, J., Fan, Y., and Zhang, S. (2021) Layered double hydroxide modified biochar combined with sodium alginate: a powerful biomaterial for enhancing bioreactor performance to remove nitrate. *Bioresource Technology* 323, 124630. DOI: 101016/j.biortech.2020.124630

Zhou, J., Hong, W., Feng, J., Song, L., Li, X. *et al.* (2023) Effects of applying peanut shell and its biochar on the microbial activity and community structure of dryland red soil. *Heliyon* 9(2), e12604. DOI: 101016/j.heliyon.2022.e12604

3 Synergies of Microbial Inoculants and Agrifood Waste in Sustainable Farming

Manju M. Gupta*
Sri Aurobindo College, University of Delhi, Delhi, India

Abstract

Synergy—the cooperative interaction of multiple components yielding outcomes greater than the sum of their parts—is the theme of this chapter: integrating microbial inoculants with agrifood waste-derived amendments can measurably enhance soil fertility, crop nutrition, and sustainability. The chapter reviews major inoculant classes and their mechanisms: nitrogen fixers (rhizobia, *Azotobacter*, *Azospirillum*, cyanobacteria/*Azolla*), phosphorus-solubilizing microorganisms (bacteria and fungi, including *Penicillium*), arbuscular mycorrhizal fungi that extend root foraging and improve soil aggregation via glomalin, and plant-growth-promoting/biocontrol taxa (e.g. *Trichoderma*, *Bacillus*, *Pseudomonas*) that also mobilize micronutrients. It further discusses that thoughtfully designed consortia—combining complementary functions such as N fixation, P solubilization, lignocellulose degradation, and pathogen suppression—outperform single strains, particularly when paired with carbon- and nutrient-rich agro-waste substrates. It also highlights commercialization status, exemplar products, and circular-economy pathways for nutrient recovery (e.g. digestate, biochar, P recovery), alongside constraints in cost, quality assurance, and policy.

3.1 Introduction

Synergy, originating from the Greek word *synergia* ("cooperation"), embodies the idea of the joint action of multiple elements. Together they can achieve results exceeding the sum of their separate effects, emphasizing the strength of collaboration over isolation (Northouse, 2018). In the realm of microbial inoculants and agrifood waste-derived nutrients, the principle of synergy highlights the benefits of their integrated use. Agrifood waste offers a plentiful reservoir of nutrients required by the plant, which the microbial allies transform into simpler, usable forms through biodegradation, nutrient release, and uptake (Fig. 3.1). Microbial inoculants unlock plant-accessible forms of phosphorus, nitrogen, and micronutrients through mechanisms like nitrogen fixation, phosphate dissolution, and the enzymatic degradation of organic compounds (Sharma *et al.*, 2013). This natural partnership shows a clear and effective strategy to improve soil fertility.

Microbial inoculants and biofertilizers are not nutrient stores; rather, they help in the uptake of nutrients (see Chapter 1). Microbial inoculants and biofertilizers differ from compost, and the compost made from agrifood waste is

*Corresponding author: mbansalsac@outlook.com

DOI: 10.1079/9781836991021.0003

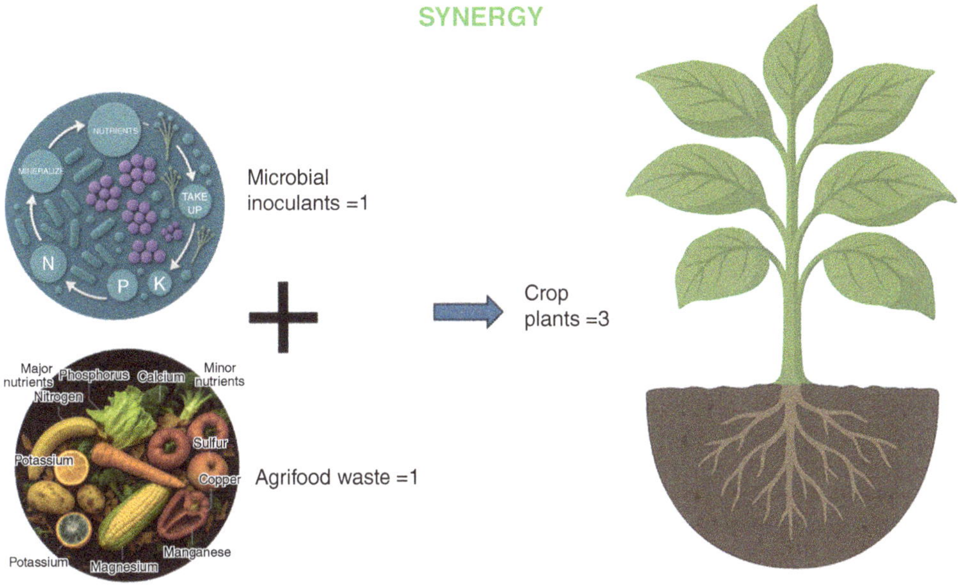

Fig. 3.1. Concept of synergy between microbial inoculants and agrifood waste. Figure author's own.

not included in biofertilizers. However, both composted agrifood waste and microbial inoculants are essential for sustainable farming. The former supplies essential nutrients directly to plants. The latter facilitates the breakdown of complex organic matter into simpler molecules, releases bound minerals, and enhances nutrient uptake by crops. Thus, together they offer a sustainable alternative to chemical fertilizers, which often lead to nutrient runoff and environmental degradation. Most microorganisms classified as biofertilizers or microbial inoculants have been commercialized and are in widespread use (Sharma *et al.*, 2013). Recently, O'Callaghan *et al.* (2022) critically reviewed the limitations of and opportunities offered by microbial inoculants.

Together, composted agrifood waste and biofertilizers reduce reliance on synthetic fertilizers, promote a circular economy, and support long-term soil health, making biofertilizers a cornerstone of eco-friendly farming practices. Several microorganisms are known to perform both the abovementioned functions; for example, *Trichoderma harzianum* facilitates both the uptake and degradation of nutrients from compost. Adding *Trichoderma harzianum* to compost made from agrifood waste speeds up the breakdown of complex organic materials. This results in simpler compounds like sugars and amino acids. *Trichoderma*'s mycoparasitic and plant growth-promoting activities boost nutrient uptake. It produces enzymes and siderophores for solubilizing micronutrients such as iron and zinc reported in crops like tomatoes and wheat (Bonanomi *et al.*, 2007). This dual action of *Trichoderma* improves soil fertility, supports root development, and exemplifies how microbial inoculants can maximize the benefits of compost in sustainable agriculture.

Microbial consortia are groups of two or more microbial species that live and interact within a shared environment. The outcome of their interactions depends on the purpose for which they are combined. They may be natural or synthetic. For example, plant growth-promoting microbial consortia include bacteria and fungi, which interact to increase the growth of the plant. Microbial consortia are essential for breaking down agrifood waste and need to be specially designed for this purpose. A microbial consortium in agrifood waste composting might feature specialized bacteria and fungi that target components like cellulose and lignin. Their interaction should streamline the decomposition process in such a way as to yield high-quality compost more rapidly than single-strain methods (Bonanomi *et al.*, 2007). Their synergistic interactions accelerate the decomposition and release of complex organic matter into plant nutrients, which are readily absorbed by plants (Nanda *et al.*, 2025). This chapter investigates the primary categories of commercialized microbial inoculants—nitrogen-fixing bacteria, phosphate-solubilizing microorganisms, plant growth-promoting microorganisms, and microbial consortia. It further details the specific microorganisms, their commercial names, and the mechanisms they employ to improve soil fertility. It also analyzes their connection with agrifood waste and highlights recent progress in their synergistic applications, reinforcing their value in sustainable farming practices.

3.2 Nitrogen Nutrition

Nitrogen (N) nutrition is essential for plant growth, serving as a key component of proteins, chlorophyll, and nucleic acids. The optimal crop yields often requires 100–200 kg N/ha depending on soil and crop type (Harris *et al.*, 2023). When there is not enough nitrogen, both crop yield and the amount of nitrogen in the grains are reduced (Lal, 2020). Although nitrogen makes up about 78% of the Earth's atmosphere, plants cannot directly use this form. Nitrogen fixation is the biological or chemical process that converts inert atmospheric nitrogen (N_2) into ammonia (NH_3) or other usable forms that plants can absorb and use for growth. In the biological context, this process is primarily carried out by certain microorganisms, such as nitrogen-fixing bacteria like *Rhizobium* and *Azotobacter*, which possess the enzyme nitrogenase. Nitrogen fixation by soil microbes can provide 300–400 kg N/ha/year (Daniel *et al.* 2022) and increase crop yield by 10–50% (Kawalekar, 2013). Exudates from plant roots are released into the soil, attracting bacteria and facilitating colonization and nitrogen fixation inside the plant rhizosphere (Bansal and Mukerji, 1996; Bansal *et al.*, 2000). These bacteria form symbiotic or asymbiotic associations with plant roots, facilitating nitrogen fixation, improving nutrient availability, and ultimately enhancing plant growth and soil fertility. Such organisms are used as microbial inoculants to fix atmospheric nitrogen and are commercially available for agricultural application. The main groups of nitrogen-fixing microbial inoculants, along with their commercial names, functions, and associated benefits, are summarized below.

3.2.1 The rhizobial inoculants

Rhizobial inoculants are the most common and widely used biofertilizers around the world. These products contain nitrogen-fixing bacteria from the *Rhizobium* group, which live in symbiosis with legume plant roots. *Rhizobium* inoculation is a well-established practice and is often considered a major success in applied soil microbiology (Catroux *et al.*, 2001). It is estimated that approximately 2000 t of inoculants are produced worldwide annually, a quantity sufficient to inoculate 20 million hectares of legumes. The largest single producers are in the USA, with an annual production of 1000 t (Singleton *et al.*, 1997).

Rhizobium, nitrogen-fixing bacteria, form symbiotic relationships with leguminous plants. Residing in root nodules, they use the enzyme nitrogenase to convert atmospheric nitrogen into ammonia, supplying the plant with nitrogen for protein synthesis. In exchange, the plant provides the bacteria with nutrients, such as carbohydrates. Rhizobial inoculants can fix 50–100 kg of nitrogen per hectare annually, reducing the need for synthetic fertilizers, enhancing soil fertility, and improving crop yields in sustainable agriculture systems (Gupta, 2024). Soybean inoculation with nitrogen-fixing *Bradyrhizobium* spp. is done in Brazil. This could meet the nitrogen demand of the crop, eliminating the need for chemical nitrogen fertilizers, which would save US$13 billion per year, in addition to reducing emissions of greenhouse gases (Zilli *et al.*, 2021). *Rhizobium*-based biofertilizers are widely used not only in Brazil but globally, leveraging the bacteria's symbiotic relationship with leguminous plants to fix atmospheric nitrogen into ammonia, enhancing plant growth and soil fertility while reducing the need for chemical fertilizers.

Rhizobial inoculants encompass not only the genus *Rhizobium* but also other nitrogen-fixing bacteria within the family Rhizobiaceae and related genera in the order Rhizobiales (Gupta, 2024). These bacteria are chosen for their ability to form symbiotic relationships with specific leguminous plants. For instance, *Rhizobium leguminosarum* is highly effective for legumes like peas, lentils, and vetches, forming nitrogen-fixing nodules in temperate soils. *Rhizobium etli* is specialized for common beans, enhancing their nitrogen uptake. Similarly, *Rhizobium phaseoli*, a close relative of *R. etli*, is used to boost bean crop yields. *Rhizobium tropici*, suited to tropical and subtropical climates, nodulates beans and other legumes under challenging conditions like high temperatures and acidic soils.

Various *Rhizobium* strains are commercially available, tailored to specific host plants or environmental conditions, ensuring optimal nitrogen fixation and crop productivity. These rhizobial species are critical components of biofertilizer products, selected for their compatibility with specific legume crops and their ability to form effective root nodules that convert atmospheric nitrogen into forms usable by plants. The choice of strain for inoculation depends on the legume species being grown and local soil conditions to ensure efficient symbiotic performance (Harris *et al.*, 2023). However, they differ in growth rate, genomic characteristics, preferred host plants, and their taxonomic classification. For example, *Bradyrhizobium japonicum*, a slow-growing species, is commonly used for soybean cultivation and can fix substantial amounts of nitrogen—up to 100 kg/ha—in a wide range of soil types (O'Callaghan *et al.*, 2022). *Bradyrhizobium elkanii* is notable for its tolerance to acidic soils and its high nitrogen fixation potential, making it useful in challenging environments. *Sinorhizobium meliloti* is primarily associated with alfalfa and sweet clover. It performs well in alkaline soils and arid regions. *Mesorhizobium loti* forms symbiotic relationships with legumes such as lotus and trefoil and is well suited for pasture systems in diverse climates. *Mesorhizobium ciceri* is specific to chickpeas and is commonly used in semi-arid agricultural systems to improve nitrogen availability and crop yield.

Some common rhizobial inoculants include HiStick (with *Bradyrhizobium japonicum* for soybeans), Nitragin Gold (for soybeans), Cell-Tech (with *Sinorhizobium meliloti* for alfalfa and clover), and TagTeam (combining *Mesorhizobium ciceri* and phosphorus-solubilizing fungi for chickpeas) (Table 3.1). These products are widely used in sustainable agriculture to reduce dependence on chemical nitrogen fertilizers and enhance soil health. Two main situations where inoculation is specifically useful are (i) when there are no indigenous strains of the required rhizobia in soil, and (ii) when the level of the indigenous population is low (O'Callaghan *et al.* 2022). Rhizobia may also be absent due to severe detrimental soil conditions, especially soil pH. Acidic soils are generally free of *Sinorhizobium meliloti*, the symbiont of alfalfa, and basic soils are free of *Bradyrhizobium* sp. (Lupinus), the symbiont of lupins (Amarger, 1980). Other soil conditions like high temperature, dryness and salinity can also be detrimental to the occurrence of the required rhizobial populations.

3.2.2 The *Azotobacter* inoculants

Azotobacter, a free-living, aerobic, nitrogen-fixing bacterium commonly found in neutral to alkaline soils, benefits non-leguminous crops like cereals, vegetables, and cotton. These bacterial species, such as *Azotobacter chroococcum*, fix atmospheric nitrogen (up to 20–30 kg/ha/year) and produce growth-promoting substances like gibberellins and auxins, which stimulate root development and plant growth. *Azotobacter* also improves soil health by secreting polysaccharides that enhance soil aggregation and water retention, making it a valuable biofertilizer for improving fertility in diverse cropping systems.

Azotobacter inoculants thrive in soils enriched with agrifood waste, such as straw or vegetable trimmings, which supply the organic carbon needed for their nitrogen-fixing activity. The decomposition of agrifood waste by other soil microbes releases nutrients that *Azotobacter* utilizes, while its production of polysaccharides improves soil aggregation, enhancing the waste's beneficial effects on soil structure (Sumbul *et al.*, 2020). This synergy increases nutrient availability and water retention for crops, promoting a circular economy by transforming waste into a fertility-boosting resource.

3.2.3 The *Azospirillum* inoculants

Strains of *Azospirillum*, such as *Azospirillum brasilense*, are associative nitrogen-fixing bacteria that colonize the root surfaces of grasses, cereals,

Table 3.1. Some commercially available nitrogen-fixing products and microorganisms. Table author's own.

Microbial group	Product name	Active microorganism(s)	Manufacturer/ supplier	Website
Rhizobium	Nitragin®	*Rhizobium* spp. (legumes: soybean, peas, etc.)	Nobbe and Hiltner	https://www.novonesis.com/en/biosolutions/animal-and-plant/plant/forages/nitragin-gold-alfalfa-and-sweet-clover
Rhizobium	Biofix®	*Rhizobium* spp. (legumes: soybean, alfalfa)	Novozymes (Brazil/Global)	https://www.novonesis.com/en/biosolutions/animal-and-plant/plant/think-biological
Rhizobia	Rhizo Liq	*Bradyrhizobium* sp.	MBFi Group	https://mbfi.co.za/
Rhizobia	Vault HP	*Bradyrhizobium* sp.	BASF	https://agriculture.basf.us/content/dam/cxm/agriculture/crop-protection/products/documents/Vault_HP_TIB2019.pdf
Rhizobia	Nitragin® Gold	*Sinorhizobium meliloti*	Novozymes (Brazil/Global)	https://www.novonesis.com/en/biosolutions/animal-and-plant/plant/forages/nitragin-gold-alfalfa-and-sweet-clover
Azospirillum	MicroAZ-IF Liquid	*Azospirillu A* sp.	Terramax	https://terramaxag.com/product/corn/microaz-if-liquid/
Azotobacter	Azotobacter Biofertilizer	*Azotobacter* spp.	Indian Farmers Fertiliser Cooperative Limited (IFFCO)	https://www.iffco.in
Rhizobia	HiStick	*Bradyrhizobium japonicum* (for soybeans)	BASF	https://www.agro.basf.co.za/en/Products/Overview/Inoculant/HiStick.html
Rhizobia	Cell-Tech	*Sinorhizobium meliloti* (for alfalfa and clover)	FMC	https://ag.fmc.com/ca/en/biologicals/cell-tech

and other non-leguminous plants like wheat and maize. Associative nitrogen-fixing bacteria live in close association with plant roots, particularly in the rhizosphere, and convert atmospheric nitrogen into ammonia, which plants can use for growth, without forming specialized structures like root nodules. Unlike symbiotic nitrogen-fixing bacteria (e.g. *Rhizobium*) and free-living bacteria like *Azotobacter*, these bacteria are loosely associated with plants, enhancing soil fertility and plant nutrition through nitrogen fixation and other growth-promoting activities (Gupta, 2024). They fix 20–40 kg of nitrogen per hectare annually and produce plant growth hormones like auxins, cytokinins, and gibberellins, which promote root elongation and increase nutrient uptake. By enhancing root biomass and soil microbial activity, *Azospirillum* improves drought tolerance and nutrient efficiency, making it an effective biofertilizer for sustainable agriculture in water-scarce regions. MicroAZ-IF Liquid™ is an example of a commercial *Azospirillum* corn inoculant, designed to fix atmospheric nitrogen, stimulate root growth, and increase crop yields.

3.2.4 Blue-green algal inoculants

Blue-green algal inoculants, also known as cyanobacteria (e.g. *Nostoc*, *Anabaena*), are photosynthetic microorganisms used primarily in wetland rice cultivation to fix atmospheric nitrogen and enrich soil fertility. Capable of fixing 20–30 kg of nitrogen per hectare per season, they thrive in flooded paddy fields, forming a green mat on the soil surface that also adds organic matter upon decomposition. Cyanobacteria release extracellular polysaccharides that improve soil structure and water retention, while their nitrogen-fixing ability reduces dependency on chemical fertilizers, supporting eco-friendly rice farming.

Blue-green algal inoculants, such as *Nostoc*, exhibit synergy with agrifood waste in rice paddies when waste like rice straw or fish processing residues is incorporated into the soil (Venkataraman, 2017). The organic matter from these wastes supports cyanobacterial growth, providing a carbon source that enhances their nitrogen fixation (20–30 kg/ha/season). In return, cyanobacteria add fixed nitrogen and extracellular polysaccharides to the soil, improving fertility and structure, while the waste decomposition contributes additional organic matter, fostering a nutrient-rich environment for rice cultivation. Widely used in India for rice paddies, it is produced by research institutions like the Indian Agricultural Research Institute (IARI), often containing *Anabaena* and *Nostoc*.

3.2.5 *Azolla-Anabaena* inoculants

Azolla-Anabaena inoculants involve a symbiotic relationship between the aquatic fern *Azolla* and the nitrogen-fixing cyanobacterium *Anabaena azollae*, commonly used in rice paddies. Azolla grows rapidly on water surfaces, and *Anabaena* within its leaf cavities fixes 40–60 kg of nitrogen per hectare per crop cycle, providing a natural nitrogen source for rice (Gupta, 2024). As a green manure, *Azolla* decomposes to release nutrients and organic matter, improving soil fertility, suppressing weeds, and reducing the need for synthetic fertilizers in wetland agriculture.

3.2.6 Agrifood waste as a source of nitrogen

Agrifood waste, such as crop stalks, vegetable trimmings, fruit peels, and livestock manure, contains 1–3% nitrogen by dry weight, making it a valuable resource for improving soil fertility. Through processes like composting or anaerobic digestion, this nitrogen is stabilized and gradually released into the soil, delivering 20–50 kg of nitrogen per hectare per season as a sustainable alternative to synthetic fertilizers. Combining agrifood waste with nitrogen-fixing microbes like *Rhizobium* boosts nitrogen availability, supporting a circular economy by recycling nutrients and reducing greenhouse gas emissions. Microalgae and cyanobacteria, with up to 70% protein content in their biomass, can also effectively recover nitrogen from waste, further enhancing nutrient recycling for agricultural systems.

3.3 Phosphorus Nutrition

Agricultural soils often contain large reserves of phosphorus, but much of it is in forms unavailable for plant uptake. Soil phosphorus exists primarily as organic phosphorus and inorganic phosphorus. Organic phosphorus is incorporated into the structures of living organisms and dead organic matter in the soil, such as plant and animal residues including agrifood waste, and it is not directly available for plant uptake. It needs to be converted into an inorganic form through a process called mineralization. Soil microorganisms play a crucial role in mineralization of organic matter, releasing inorganic phosphorus that plants can absorb. Plant-available, inorganic phosphorus is present in soil solution (see Chapter 2). This includes orthophosphate ions ($H_2PO_4^-$ and HPO_4^{2-}) dissolved in soil water and the sorbed phosphorus, which is inorganic phosphorus attached to soil particles.

Due to current issues such as scarcity of phosphorus supply, the high cost of extraction, processing, and transportation of phosphorus fertilizers, and the negative environmental impact on water quality from excess phosphorus fertilizers, there is growing attention on utilizing the phosphorus already present in soil more effectively. Soil microorganisms, including bacteria, e.g. phosphate-solubilizing microorganisms (PSMs), and fungi, e.g. arbuscular mycorrhizal fungi (AMF), are key players in the biogeochemical cycling of both inorganic and organic phosphorus in the rhizosphere. They possess the ability to convert these unavailable forms of phosphorus into readily available forms, primarily orthophosphate ions, which plants can then absorb and utilize for growth. In addition to discussing the importance of PSMs and AMFs as phosphorus-solubilizing biofertilizers, this section also discusses the synergy and nutrient storage potential of agrobiomass.

3.3.1 Phosphate-solubilizing microorganisms

PSMs, a group of organisms composed of actinobacteria, bacteria such as *Bacillus* and

Pseudomonas, and fungi, arbuscular mycorrhizae, and cyanobacteria, enhance soil fertility by converting insoluble phosphorus compounds into plant-available forms through the production of organic acids like citric and gluconic acid (Richardson, 2001; Richardson and Simpson, 2011). These microorganisms can solubilize 30–50 kg of phosphorus per hectare, addressing phosphorus deficiency in soils where it is often fixed and unavailable (Gupta, 2024). PSMs also produce growth-promoting substances and improve soil microbial activity, making them effective biofertilizers for a wide range of crops, including cereals and vegetables.

More progress has been made with phosphorus solubilizing fungi. For example, jumpStart® (Bayer Crop Science) is a biofertilizer containing *Penicillium bilaii*, recommended for wheat and canola, with Canadian wheat growers reporting average yield increases of ~6%, and up to 66% in some cases (Harvey *et al.*, 2009), though other studies suggest more modest benefits (Karamanos *et al.*, 2010).

PSMs also work with agrifood waste by utilizing the organic acids and carbon sources in residues such as fruit peels or manure to enhance their activity (Kim *et al.*, 1998). PSMs solubilize phosphorus from soil and waste-bound phosphates, making it available to plants, and the waste provides a substrate for microbial growth. This synergy improves phosphorus availability in nutrient-poor soils, reduces reliance on chemical fertilizers, and recycles agrifood waste into a valuable soil amendment (Hart *et al.*, 2022). When it comes to agrifood waste, animal residues are generally considered the richest sources of phosphorus (Hart *et al.*, 2022). Specifically, animal bones are rich in hydroxyapatite, a mineral containing calcium and phosphate.

Several PSMs are commercially available and are widely incorporated into biofertilizer formulations due to their ability to enhance nutrient availability and support plant growth. These formulations often include specific strains marketed under various commercial or brand names. Examples of such microorganisms and their corresponding products are provided in Table 3.2, highlighting their practical application in sustainable agriculture. A long list of microbes has been reported with functions of PSMs; however, not many are in commercial production (Raymond *et al.* 2021; Ramos Cabrera *et al.*, 2024). According to the National Institutes of Health (NIH), the high cost of production and competition from indigenous soil microbes are the main hurdles hindering the widespread adoption of PSMs (Raymond *et al.* 2021). PSMs also have the potential to enhance phosphate-induced immobilization of metals to remediate contaminated soil (Yuan *et al.*, 2017). Phosphate compounds are capable of immobilizing heavy metals, especially lead (Pb), in contaminated environments through phosphate-heavy metal precipitation. However, most phosphorus compounds are not readily soluble in soils, so it is not readily used for metal immobilization (Yuan *et al.*, 2017). But PSM can help in heavy-metal remediation to an extent this way.

3.3.2 Arbuscular mycorrhizal fungi

AMF form mutualistic relationships with the roots of most plants, including key crops like maize, wheat, and legumes (Smith and Read, 2010). By creating extensive hyphal networks, AMF significantly enhance the uptake of phosphorus, water, and micronutrients such as zinc and copper, often contributing the equivalent of 50–100 kg of phosphorus per hectare (Gupta, 2024). As natural biofertilizers, AMF supply plants with essential nutrients and protect against pathogens in exchange for photosynthetic products. Additionally, AMF improve soil structure through the production of glomalin, reducing reliance on chemical phosphorus fertilizers and promoting sustainable crop production (Rillig and Mummey, 2006; Gupta and Abbott, 2021).

AMF are established as one of the most widespread and ecologically significant symbioses with terrestrial plants, fundamentally shaping nutrient cycling and ecosystem dynamics (Smith and Read, 2010; Willis *et al.*, 2013). This mutualistic relationship depends on a finely orchestrated interaction of specialized fungal structures designed for efficient nutrient absorption and strategic resource storage. Central to this exchange are the arbuscules, intricate tree-like fungal formations within the plant root cortical cells, which serve as the primary interface for bidirectional nutrient transfer (Fig. 3.2; Luginbuehl and Oldroyd, 2017). The fungus delivers essential nutrients (phosphorus, nitrogen, micronutrients) to the plant across the periarbuscular membrane. In return, it receives carbon compounds

Table 3.2. Phosphorus-solubilizing microorganisms with their commercial names. Table author's own.

Microbial group	Species/product name	Active microorganism(s)	Manufacturer/supplier	Function	Website
Bacillus and *Pseudomaons*	Katyayani Phosphate Solubilizing Bacteria Biofertilizer	*Bacillus* and *Pseudomonas* spp. (CFU: minimum 2 x 10^8 per ml)	Katayani Orgainics	Converts insoluble phosphorus to plant-usable form	https://www.katyayaniorganics.com/
Bacillus spp.	CBF	*Bacillus mucilaginosus*, *B. subtilis*	China Bio-Fertilizer AG	Biofertilizer, nutrient solubilization	
Bacillus spp.	Symbion van plus	*Bacillus megaterium*	T. Stanes and Co. Ltd	Phosphate solubilization	https://www.tstanes.com/domestic-crop-care/bio-fertilizers/symbion-vam-plus
Azotobacter sp. and *Bacillus* spp.	Life Force Bio-P	*Azotobacter* sp. and *Bacillus subtilis*	Microbial Solutions (Psty) Ltd	Biofertilizer, nutrient cycling	https://www.microbial-solutions.com/
Bacillus spp.	InomixR	*Bacillus polymyxa*, *B. subtilis*	Lab (Labiotech)	Biofertilizer, nutrient solubilization	
Penicillium spp.	Penicillium bilaii	*Penicillium bilaii*	Novobac	Phosphate solubilization and a pet-safe fungicide	https://www.novobac.com

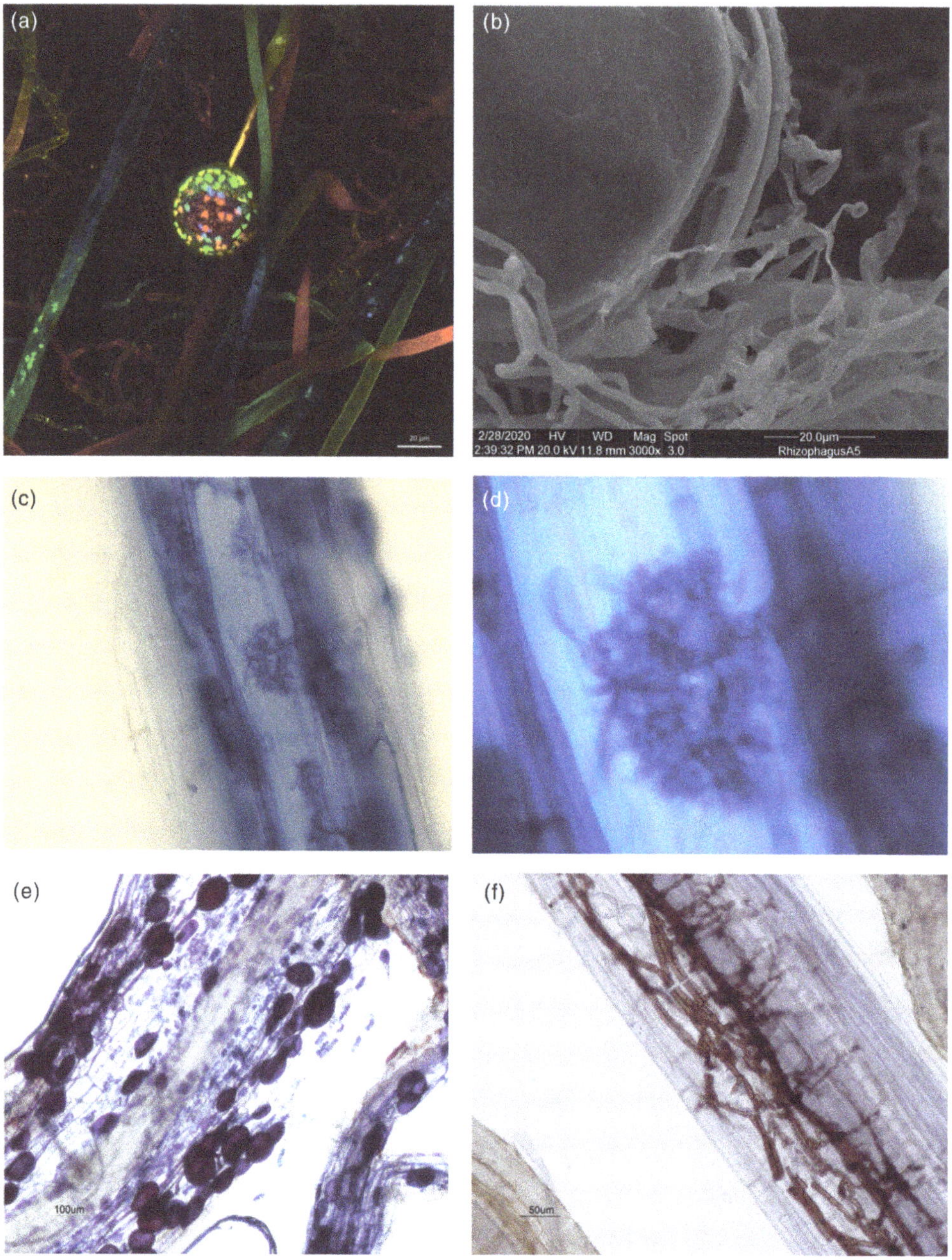

Fig. 3.2. Clockwise from top left. (a) AMF spores: Confocal microscopy visualizing nuclei within spores and hyphae of the species *Rhizophagus irregularis* (Pseudo-coloration has been applied for depth recognition). (b) Scanning electron microscopy showing hyphae and spore of *R. irregularis*. (c) Brightfield microscopy visualizing intraradical colonization of the species *R. irregularis* in flax roots (AM fungal tissue was stained using trypan blue). (d) An arbuscule enlarged. (e) Root clearing preparation under light microscopy showing extensive hyphae and vesicles (stained with trypan blue). (f) Dark septate hyphae inside the root. Photo credits: (a–d) Vasilis Kokkoris @Amsterdam Institute for Life and Environment (A-LIFE), Vrije Universiteit Amsterdam. (e–f) Sedona Spun@ Northern Arizona University, USA.

(mostly lipids) from the plant (Smith and Read, 2010). Beyond this active exchange, AMF exhibit remarkable capacities for nutrient reserve, ensuring fungal persistence and sustained plant benefit. Vesicles are thick-walled, lipid-rich structures (Fig. 3.2). They store carbon (as triacylglycerides) and phosphorus, buffering against nutrient shortages (Olsson *et al.*, 2011). Fungal hyphae form an extensive network in the soil. This network transports nutrients and stores them temporarily, efficiently distributing resources (Rillig and Mummey, 2006; Gupta and Abbott, 2021). AM fungal spores, crucial for dispersal and survival, are nutrient-dense. They contain lipids and carbohydrates like glycogen and trehalose, essential for germination and initial growth before finding a host. Dark septate endophytic fungi (DSE) improve host plant growth and nutrient uptake, similar to mycorrhizal symbiosis. However, the nature of their relationship remains unclear, potentially ranging from mutualistic to parasitic (Ruotsalainen *et al.* 2022). DSEs are typically defined as ascomycetous fungi that form melanized septate hyphae and, sometimes, micro-sclerotia within living roots (Fig. 3.2f). They are very often confused with AMF.

Several internationally recognized companies—such as Mycorrhizal Applications, Valent USA, Symborg, Koppert, PremierTech, and GroundWork BioAG—have commercialized AMF inoculants and offer them under brand names like *MycoApply*®, *MycoUp, Panoramix,* and *Rootella* (Table 3.3). These products, often containing well-characterized species like *Rhizophagus irregularis, Funneliformis mosseae,* and *Claroideoglomus etunicatum*, are widely marketed

Table 3.3. Commercial AMF inoculants by global biofertilizer suppliers. Table author's own.

Company (region)	Product	Form	AMF species	Website
Ecological Resources / Oikos (USA/Chile)	Oiko-Rhiza-E	powder	*Rhizophagus irregularis, Funneliformis mosseae, Glomus deserticola, Rhizophagus clarus*	https://oikos.cl/
Mycorrhizal Applications (USA/Chile)	MycoApply® EndoMaxx	granular	*Rhizophagus aggregatum, Rhizophagus irregularis, Funneliformis mosseae, Claroideoglomus etunicatum*	mycorrhizae.com
Valent USA Agricultural Products (USA)	MycoApply® EndoMaxx	powder	*Rhizophagus aggregatum, Rhizophagus irregularis, Funneliformis mosseae, Claroideoglomus etunicatum*	valentbiosciences.com
Symborg Inc. (Spain)	MycoUp Activ	powder	*Rhizophagus iranicum*	symborg.com
PremierTech (Canada)	Activ field crops granular	granular	*Rhizophagus irregularis*	premiertech.com
Sustane Natural Fertilizer (USA)	Sustane 3-7_2 w/ Mycorrhizae and Humates	granular	*Rhizophagus aggregatum, Claroideoglomus etunicatum, Glomus deserticola, Rhizophagus clarus*	sustane.com
JH Biotech (USA)	MYCORMAX Biological Inoculum	powder	*Rhizophagus irregularis, Funneliformis mosseae*	jhbiotech.com
Plant Health Care (Mexico)	PHC VAM.PWI	powder	*Rhizophagus irregularis, Claroideoglomus etunicatum*	https://phcmexico.com.mx/ft-vam-pwi/
Vegalab Inc. (USA)	Vegalab MYCO BIOBOOST	powder	*Rhizophagus irregularis*	vegalab.com
GroundWork BioAG (Israel)	Rootella G	granular	*Rhizophagus irregularis*	groundworkbioag.com
Agronutrition (France)	CONNECTIS	liquid	*Rhizophagus irregularis*	agronutrition.com

as sustainable solutions for improving plant nutrient uptake, especially phosphorus. However, despite their proven efficacy in controlled environments and pot trials, their performance under real field conditions remains inconsistent (Gupta and Abbott, 2021). Factors such as native soil microbial competition, environmental variability, crop genotype, and inoculant formulation often diminish their effectiveness (Gupta and Abbott, 2021). This gap between laboratory success and field application continues to challenge the broader adoption of AMF-based biofertilizers in mainstream agriculture.

Despite the gap in field success and impact of AMF being reported as context dependent, the market for AMF inoculants is expanding consistently, with commercial products primarily utilizing strains from the Glomeraceae family, such as *Rhizophagus*, *Funneliformis*, and *Claroideoglomus* (Basiru and Hijri, 2022). A study by Basiru *et al.* (2020) analyzing 68 mycorrhizal products from 28 manufacturers across Europe, America, and Asia revealed that 90% of these products are solid formulations—65% in powder form and 25% granular—while only 10% are liquid. Research indicates that single-species AMF inoculations often result in greater shoot biomass compared to multi-species inoculations (Berruti *et al.*, 2016). For instance, Gosling *et al.* (2016) found that under controlled greenhouse conditions with a single stress factor, AMF optimized for that stress maximize plant benefits, whereas diverse AMF communities are more effective in complex field environments (Wagg *et al.*, 2015). This suggests that the specific composition of AMF species, rather than their diversity, primarily drives functional outcomes.

AMF also influence soil microorganisms in the mycorrhizosphere, a distinct soil zone surrounding their extraradical mycelium (Gupta *et al.*, 2018). Interactions between AMF and soil microbes can be positive, negative (Albertsen *et al.* 2006), or neutral (Olsson *et al.*, 1996), affecting microbial biomass and the growth of specific taxa (Marschner and Timonen, 2006). Certain bacterial species exhibit specificity with particular AMF, fostering indirect synergistic effects that enhance plant growth, nutrient acquisition, and root branching (Marschner and Timonen, 2006). Developing inoculum from native AMF is said to be the most effective, environmentally friendly, and sustainable method of inoculation with these fungi (Hart *et al.*, 2018; Gupta and Abbott, 2021).

While AMF inoculants containing multiple strains can produce additive or synergistic effects on crops, increasing AMF diversity may not yield additional benefits if the strains perform redundant functions. Combining AMF with other bioinoculants, such as plant growth-promoting rhizobacteria, nitrogen-fixing bacteria, or organic compounds like humic acid, can enhance plant responses (Gao *et al.*, 2020). However, few commercial products incorporate such complex formulations (Basiru *et al.*, 2020). The use of AMF in microbial consortia for inoculation is explored in greater detail in the upcoming section.

3.3.3 Agrifood waste as a source of phosphorus

Agrifood waste represents a valuable resource for phosphorus recovery, offering a sustainable alternative to traditional phosphorus sources and contributing to a circular economy in agriculture. Agrifood waste contains considerable amounts of phosphorus, especially in forms like orthophosphate, phytic acid, and phosphate diesters (Xie *et al.*, 2023). Unlike mined phosphate rock, which is a finite resource, agrifood waste is continuously generated, making it a renewable source of phosphorus. Recycling phosphorus from agrifood waste prevents nutrient runoff, a major cause of water pollution and eutrophication, protecting ecosystems (Mokjatturas *et al.*, 2025). Recovering phosphorus from agrifood waste can reduce waste management costs and generate revenue by producing marketable products like fertilizers and animal feed. Puraloop® (https://www.icl-group.com) is such an innovative fertilizer created by recovering phosphorus from organic waste streams. It represents a forward-thinking solution to the global phosphorus challenge and, by transforming waste into a valuable resource for agriculture, is leading the way in sustainable development in line with the principles of the circular economy.

Despite this potential, phosphorus recovery from agrifood waste faces several challenges. Recovered phosphorus generally cannot compete economically with the relatively low cost of mined phosphate rock. Challenges also include issues with the quality of recovered phosphorus products, the need for further research on optimizing recovery processes, and the importance of implementing supportive policies and raising public awareness (Xie *et al.*, 2023). Supportive policies and greater public awareness are also critical to overcoming these challenges and promoting sustainable phosphorus management.

3.4 Plant Growth Promotion and Protection against Soil Pathogens

The concept of plant growth promotion and pathogen protection involves enhancing plant development through microbial inoculation, which boosts growth and shields plants from diseases. These microbes operate not only in soil but also in agrifood waste substrates. Plant growth-promoting rhizobacteria (PGPR), a diverse bacterial group, colonize plant roots and foster growth through mechanisms such as producing phytohormones (e.g. auxins, cytokinins, gibberellins), solubilizing nutrients like phosphorus and iron, and suppressing soil-borne pathogens via antibiotics or siderophores. Notable PGPR include bacteria like *Pseudomonas fluorescens*, *Bacillus subtilis*, *Azospirillum brasilense*, and *Rhizobium leguminosarum*, as well as fungi like *Trichoderma harzianum* and actinomycetes such as *Streptomyces griseus* and *Frankia* spp.

These microbes can boost crop yields by 10–20% and enhance stress tolerance, making them effective biofertilizers. Concurrently, these microbes combat soil-borne pathogens (e.g. *Fusarium oxysporum*, *Rhizoctonia solani*, *Pythium ultimum*) through mechanisms like resource competition, antibiotic synthesis, and induced systemic resistance, reducing disease incidence by 20–40% in crops such as tomatoes, wheat, and rice. In South America, over 100 products use *Azospirillum* bacteria, mainly for wheat and maize, though they're suggested for 16 different crops (O'Callaghan *et al.*, 2022). The yield benefits vary due to factors like nitrogen fertilizer use, plant type, and the specific *Azospirillum* strain (Veresoglou and Menexes, 2010). While *Azospirillum* can't provide all the nitrogen needed for crops like cereals and maize, it can cut fertilizer use by 25–50% (Santos *et al.*, 2019).

Many microorganisms that promote plant growth also serve as biocontrol agents, making it challenging to distinguish the mechanisms driving enhanced growth, as some inoculants likely provide multiple benefits. For instance, *Trichoderma* species boost plant growth through various means, such as controlling soil-borne pathogens, strengthening plant resistance to pathogens and insects, and promoting root development (Schuster and Schmoll, 2010). *Trichoderma* strains are key components in numerous agricultural products sold globally (Woo and Pepe, 2018). Similarly, *Pseudomonas chlororaphis*, the active ingredient in various seed treatments like Cedomon®, Cerall®, and Cedress® for barley, wheat, and peas, respectively, enhances root growth, supports early plant establishment, and combats several plant pathogens (Lantmännen BioAgri, 2021).

Many plant growth-promoting microbial inoculants confer the growth benefit by enhancing the availability and uptake of micronutrients in soil like iron, zinc, copper, manganese, and molybdenum (Gorain *et al.*, 2022). These microorganisms improve micronutrient dynamics through various mechanisms, supporting sustainable agriculture and soil health (Gupta, 2020). Mycorrhizal fungi, such as *Rhizophagus intraradices*, extend the root system via hyphal networks, increasing the surface area for micronutrient absorption by up to tenfold. These fungi enhance uptake of zinc, copper, and manganese from soil particles and organic matter, including compost, delivering 20–50% more micronutrients to plants like maize and wheat compared to non-mycorrhizal roots (Ibiang *et al.*, 2018). Another example of two commercial biofertilizers used for micronutrient nutrition are Zn Sol B® (*Thiobacillus novellus*) and Mn Sol B® (*Penicillium citrinum*), which are used for zinc and manganese mobilization, respectively (https://agrilife.in/subcat.php). Silicate-solubilizing bacteria, such as *Bacillus mucilaginosus*, enhance soil fertility by breaking down silicate minerals in the soil, releasing potassium, silicon, and other micronutrients for plant uptake (Sharma *et al.*, 2023). They produce organic acids and polysaccharides that weather silicate rocks, making 20–40 kg of potassium per hectare available annually, which is crucial in potassium-deficient soils. Silicate solubilizing bacteria also improve plant resistance to pests and diseases.

When combined with agrifood waste, microbial inoculants like *Trichoderma harzianum* degrade organic matter, releasing bound micronutrients such as zink and iron. This process can increase micronutrient availability by 10–20% in compost-amended soils, as the microbes unlock nutrients trapped in lignocellulosic structures (Ali *et al.*, 2022). This synergy boosts soil health, minimizes reliance on chemical fertilizers, and converts agrifood waste into a valuable asset for sustainable crop production across diverse crops such as tomatoes, rice, and potatoes.

There are large numbers of commercially available products representing a diverse array of microbial inoculants used in sustainable agriculture for disease suppression, nutrient cycling, and overall plant health enhancement. The listed biocontrol agents, particularly strains of *Trichoderma* and *Bacillus*, are widely incorporated into integrated pest management strategies (Table 3.4). Additionally, AMF such as *Rhizophagus irregularis* and *Funneliformis mosseae* offer valuable symbiotic benefits, though their field effectiveness remains variable and often depends on environmental compatibility and formulation quality.

In conclusion, microbial inoculants play a pivotal role in sustainable agriculture by promoting plant growth and protecting against pathogens. Their integration with agrifood waste enhances nutrient availability, fostering resilient and eco-friendly crop production systems.

3.5 Synergies of Microbial Consortia as Biofertilizer

Microbial consortia are structured communities of two or more types of microorganisms (bacteria, fungi, archaea, etc.) that work together. They perform various biological roles, including nutrient cycling and pathogen suppression. These communities are often strategically formulated, with each member contributing a distinct function, leading to greater overall effectiveness compared to the use of single-strain inoculants. For instance, combining *Rhizobium*, a nitrogen-fixer, with phosphate-solubilizing *Bacillus* can significantly improve soil fertility. Similarly, pairing *Trichoderma* with *Pseudomonas* can enhance both plant growth and disease resistance. Such synergies can help fulfill the comprehensive nutrient requirements of a particular soil environment and often deliver enhanced results (O'Callaghan *et al.*, 2022).

The phenomenon of microbial consortia is derived from the study of native microbial consortia. *Arabidopsis thaliana* root scanning electron microscopy images (Fig. 3.3; Hassani *et al.*, 2018) show a dense, diverse community of microorganisms on the root surface. These included biofilm-forming bacteria, filamentous fungal or Oomycetes hyphae, and a variety of protists—some likely from the Bacillariophyceae class—interwoven around root hairs and the primary root axis. Spores, bacterial filaments, and morphologically distinct bacterial cells contribute to a rich microbial landscape that suggests cooperative and competitive interactions. This highly structured and multi-species colonization hints at a naturally evolved microbial network, potentially playing key roles in nutrient exchange, root protection, and signalling within the rhizosphere.

The same concept is applied in designing microbial consortia in industries. Like other biofertilizers, microbial consortia are commonly blended with carriers such as compost or peat to ensure stability and ease of application. These consortia are engineered to maximize nutrient uptake and encourage sustainable farming practices. Their microbial composition may vary—from minimal yet effective blends to complex formulations. For example, Sigma Bio Liquid (Sigma AgriScience, 2021) comprises four bacterial species, four mycorrhizal fungi, and two *Trichoderma* species. Another product, BioGro Plus, includes a compact blend of three to four microorganisms—namely *Bacillus subtilis*, *Trichoderma harzianum*, and *Pseudomonas fluorescens*—geared toward nutrient cycling and plant growth promotion (https://biogro.com/). The synergy among these microbes supports nitrogen fixation, phosphorus solubilization, and pathogen resistance, making the product ideal for crops such as tomatoes and wheat.

In many cases, co-inoculation has shown improved outcomes over the use of individual strains. Wheat grown in phosphorus-deficient soil with both *Penicillium bilaiae* and *Bacillus simplex* showed increased phosphorus uptake compared to using either microbe individually (Hansen *et al.*, 2020). One recent field trial compared various inoculant groups—plant growth-promoting rhizobacteria (PGPR), rhizobia, AMF, and their combinations—on faba bean and wheat. The group that received all three microbial

Table 3.4. Commercial microorganisms used for biological control of plant diseases. Table author's own.

Microorganism group	Product name(s)	Active strain(s)	Company	Application type	Website
Bacillus spp.	Valcure, Revitalize, Amylo-X WG, Garden Friendly Fungicide	*Bacillus amyloliquefaciens* D747	Certis Belchim	Biofungicide	certisbelchim.com
Bacillus spp.	Taegro, RhizoVital 42	*Bacillus amyloliquefaciens* FZB42	Novonesis	Biocontrol	novonesis.com
Trichoderma spp.	Remedier®, Tenet®	*Trichoderma asperellum*, *T. gamsii*	Blacksmith BioScience	Biocontrol	blacksmithbio.com
Trichoderma spp.	Tusal®	*Trichoderma asperellum*, *T. atroviride*	Prochemica	Biocontrol	prochemica.com
Trichoderma spp.	T34 Biocontrol®	*Trichoderma asperellum*	Arbico Organics	Biocontrol	arbico-organics.com
Trichoderma spp.	Xedavir®	*Trichoderma asperellum*	Perfarelalbero	Biocontrol	perfarelalbero.it
Trichoderma spp.	Tri-Soil®	*Trichoderma atroviride*	Agrauxine	Biocontrol	agrauxine.com
AMF	MycoGuard	*Rhizophagus irregularis*, *Funneliformis mosseae*	Gardenshop	Biocontrol	gardenshop.co.za
Trichoderma spp.	Avengelus®	*Trichoderma atrobrunneum*	Sorbus Intl	Biocontrol	sorbus-intl.co.uk
Trichoderma spp.	Vintec®	*Trichoderma atroviride*	Bi-PA	Biocontrol	bi-pa.com
Trichoderma spp.	Mikro-Roots	*Trichoderma* spp.	Mikrobs	Biofertilizer	mikrobs.com

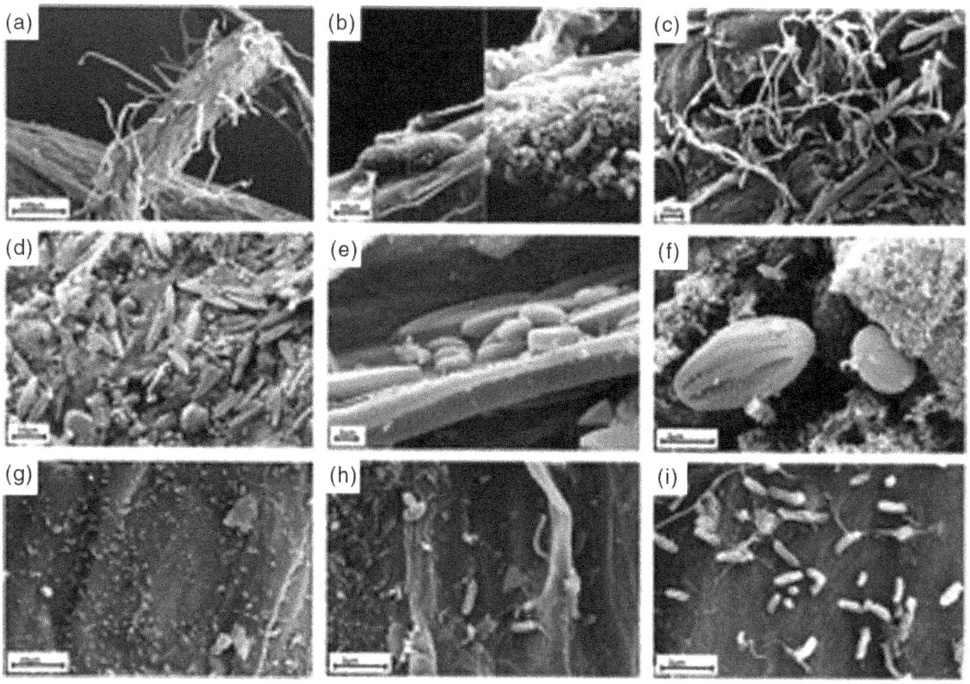

Fig. 3.3. Microbial consortia naturally formed on the roots of *Arabidopsis thaliana*. Scanning electron microscopy pictures of root surfaces from natural *A. thaliana* populations showing the complex microbial networks formed on roots. (a) Overview of an *A. thaliana* root (primary root) with numerous root hairs. (b) Biofilm-forming bacteria. (c) Fungal or oomycete hyphae surrounding the root surface. (d) Primary root densely covered by spores and protists. (e, f) Protists, most likely belonging to the Bacillariophyceae class. (g) Bacteria and bacterial filaments. (h, i) Different bacterial individuals showing great varieties of shapes and morphological features. Reproduced from Hassani *et al*., 2018. Figure used under licence CC BY-SA 4.0.

types exhibited significant increases in plant biomass and nutrient content compared to both uninoculated and singly inoculated controls (Raklami *et al*., 2019).

Microbial communities often function synergistically (Barea *et al*., 2005). However, the selection of organisms in commercial products and the consistency of their cooperation remain unclear. There is limited published evidence that quantitatively assesses the benefits of specific combinations of well-characterized micro-organisms (O'Callaghan *et al*., 2022). A key goal in biofertilizer technology is developing polymicrobial inoculants with multiple benefits, usable across various crops and regions (Reddy and Saravanan, 2013). However, while many such products are marketed for use across a wide crop range, scientific validation for their broad-spectrum efficacy is still limited.

One practical example of a microbial consortium used for nutrient recovery from agrifood waste involves the combination of *Bacillus subtilis* and *Trichoderma harzianum* (Fig. 3.4; Vurukonda *et al*., 2024). This mixture effectively breaks down waste materials like potato peels, sugarcane bagasse, and vegetable scraps, releasing vital nutrients such as nitrogen, phosphorus, and potassium. *Bacillus subtilis* breaks down proteins and complex carbohydrates. *Trichoderma harzianum* facilitates the decomposition of cellulose and lignin while providing disease suppression, and *Aspergillus niger* helps solubilize phosphorus and other micronutrients. Together, they speed up composting processes and convert organic waste into nutrient-rich biofertilizer. Research shows that this approach enhances nutrient availability for crops like maize and rice, promoting sustainable nutrient recycling from agricultural residues.

Together, these insights highlight the vast potential of microbial consortia in advancing sustainable agriculture and nutrient recycling. By drawing inspiration from natural root-associated

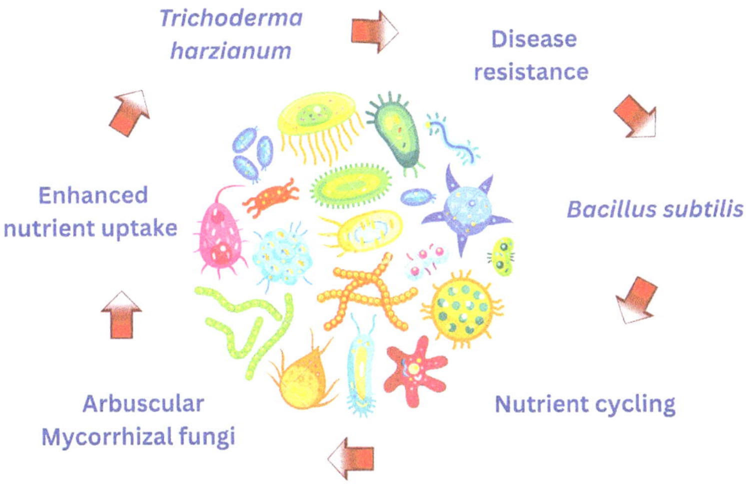

Fig. 3.4. A conceptual microbial consortium structure and function. Figure author's own.

communities and strategically combining functionally diverse microorganisms, researchers and practitioners can develop targeted, synergistic biofertilizers. However, realizing their full potential requires more robust field validation, careful strain selection, and a deeper understanding of microbe–microbe and plant–microbe interactions. As research progresses, microbial consortia are poised to become a cornerstone of ecologically resilient and resource-efficient farming systems.

3.6 Conclusions

The integration of microbial inoculants with agrifood waste represents a transformative approach to sustainable agriculture. This synergy not only enhances the decomposition and nutrient-releasing potential of organic residues but also improves nutrient availability, soil structure, and crop productivity. Microbial inoculants—whether used individually or in consortia—play critical roles in nitrogen fixation, phosphorus solubilization, micronutrient mobilization, and disease suppression. Their cooperation with composted agrifood waste ensures a more efficient and environmentally friendly use of resources, aligning with the principles of the circular economy. As agriculture continues to face the challenges of soil degradation, climate change, and diminishing fertilizer resources, embracing such synergistic strategies can provide scalable, low-impact solutions for food security and ecological sustainability. Moving forward, greater focus on refining microbial consortia, enhancing commercialization, and promoting farmer education will be essential to maximize the impact of this powerful partnership.

References

Albertsen, A., Ravnskov, S., Green, H., Jensen, D., and Larsen, J. (2006) Interactions between the external mycelium of the mycorrhizal fungus glomus intraradices and other soil microorganisms as affected by organic matter. *Soil Biology and Biochemistry* 38(5), 1008–1014. DOI: 10.1016/j.soilbio.2005.08.015

Ali, I., Khan, A., Ali, A., Ullah, Z., Dai, D.-Q. *et al.* (2022) Iron and zinc micronutrients and soil inoculation of *Trichoderma harzianum* enhance wheat grain quality and yield. *Frontiers in Plant Science* 13, 960948. DOI: 10.3389/fpls.2022.960948

Amarger, N. (1980) Aspect microbiologique de la culture des legumineuses. *Le Selectionneur francais* 28, 61–66.

Bansal, M. and Mukerji, K.G. (1996) Root exudates in rhizosphere biology. In: Mukerji, K.G., Singh V.P., and Dwivedi, S. (eds) *Concepts in Applied Microbiology and Biotechnology*. Aditya Books, New Delhi, pp. 98–120.

Bansal, M., Chamola, B.P., Sarwar, N., and Mukerji, K.G. (2000) Mycorrhizosphere: interaction between rhizosphere microflora and VAM fungi. In: Mukerji, K.G., Chamola, B.P., and Singh, J. (eds) *Mycorrhizal Biology*. Springer, New York. pp. 143–152.

Barea, J.M., Pozo, M.J., Azcón, R., and Azcón-Aguilar, C. (2005) Microbial co-operation in the rhizosphere. *Journal of Experimental Botany* 56(417), 1761–1778. DOI: 10.1093/jxb/eri197

Basiru, S. and Hijri, M. (2022) The potential applications of commercial arbuscular mycorrhizal fungal inoculants and their ecological consequences. *Microorganisms* 10(10), 1897. DOI: 10.3390/microorganisms10101897

Basiru, S., Mwanza, H.P., and Hijri, M. (2020) Analysis of arbuscular mycorrhizal fungal inoculant benchmarks. *Microorganisms* 9(1), 81. DOI: 10.3390/microorganisms9010081

Berruti, A., Lumini, E., Balestrini, R., and Bianciotto, V. (2016) Arbuscular mycorrhizal fungi as natural biofertilizers: let's benefit from past successes. *Frontiers in Microbiology* 6, 1559. DOI: 10.3389/fmicb.2015.01559

Bonanomi, G., Antignani, V., Pane, C., and Scala, F. (2007) Suppressiveness of 18 composts against 7 pathogens and 3 seedlings diseases. *Soil Biology and Biochemistry* 39(9), 2303–2314. DOI: /10.1016/j.soilbio.2007.03.003

Catroux, G., Hartmann, A., and Revellin, C. (2001) Trends in rhizobial inoculant production and use. *Plant and Soil* 230(1), 21–30. DOI: 10.1023/A:1004777115628

Daniel, A.I., Fadaka, A.O., Gokul, A., Bakare, O.O., Aina, O. *et al.* (2022) Biofertilizer: the future of food security and food safety. *Microorganisms* 10(6), 1220. DOI: 10.3390/microorganisms10061220

Gao, C., El-Sawah, A.M., Ali, D.F.I., Alhaj Hamoud, Y., Shaghaleh, H. *et al.* (2020) The integration of bio and organic fertilizers improve plant growth, grain yield, quality and metabolism of hybrid maize (*Zea mays* L.). *Agronomy* 10(3), 319. DOI: 10.3390/agronomy10030319

Gorain, B., Paul, S., and Parihar, M. (2022) Role of soil microbes in micronutrient solubilization. In: Gupta, V.K. (ed.) *New and Future Developments in Microbial Biotechnology and Bioengineering*. Elsevier, Amsterdam, pp. 131–150.

Gosling, P., Jones, J., and Bending, G.D. (2016) Evidence for functional redundancy in arbuscular mycorrhizal fungi and implications for agroecosystem management. *Mycorrhiza* 26(1), 77–83. DOI: 10.1007/s00572-015-0651-6

Gupta, M.M. (2020). Arbuscular mycorrhizal fungi: the potential soil health indicators. In: Giri, B. and Varma, A. (eds) *Soil Health*. Springer, Cham, Switzerland, pp. 183–195.

Gupta, M.M. (2024) *Biofertilizers NEP Curricula*. Swaraj Prakashan, New Delhi.

Gupta, M.M. and Abbott, L.K. (2021) Exploring economic assessment of the arbuscular mycorrhizal symbiosis. *Symbiosis* 83(2), 143–152. DOI: 10.1007/s13199-020-00738-0

Gupta, M.M., Aggarwal, A., and Asha, A. (2018) From mycorrhizosphere to rhizosphere microbiome: the paradigm shift. In: Giri, B., Prasad, R., and Varma, A. (eds) *Root Biology*. Springer, Cham, Switzerland, pp. 487–500.

Hansen, V., Bonnichsen, L., Nunes, I., Sexlinger, K., Lopez, S.R. *et al.* (2020) Seed inoculation with *Penicillium bilaiae* and *Bacillus simplex* affects the nutrient status of winter wheat. *Biology and Fertility of Soils* 56(1), 97–109. DOI: 10.1007/s00374-019-01401-7

Harris, M., Hussain, T., Tauseef, A., Khan, A., and Khan, A.A. (2023) Application of microbial inoculants as an alternative to chemical products for decomposition of organic wastes. In: Sharma, V.K., Kumar, A., Passarini, M.R.Z., Parmar, S., and Singh, V.K. (eds) *Microbial Inoculants*. Academic Press, London, pp. 29–52.

Hart, A., Ebiundu, K., Peretomode, E., Onyeaka, H., Nwabor, O.F. *et al.* (2022) Value-added materials recovered from waste bone biomass: technologies and applications. *RSC Advances* 12(34), 22302–22330. DOI: 10.1039/d2ra03557j

Hart, M.M., Antunes, P.M., Chaudhary, V.B., and Abbott, L.K. (2018) Fungal inoculants in the field: is the reward greater than the risk? *Functional Ecology* 32(1), 126–135. DOI: 10.1111/1365-2435.12976

Harvey, P.R., Warren, R.A., and Wakelin, S. (2009) Potential to improve root access to phosphorus: the role of non-symbiotic microbial inoculants in the rhizosphere. *Crop and Pasture Science* 60(2), 144. DOI: 10.1071/CP08084

Hassani, M.A., Durán, P., and Hacquard, S. (2018) Microbial interactions within the plant holobiont. *Microbiome* 6(1), 58. DOI: 10.1186/s40168-018-0445-0

Ibiang, Y.B., Innami, H., and Sakamoto, K. (2018) effect of excess zinc and arbuscular mycorrhizal fungus on bioproduction and trace element nutrition of tomato (*Solanum lycopersicum* L. cv. Micro-Tom). *Soil Science and Plant Nutrition* 64(3), 342–351. DOI: 10.1080/00380768.2018.1425103

Karamanos, R.E., Flore, N.A., and Harapiak, J.T. (2010) Revisiting use of *Penicillium bilaii* with phosphorus fertilization of hard red spring wheat. *Canadian Journal of Plant Science* 90(3), 265–277. DOI: 10.4141/CJPS09123

Kawalekar, J.S. (2013) Role of biofertilizers and biopesticides for sustainable agriculture. *Journal of Bio Innovation* 2, 73–78.

Kim, K.Y., Jordan, D., and McDonald, G.A. (1998) Enterobacter agglomerans, phosphate solubilizing bacteria, and microbial activity in soil: effect of carbon sources. *Soil Biology and Biochemistry* 30(8–9), 995–1003. DOI: 10.1016/S0038-0717(98)00007-8

Lal, R. (2020) Soil organic matter content and crop yield. *Journal of Soil and Water Conservation* 75(2), 27A–32A. DOI: 10.2489/jswc.75.2.27A

Lantmännen BioAgri (2021) Product information for Cedoman®, Cerall®, and Cedess® biological seed treatments. Available at: https://www.lantmannenbioagri.com (accessed September 5, 2025).

Luginbuehl, L.H. and Oldroyd, G.E.D. (2017) Understanding the arbuscule at the heart of endomycorrhizal symbioses in plants. *Current Biology* 27(17), R952–R963. DOI: 10.1016/j.cub.2017.06.042

Marschner, P. and Timonen, S. (2006) Bacterial community composition and activity in rhizosphere of roots colonized by arbuscular mycorrhizal fungi. In: Mukerji, K.G., Manoharachary, C. and Singh, J. (eds) *Microbial Activity in the Rhizosphere*. Springer, Heidelberg, Germany, pp. 139–154.

Mokjatturas, S., Chinwetkitvanich, S., Patthanaissaranukool, W., Polprasert, C., and Polprasert, S. (2025) Phosphorus mass flows and economic benefits of food waste management: the case study of selected retail and wholesale fresh markets in Thailand. *Clean Technologies and Environmental Policy* 27(1), 219–233. DOI: 10.1007/s10098-024-02847-6

Nanda, K., Singh, V., Kumar, S., Sharma, P., and Singh, S.P. (2025) Waste mitigation through synergistic solutions with plants and microbes. In: Ren, J. (ed.) *Waste-to-Energy*. Elsevier, Amsterdam, pp. 163–193.

Northouse, P.G. (2018) *Leadership: Theory and Practice*, 8th edn. SAGE Publications, Thousand Oaks, California.

O'Callaghan, M., Ballard, R.A., and Wright, D. (2022) Soil microbial inoculants for sustainable agriculture: limitations and opportunities. *Soil Use and Management* 38(3), 1340–1369. DOI: 10.1111/sum.12811

Olsson, P.A., Bååth, E., Jakobsen, I., and Söderström, B. (1996) Soil bacteria respond to presence of roots but not to mycelium of arbuscular mycorrhizal fungi. *Soil Biology and Biochemistry* 28(4–5), 463–470. DOI: 10.1016/0038-0717(96)00011-9

Olsson, P.A., Hammer, E.C., Pallon, J., Van Aarle, I.M., and Wallander, H. (2011) Elemental composition in vesicles of an arbuscular mycorrhizal fungus, as revealed by PIXE analysis. *Fungal Biology* 115(7), 643–648. DOI: 10.1016/j.funbio.2011.03.008

Raklami, A., Bechtaoui, N., Tahiri, A.-I., Anli, M., Meddich, A. *et al.* (2019) Use of rhizobacteria and mycorrhizae consortium in the open field as a strategy for improving crop nutrition, productivity and soil fertility. *Frontiers in Microbiology* 10, 1106. DOI: 10.3389/fmicb.2019.01106

Ramos Cabrera, E.V., Delgado Espinosa, Z.Y., and Solis Pino, A.F. (2024) Use of phosphorus-solubilizing microorganisms as a biotechnological alternative: a review. *Microorganisms* 12(8), 1591. DOI: 10.3390/microorganisms12081591

Raymond, N.S., Gómez-Muñoz, B., van der Bom, F.J., Nybroe, O., Jensen, L.S., Müller-Stöver, D.S., Oberson, A., and Richardson, A.E. (2021) Phosphate-solubilising microorganisms for improved crop productivity: a critical assessment. *New Phytologist* 229(3),1268–1277. DOI: 10.1111/nph.16924

Reddy, C.A. and Saravanan, R.S. (2013) Polymicrobial multi-functional approach for enhancement of crop productivity. *Advances in Applied Microbiology* 82, 53–113. DOI: 10.1016/B978-0-12-407679-2.00003-X

Richardson, A.E. (2001) Prospects for using soil microorganisms to improve the acquisition of phosphorus by plants. *Functional Plant Biology* 28(9), 897. DOI: 10.1071/PP01093

Richardson, A.E. and Simpson, R.J. (2011) Soil microorganisms mediating phosphorus availability. *Plant Physiology* 156, 989–996. DOI: 10.1104/pp.111.175448

Rillig, M.C. and Mummey, D.L. (2006) Mycorrhizas and soil structure. *New Phytologist* 171(1), 41–53. DOI: 10.1111/j.1469-8137.2006.01750.x

Ruotsalainen, A.L., Kauppinen, M., Wäli, P.R., Saikkonen, K., Helander, M., and Tuomi, J. (2022) Dark septate endophytes: mutualism from by-products? *Trends in Plant Science* 27(3), 247–254. DOI: 10.1016/j.tplants.2021.10.001

Santos, M.S., Nogueira, M.A., and Hungria, M. (2019) Microbial inoculants: reviewing the past, discussing the present and previewing an outstanding future for the use of beneficial bacteria in agriculture. *AMB Express* 9(1), 205. DOI: 10.1186/s13568-019-0932-0

Schuster, A. and Schmoll, M. (2010) Biology and biotechnology of *Trichoderma*. *Applied Microbiology and Biotechnology* 87(3), 787–799. DOI: 10.1007/s00253-010-2632-1

Sharma, B., Kumawat, K.C., Tiwari, S., Kumar, A., Dar, R.A. *et al.* (2023) Silicon and plant nutrition—dynamics, mechanisms of transport and role of silicon solubilizer microbiomes in sustainable agriculture: a review. *Pedosphere* 33(4), 534–555. DOI: 10.1016/j.pedsph.2022.11.004

Sharma, S.B., Sayyed, R.Z., Trivedi, M.H., and Gobi, T.A. (2013) Phosphate solubilizing microbes: sustainable approach for managing phosphorus deficiency in agricultural soils. *SpringerPlus* 2, 587. DOI: 10.1186/2193-1801-2-587

Sigma AgriScience, S. (2021) Product information for Sigma Bio. Available at: https://sigmaagriculture.com/catalog/product/sigma-bio/ (accessed September 5, 2025).

Singleton, P.W., Boonkerd, N., Carr, T.J., and Thompson, J.A. (1997) Technical and market constraints limiting legume inoculant use in Asia. In: Rupela, O.P., Johansen, C., and Herridge, D.F. (eds) *Extending Nitrogen Fixation Research to Farmers' Fields*. ICRISAT, Patancheru, India, pp. 17–38.

Smith, S.E. and Read, D.J. (2010) *Mycorrhizal Symbiosis*. Academic Press, London.

Sumbul, A., Ansari, R.A., Rizvi, R., and Mahmood, I. (2020) Azotobacter: a potential bio-fertilizer for soil and plant health management. *Saudi Journal of Biological Sciences* 27(12), 3634–3640. DOI: 10.1016/j.sjbs.2020.08.004

Venkataraman, L.V. (2017) Blue-green algae as biofertilizer. In: Richmond, A. (ed.) *Handbook of Microalgal Mass Culture*. CRC Press, Boca Raton, Florida, pp. 455–472.

Veresoglou, S.D. and Menexes, G. (2010) Impact of inoculation with *Azospirillum* spp. on growth properties and seed yield of wheat: a meta-analysis of studies in the ISI Web of Science from 1981 to 2008. *Plant and Soil* 337(1–2), 469–480. DOI: 10.1007/s11104-010-0543-7

Vurukonda, S.S.K.P., Fotopoulos, V., and Saeid, A. (2024) Production of a rich fertilizer base for plants from waste organic residues by microbial formulation technology. *Microorganisms* 12(3), 541. DOI: 10.3390/microorganisms12030541

Wagg, C., Barendregt, C., Jansa, J., and van der Heijden, M.G.A. (2015) Complementarity in both plant and mycorrhizal fungal communities are not necessarily increased by diversity in the other. *Journal of Ecology* 103(5), 1233–1244. DOI: 10.1111/1365-2745.12452

Willis, A., Rodrigues, B.F., and Harris, P.J.C. (2013) The ecology of arbuscular mycorrhizal fungi. *Critical Reviews in Plant Sciences* 32(1), 1–20. DOI: 10.1080/07352689.2012.683375

Woo, S.L. and Pepe, O. (2018) Microbial consortia: promising probiotics as plant biostimulants for sustainable agriculture. *Frontiers in Plant Science* 9, 1801. DOI: 10.3389/fpls.2018.01801

Xie, S., Tran, H.T., Pu, M., and Zhang, T. (2023) Transformation characteristics of organic matter and phosphorus in composting processes of agricultural organic waste: research trends. *Materials Science for Energy Technologies* 6, 331–342. DOI: 10.1016/j.mset.2023.02.006

Yuan, Z., Yi, H., Wang, T., Zhang, Y., Zhu, X. *et al.* (2017) Application of phosphate solubilizing bacteria in immobilization of Pb and Cd in soil. *Environmental Science and Pollution Research* 24(27), 21877–21884. DOI: 10.1007/s11356-017-9832-5

Zilli, J.É., Pacheco, R.S., Gianluppi, V., Smiderle, O.J., Urquiaga, S. *et al.* (2021) Biological N2 fixation and yield performance of soybean inoculated with bradyrhizobium. *Nutrient Cycling in Agroecosystems* 119(3), 323–336. DOI: 10.1007/s10705-021-10128-7

4 Mass Multiplication of Microbial Inoculants

Anirudh Sharma*
Department of Biotechnology, Jaypee Institute of Information Technology, Noida, Uttar Pradesh, India

Abstract

Microbial inoculants play a pivotal role in modern sustainable agriculture, enhancing soil fertility, plant health, and productivity while reducing dependency on chemical fertilizers. This chapter peeks into the mass multiplication techniques of key microbial inoculants, addressing both their biological characteristics and industrial-scale production challenges. Beginning with an overview of microbial inoculants, we classify key types, explore their mechanisms of action, and evaluate their benefits in agricultural ecosystems. The chapter systematically examines rhizobial, *Azotobacter*, and *Azospirillum* inoculants, detailing optimized culturing techniques, quality control measures, and field applications. A special focus is given to phosphate-solubilizing bacteria and cyanobacterial inoculants, highlighting their roles in nutrient cycling and soil enhancement. *Frankia* inoculants are explored for their nitrogen-fixing capabilities, particularly in reclaiming degraded lands. Arbuscular mycorrhizal fungi are discussed in the context of large-scale propagation, plant compatibility, and application strategies. Advancements in mass production technologies, including fermentation approaches and carrier formulations, are examined alongside challenges such as microbial viability, storage limitations, and strain specificity. The chapter further considers innovative strategies for enhanced inoculant effectiveness, including genetic engineering, synergistic microbial combinations, and integration with agrifood waste for sustainable bioprocessing.

4.1 Introduction

Agriculture faces unprecedented challenges in the current era, driven by factors such as land degradation, climate change, declining soil fertility, pest-driven losses, and excessive dependence on chemical inputs (Seetharaman *et al.*, 2025). A significant concern is soil degradation, exacerbated by intensive farming practices and excessive synthetic fertilizer use, which disrupt natural microbial communities and lead to reduced nutrient availability. According to estimates by the Food and Agriculture Organization (FAO), up to 33% of the world's arable land is already degraded (Zorn *et al.*, 2013), affecting global food security and sustainable crop yields. Other than this, most of the other areas in the world are not arable owing to various factors related to edaphic factors, population expansion, being unfertile fallow land, or being affected by climatic factors, deprivation of irrigation water sources, and many other factors (Mudgal *et al.*, 2025). Economic losses due to soil deterioration and inefficient fertilization methods are also alarming. Studies indicate that global agricultural losses due to soil depletion, salinization,

*Corresponding author: anirudhsharma172@gmail.com

DOI: 10.1079/9781836991021.0004

and declining fertility exceed US$400 billion annually (Goss *et al.*, 2017; Kopittke *et al.*, 2019). Furthermore, excessive reliance on agrochemicals contributes to environmental pollution, groundwater contamination, and biodiversity loss (Sharma *et al.*, 2021; Gnanaprakasam *et al.*, 2022; Balu *et al.*, 2024; Kashyap *et al.*, 2024; Harika *et al.*, 2025; Mudgal *et al.*, 2025; Seetharaman *et al.*, 2025), raising concerns about the sustainability of current agricultural models. The growing demand for high-yield crops (Checco *et al.*, 2023; Kaur *et al.*, 2025) and the pressure of meeting food supply expectations necessitate alternative solutions that not only maintain productivity but also restore ecological balance (Kumar *et al.*, 2022).

To curb these problems, innovative biotechnological approaches, including precision agriculture (Pedersen *et al.*, 2017), biofertilization (Singh *et al.*, 2018), and soil microbiome restoration (Kaur *et al.*, 2022; Faskhutdinova *et al.*, 2024; Serazetdinova *et al.*, 2025), are gaining traction. Researchers have identified beneficial soil microbes that enhance nutrient uptake, boost plant immunity, and improve soil structure. Microbial inoculants are at the forefront of sustainable agricultural practices, offering an eco-friendly alternative to synthetic fertilizers. These beneficial microorganisms—such as nitrogen-fixing bacteria, phosphate-solubilizing bacteria, cyanobacteria, and mycorrhizal fungi—enhance nutrient availability, improve soil health, and boost plant resilience against environmental stresses. Their ability to restore soil microbial diversity and promote sustainable crop production makes them vital in mitigating the negative impacts of intensive farming practices. By fostering nutrient cycling, reducing dependence on chemical inputs, and improving water retention, microbial inoculants play a pivotal role in enhancing productivity while maintaining environmental integrity. Nitrogen-fixing bacteria, phosphate-solubilizing microbes, mycorrhizal fungi, and plant growth-promoting rhizobacteria (PGPR) have emerged as biological alternatives to conventional fertilizers. These inoculants reduce environmental impact while enhancing crop resilience to abiotic and biotic stress factors (Singh *et al.*, 2021; Mudgal *et al.*, 2022; Kaur *et al.*, 2023; Parashar *et al.*, 2023; Dhar *et al.*, 2024). Other than this, many of these plant growth-enhancing microbial bioinoculant candidates have been known to effectuate epigenetic and other regulatory controls by effectuating elicitor metabolites intervening and upregulating stress management in crops either directly or indirectly. Microbial growth enhancers can regulate auxin response factors (ARFs), which are key transcription factors involved in auxin-mediated gene expression and plant development (Verma *et al.*, 2022). Several beneficial microbes, including PGPR and arbuscular mycorrhizal (AM) fungi, influence ARF activity through various mechanisms (Roosjen *et al.*, 2017; Çakmakçı *et al.*, 2020; Liu *et al.*, 2024).

Recent advancements in synthetic biology, microbial engineering, and formulation technologies have enabled efficient microbial strain selection, optimizing inoculant effectiveness in diverse soil conditions. Research on biofilm formation and microbial consortia has shown promising results in sustaining microbial survival and activity within soil ecosystems, ensuring consistent agricultural benefits (Aguilar-Paredes *et al.*, 2020; Vishwakarma *et al.*, 2020; Chaudhary *et al.*, 2023). Additionally, integration of agri-food waste into biofertilizer production presents a circular economy model (Aguilar-Rivera and Olvera-Vargas, 2022; Chojnacka *et al.*, 2022; Ezeorba *et al.*, 2024; Singh *et al.*, 2025), reducing waste accumulation while enhancing microbial mass multiplication efficiency. However, while microbial inoculants offer promising solutions, their mass multiplication and commercialization present technical challenges that need to be addressed for broader adoption (Chaudhary and Shukla, 2020). Their growth media composition significantly influences microbial viability, requiring optimized formulations enriched with carbon sources, minerals, and buffering agents to sustain metabolic activity (Huang *et al.*, 2023). Several key challenges complicate the widespread implementation of these beneficial microorganisms. Nutrient requirements and media optimization pose significant hurdles, as different inoculants require specific media formulations, which directly impacts production scalability and cost-effectiveness (O'Callaghan *et al.*, 2022; Díaz-Rodríguez *et al.*, 2025). Survival during storage represents another critical challenge, with many microbial inoculants exhibiting reduced viability over time, necessitating improved storage techniques such as lyophilization, encapsulation, and liquid suspensions to maintain their effectiveness (Berninger *et al.*, 2018; Balla *et al.*, 2022; Rojas-Sánchez *et al.*, 2022; Rojas-Padilla *et al.*, 2024). Strain specificity and adaptability issues further complicate field applications, as certain inoculants that perform well under controlled laboratory conditions often fail in real-world agricultural settings due to soil variability and environmental fluctuations (Zhu *et al.*, 2023). Additionally, carrier material selection remains problematic, with peat, vermiculite, lignite-based carriers, and biochar being commonly used options, yet finding cost-effective and eco-friendly carriers that maintain microbial viability while being economically viable continues to challenge researchers and manufacturers in the field (Egamberdieva *et al.*, 2018; Aloo *et al.*, 2022; Gupta *et al.*, 2022; Bolan *et al.*, 2023; Sivaram *et al.*, 2023). The successful application of microbial inoculants requires precise strain selection, formulation improvement, and field adaptability testing (Chaudhary *et al.*, 2020). Laboratory-optimized strains must undergo compatibility testing with specific soil types, crop varieties, and climatic conditions before large-scale agricultural adoption. Furthermore, farmers need technical training to ensure appropriate inoculant application. Incorrect dosing, poor seed coating methods, or suboptimal field inoculation techniques may result in inconsistent microbial performance (Bhowmick, 2018; Rocha *et al.*, 2019; O'Callaghan *et al.*, 2022; Chemla *et al.*, 2025). Thus, field-scale validation studies and customized application protocols must accompany microbial product commercialization efforts (Salomon *et al.*, 2022; Papin, 2024). The commercialization of microbial inoculants holds immense potential but requires overcoming regulatory, economic, and technological barriers. A significant advancement in biofertilizer production is the integration of agrifood waste as a substrate for microbial inoculant cultivation (Kiruba N *et al.*, 2022). Organic residues from food processing, crop remains, and dairy industry by-products serve as nutrient-rich sources for microbial growth, contributing to a circular economy approach in agriculture (Rafiq *et al.*, 2023). This integration not only reduces waste accumulation but also enhances the efficiency and cost-effectiveness of inoculant mass multiplication. The use of agrifood waste as a fermentation medium supports microbial viability, improves inoculant efficacy, and aligns with sustainable waste management strategies (Kieliszek *et al.*, 2020; Parajuli *et al.*, 2022). Additionally, repurposing agricultural by-products for biofertilizer development minimizes environmental pollution and strengthens the link between agriculture and biotechnology-driven solutions for soil restoration and plant health improvement.

This chapter aims to provide a comprehensive overview of mass multiplication techniques for microbial inoculants, focusing on cultivation methods, formulation strategies, and application approaches. The discussion extends to the selection of microbial strains, optimization of fermentation processes, and advancements in biofertilizer production using agrifood waste as a substrate. Furthermore, the chapter addresses challenges such as microbial viability, scalability, and field applicability, offering insights into innovative solutions that enhance the performance and sustainability of microbial inoculants. By integrating scientific research with practical applications, this chapter serves as a valuable resource for researchers, agronomists, and industry professionals seeking to leverage microbial technologies for sustainable agriculture and environmental conservation.

4.2 Microbial Inoculants and Agrifood Waste Synergy

4.2.1 Types of microbial inoculants and their functions

Microbial inoculants comprise a diverse array of beneficial microorganisms that enhance plant growth, soil fertility, and nutrient availability through biological interactions. These microbes can be classified based on their functional roles: nitrogen-fixing bacteria, such as rhizobia (*Rhizobium* spp.), which form symbiotic relationships with legumes (Table 4.1; Schütz *et al.*, 2018; Sammauria *et al.*, 2020; Shahwar *et al.*, 2023), and free-living nitrogen fixers like *Azotobacter* and *Azospirillum*, which improve nitrogen availability in non-leguminous crops; phosphate-solubilizing microorganisms, including phosphate-solubilizing bacteria (PSB; *Bacillus*, *Pseudomonas*, *Serratia* spp.), which release organic acids to mobilize phosphate, and AM fungi, which facilitate

Table 4.1. Classification of microbial inoculants. Tabulated key aspects of bioinoculant types, their biological functions, practical target applications, and notes on production processes or challenges. Table author's own.

Inoculant type	Primary functions / mechanisms	Target crops / environments	Key benefits and applications	Comments / production notes
Rhizobial inoculants	Nitrogen fixation via symbiotic nodule formation; secretion of Nod factors; organic acid production	Leguminous crops (e.g. soybean, chickpea, lentils)	Enhances nitrogen availability; lowers synthetic fertilizer dependency; promotes robust plant growth	Requires strict host compatibility; production can utilize agro-industrial waste (e.g. molasses, whey) via liquid or solid-state fermentation methods
***Azotobacter* inoculants**	Free-living nitrogen fixation; production of phytohormones (IAA, gibberellins); siderophore secretion; exopolysaccharide (EPS) production	Non-leguminous crops (e.g. wheat, maize)	Improves soil fertility; stimulates root growth; enhances nutrient uptake	Suitable for nutrient-poor soils; media can include waste by-products (molasses, dairy wastes)
***Azospirillum* inoculants**	Nitrogen fixation; secretion of plant growth hormones (e.g. IAA); enhancement of root architecture	Cereals and grasses (e.g. rice, wheat, maize)	Boosts root development; increases water and nutrient absorption; alleviates stress	Often co-formulated with other biofertilizers; scalable production using agro-residues like wheat bran and rice husk
Phosphate-solubilizing bacteria (PSB)	Production of organic acids (gluconic, citric acids) to solubilize bound phosphates; enzymatic hydrolysis (phosphatases) for releasing phosphorus	Various crops including vegetables and cereals	Enhances phosphorus uptake; reduces dependency on mineral phosphate fertilizers	Incorporates food waste derivatives (e.g. fruit pulp waste) as media; optimization of fermentation conditions is critical
Cyanobacterial inoculants	Nitrogen fixation; biomass production; oxygenation; production of bioactive compounds	Paddy fields; degraded, nutrient-depleted soils	Improves soil fertility and water quality; contributes organic matter through biomass deposition	Cultivated in outdoor raceway ponds using agro-effluents and organic leachates; harvesting, packaging, and maintaining viability are key challenges
***Frankia* inoculants**	Formation of actinorhizal nodules; nitrogen fixation in non-leguminous plants; secretion of organic acids; enhancement of soil organic matter	Actinorhizal plants (e.g. *Alnus*, *Casuarina*)	Promotes soil restoration and stabilization; effective for reclamation of degraded lands	Cultivation can be challenging due to slow growth; innovative waste-supported strategies (e.g. compost leachates) are applied to simulate natural conditions
Arbuscular mycorrhizal (AM) fungi	Extensive hyphal network formation to improve nutrient (especially phosphorus) and water uptake; production of glomalin to enhance soil structure; stress tolerance	Wide range of crops, particularly in nutrient-poor soils	Enhances plant resilience under stress; increases water absorption; improves soil structure and fertility	Requires a host plant for optimum proliferation; production often uses composted agro-waste as substrates; advanced mass multiplication and encapsulation techniques ensure viability

phosphorus uptake through extensive root colonization (Kiruba N *et al.*, 2022); biocontrol and PGPRs, such as *Bacillus* and *Pseudomonas* spp., which produce antimicrobial compounds and enhance disease resistance, along with *Trichoderma* spp., which suppresses fungal pathogens and strengthens plant resilience; and cyanobacteria and biofertilizer-enhancing microbes, including *Anabaena* and *Nostoc* spp., which contribute to nitrogen fixation in aquatic ecosystems, and *Frankia* spp., which establishes actinorhizal symbioses to aid nitrogen fixation in non-leguminous trees and shrubs (Shahwar *et al.*, 2023). Collectively, these microbial inoculants (Fig. 4.1) function as natural biofertilizers, improving nutrient cycling while reducing reliance on synthetic chemical fertilizers, thereby supporting sustainable agricultural practices.

4.2.2 Agrifood waste as a resource for microbial growth media

Food waste is a widespread issue that spans the entire food supply chain, beginning with agricultural production and continuing through distribution, consumption, and ultimately disposal

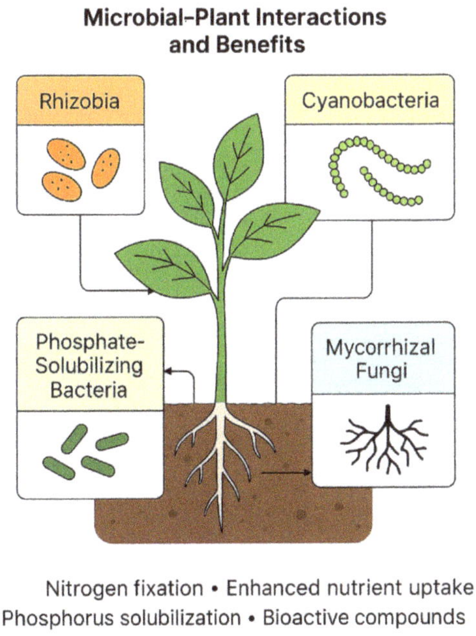

Fig. 4.1. Various beneficial plant–microbe interactions. Figure author's own.

in landfills. Each year, more than 30% of food is lost or wasted—a staggering figure, especially given the prevalence of global hunger (Muth *et al.*, 2019; Read *et al.*, 2020; Nicastro *et al.*, 2021). Beyond being an inefficiency, food waste also represents a serious social equity concern. Perfectly edible food could be redirected to support food recovery initiatives, ensuring it serves its intended purpose rather than being discarded. The environmental consequences of food waste are equally significant as wasted food contributes substantially to greenhouse gas (GHG) emissions and the depletion of natural resources (Wunderlich and Martinez, 2018; Munesue and Masui, 2019; Bhatia *et al.*, 2023). Addressing this issue through better food storage practices and waste reduction strategies can help mitigate climate change, enhance food security, and promote more sustainable food systems. In the UK alone, approximately 9.5 million tonnes of food are wasted annually (Weis *et al.*, 2021; Eaton *et al.*, 2022), complicating efforts to market misshapen produce, redistribute surplus food to those in need, and minimize environmental damage through improved post-harvest handling. Fortunately, efforts to combat food waste are gaining momentum. Many corporations, startups, and nonprofit organizations recognize the scale of this challenge and are actively working toward solutions. With billions of dollars and countless meals lost each year, initiatives such as food waste awareness campaigns, consumer education programs, and donation-driven fundraisers are helping to recover the value of wasted food while supporting charitable causes. Agrifood waste represents an underutilized yet valuable resource for microbial growth media in biofertilizer production. Many industrial and agricultural by-products contain organic carbon, essential minerals, and structural compounds that can sustain microbial proliferation during inoculant mass multiplication. Then, successful production requires stringent aseptic conditions and precise monitoring throughout the manufacturing process. Scaling up biofertilizer production demands a thorough understanding of fermentation technologies and control parameters to ensure consistency and effectiveness. Bagga and group have highlighted in a recent review the mass production techniques, leveraging both solid and liquid fermentation to cultivate biofertilizer strains using nutrients sourced from domestic and industrial wastewater (Bagga *et al.*, 2024). These methods not only improve microbial proliferation but also facilitate resource recovery, aligning with circular economy principles. Sustainable biofertilizer production ultimately depends on optimizing formulations and developing efficient carrier systems to enhance microbial viability and field efficacy. Agrifood waste substrates offer valuable nutrients for microbial growth (Table 4.2), enhancing biofertilizer production sustainably and cost-effectively. For example, molasses and sugarcane residues serve as carbohydrate sources that facilitate bacterial fermentation, promoting the proliferation of *Azotobacter*, *Rhizobium*, and *Bacillus* strains (Gabra *et al.*, 2019; Mustafa *et al.*, 2023; Stephen *et al.*, 2024). Dairy industry by-products, such as whey and lactose-rich residues, improve microbial viability, benefiting formulations of *Pseudomonas* and *Lactobacillus* (Martin *et al.*, 2021; Awasthi *et al.*, 2022; Usmani *et al.*, 2022; Malos *et al.*, 2025). Fruit and vegetable processing waste, abundant in vitamins and organic acids, supports PSB and AM fungi, while spent brewery and distillery waste supplies essential nitrogen and trace minerals, aiding the propagation of *Azospirillum* and *Frankia* (Kundu *et al.*, 2024). Additionally, legume hulls and agricultural biomass provide lignin and fiber sources necessary for fungal inoculants like *Trichoderma* and *Mycorrhizae* (Masquelier *et al.*, 2022; Sharma *et al.*, 2024). By repurposing these agrifood waste materials as microbial growth media, biofertilizer production becomes more efficient, reducing dependency on synthetic fermentation substrates while contributing to circular economy principles.

4.2.3 Environmental and economic advantages of waste-based inoculant production

Utilizing agrifood waste for microbial inoculant production offers both environmental sustainability and economic feasibility, making it a transformative approach in agricultural biotechnology. On the environmental front, repurposing waste minimizes landfill accumulation, thereby reducing greenhouse gas emissions. Additionally, lowering chemical inputs helps mitigate soil and water pollution, while enhanced carbon sequestration through microbial activity contributes to soil health restoration. Integrating circular economy principles ensures efficient recycling of agricultural by-products, further strengthening sustainable agricultural practices. Economically, agrifood waste serves as a cost-effective substrate, significantly reducing microbial production expenses. The nutrient-rich fermentation media enhance inoculant shelf life and viability, supporting commercial biofertilizer expansion and creating new employment opportunities in the sector. Moreover, decreasing reliance on synthetic fertilizers leads to long-term savings for farmers, encouraging wider adoption of biological alternatives. By harnessing agrifood waste as a microbial growth medium, agricultural biotechnology not only improves soil fertility but also mitigates environmental risks associated with conventional fertilizers. This synergy between microbial biotechnology and agro-waste recycling positions biofertilizer production as a crucial strategy for fostering both environmental stewardship and economic viability in sustainable agriculture.

4.3 Rhizobial Inoculants

Plant growth-promoting rhizobia are mostly found existing in the families *Rhizobiaceae*, *Phyllobacteriaceae*, and *Bradyrhizobiaceae* (Gopalakrishnan *et al.*, 2015). Rhizobial inoculants are fundamental to sustainable agriculture due to their ability to establish symbiotic relationships with leguminous plants, facilitating biological nitrogen fixation (BNF) (Fig. 4.2; Serazetdinova *et al.*, 2025). This intricate interaction begins when legume roots release flavonoids, which act as chemical signals to attract compatible *Rhizobium* species. In response, rhizobia synthesize Nod factors, initiating a cascade of molecular and physiological events that enable bacterial colonization and nodule formation. The process starts with root recognition and infection, where flavonoids stimulate rhizobia to produce Nod factors, triggering root hair curling and bacterial entry into root tissues. This step is tightly regulated by plant signaling pathways, including microRNA-mediated gene expression, which fine-tunes nodulation efficiency and nitrogen fixation capacity (Simon *et al.*, 2009; Tiwari

Table 4.2. Nutritional composition of agrifood waste substrates. A scientific comparison of agrifood waste substrates used in microbial inoculant production. Table author's own.

Waste substrate	Major nutrients	Key minerals and trace elements	Organic content (%)	Moisture content (%)	Microbial suitability and application	Potential limitations	References
Molasses (sugar industry waste)	High in sucrose, glucose, fructose	Calcium, potassium, iron, magnesium	~75–85%	~20–30%	Excellent carbon source for microbial growth; enhances nitrogen-fixing bacteria (*Rhizobium*, *Azotobacter*)	Requires pH adjustments due to high sugar content	(Gabra *et al.*, 2019; Mustafa *et al.*, 2023; Stephen *et al.*, 2024)
Whey (dairy industry by-product)	Lactose, whey proteins, amino acids	Calcium, phosphorus, zinc	~60–75%	~30–40%	Supports PSB and PGPR growth due to protein and mineral enrichment	Risk of microbial contamination if not sterilized	(Martin *et al.*, 2021; Awasthi *et al.*, 2022; Usmani *et al.*, 2022; Malos *et al.*, 2025)
Fruit pulp waste (citrus, mango, banana)	Organic acids, polyphenols, cellulose	Potassium, magnesium, iron	~50–70%	~60–80%	Enhances phosphorus solubilization; boosts acid production in PSB inoculants	High moisture may require dehydration before use	(El Barnossi *et al.*, 2022; da Silva *et al.*, 2023; Kundu *et al.*, 2024)
Spent brewery/ distillery grains	Fermentation residues, fiber, proteins	Nitrogen, sulfur, phosphorus	~50–65%	~40–55%	Ideal substrate for *Azospirillum* and *Frankia* due to nitrogen-rich profile	Fermentation residues may alter microbial metabolism	(Masquelier *et al.*, 2022; Sharma *et al.*, 2024)
Rice husk and wheat bran	Lignocellulose, fiber, hemicellulose	Silica, phosphorus, calcium	~40–55%	~10–20%	Works well as carrier material in solid-state fermentation	Requires processing for microbial accessibility	(Uwa *et al.*, 2018; Gunjal and Kapadnis, 2020; Lin *et al.*, 2023; Akashdeep *et al.*, 2024; Jabran *et al.*, 2024)
Sugarcane bagasse	Cellulose, lignin, hemicellulose	Potassium, phosphorus, trace metals	~40–60%	~30–50%	Supports solid-state fermentation for AM fungi and PSB	Requires enzymatic breakdown for microbial use	(Sidana *et al.*, 2014; Ehis-Eriakha *et al.*, 2023; Peralta *et al.*, 2024)
Vegetable and legume peels	Starch, fiber, simple sugars	Magnesium, potassium, calcium	~55–70%	~50–75%	Works as organic carbon source in mixed inoculant cultures	Rapid decomposition may lead to microbial competition	(A. Daanaa *et al.*, 2020; Naik *et al.*, 2020)

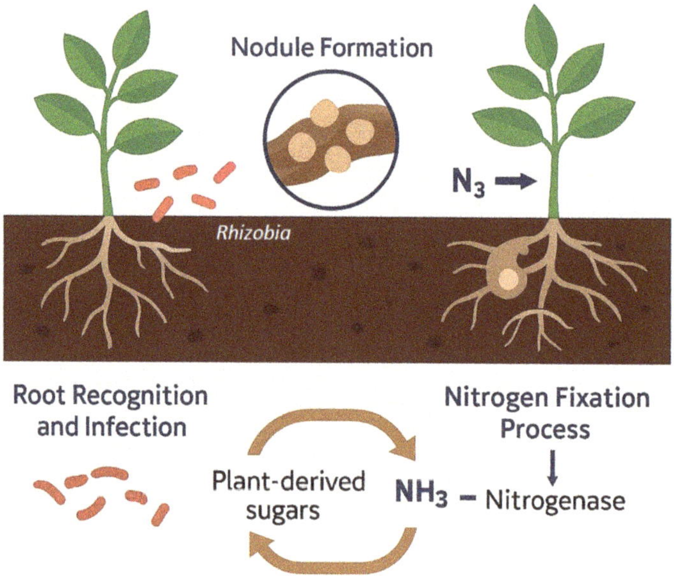

Fig. 4.2. Biological processes with bioinoculants with natural growth benefits. Figure author's own.

et al., 2021; Chopra *et al.*, 2025). Once inside the root, rhizobia induce the development of specialized nodules, creating a microenvironment that shields bacterial colonies from oxygen exposure. This protection is crucial, as nitrogenase—the enzyme responsible for nitrogen fixation—is highly oxygen-sensitive (Chiurazzi *et al.*, 2025). Environmental factors such as soil acidity, salinity, and nutrient availability significantly influence nodule development and nitrogenase activity, ultimately affecting legume productivity (Yeremko *et al.*, 2025).

Within the nodules, nitrogenase enzymatically converts atmospheric nitrogen (N_2) into ammonia (NH_3), which plants assimilate into amino acids and other essential biomolecules. This process is energetically demanding, relying on carbohydrates supplied by the host plant to sustain bacterial metabolism and nitrogen fixation. Recent advances in molecular biology, including CRISPR/Cas9 genome editing, offer promising avenues for optimizing rhizobial inoculants by enhancing nodulation efficiency and nitrogen fixation rates (Ansori *et al.*, 2023; Chen *et al.*, 2024). Additionally, research into stress-responsive miRNAs has revealed their potential in improving legume resilience under adverse environmental conditions, further strengthening the sustainability of rhizobial inoculants in agriculture (Lata *et al.*, 2019; Zhang *et al.*, 2022, 2024; Fahad *et al.*, 2025). Beyond nitrogen provision, this mutualistic exchange enhances soil fertility while reducing reliance on synthetic nitrogen fertilizers, which are associated with environmental concerns such as nitrate leaching and greenhouse gas emissions. Understanding the biochemical and genetic mechanisms governing *Rhizobium*–legume symbiosis is essential for developing improved inoculant formulations. By leveraging agro-biotechnological innovations, researchers aim to enhance nitrogen fixation efficiency, promote soil health, and reduce dependency on chemical fertilizers, thereby advancing sustainable agricultural practices globally.

Several species of rhizobia are widely recognized as effective biofertilizers and bioinoculants, particularly in enhancing nitrogen fixation in leguminous crops. Among the most notable are *Rhizobium leguminosarum*, which is commonly applied to peas, lentils, and vetch; *Sinorhizobium meliloti*, known for its efficacy in inoculating alfalfa and fenugreek; and *Bradyrhizobium japonicum*, well suited for soybean cultivation. Additionally,

Bradyrhizobium elkanii is often utilized in tropical legume systems, particularly for soybeans, while *Mesorhizobium ciceri* is specifically adapted to chickpea crops. *Mesorhizobium loti* is associated with nodulation in *Lotus* species. Notably, *Azorhizobium caulinodans* is unique in its ability to colonize stem nodules in *Sesbania* species. Finally, *Bradyrhizobium* sp. from the cowpea group demonstrates compatibility with groundnut, cowpea, and pigeon pea, expanding its utility across a range of tropical leguminous crops.

4.3.1 Rhizobial inoculants and their production using agro-industrial by-products

The application of commercial rhizobial inoculants in legume cultivation is a well-established agricultural practice aimed at enhancing nitrogen fixation (dos Santos Sousa *et al.*, 2022). Surprisingly many studies have shown that their efficiency is consistent even under abiotic stresses such as soil drought and salinity (Lindström and Mousavi, 2020; dos Santos Sousa *et al.*, 2022; Ali *et al.*, 2023). These inoculants are typically available in liquid formulations or solid preparations mixed with carrier materials. Their production involves a critical phase where large populations of rhizobial cells are cultivated before being incorporated into formulated products (Ben Rebah *et al.*, 2007). The efficiency and economic feasibility of this process are largely dictated by the cost of the growth medium and the accessibility of carrier materials such as peat for solid inoculant production. A range of agro-industrial by-products—including cheese whey and malt sprouts—contain essential growth nutrients such as nitrogen and carbon, which support rhizobial proliferation. Additionally, alternative carrier substrates, including plant compost, filter mud, and fly ash, have been investigated for their potential use in rhizobial inoculant formulations. Recent advancements have highlighted the viability of wastewater sludge, a globally abundant recyclable waste, as both a nutrient-rich growth medium and a carrier in its dehydrated form. Typically, sludge contains sufficient nutrients to sustain rhizobial growth while maintaining heavy metal concentrations within safe limits. Optimizing sludge-based inoculant production may involve pretreatment steps or nutrient supplementation to further enhance microbial viability. Studies indicate that rhizobial inoculants produced using wastewater sludge exhibit effective nodulation and nitrogen fixation capabilities in legumes, performing comparably to or better than conventional inoculants. This innovative approach presents a dual advantage, simultaneously addressing environmental concerns associated with waste disposal and offering a sustainable, cost-effective solution for biofertilizer production in modern agriculture.

In the search for cost-effective and sustainable production methods, agro-industrial by-products have proven to be excellent substrates for the cultivation of rhizobial inoculants (Ben Rebah *et al.*, 2007). These by-products offer a dual advantage: they reduce production costs and contribute to waste valorization. Ben Rebah *et al.* (2007) is a beautifully crafted literature review that explores the use of agro-industrial by-products and wastewater sludge as alternative growth media and carriers for commercial rhizobial inoculant production. A study highlights the potential of agrifood waste-derived media in supporting *Rhizobium leguminosarum* growth and exopolysaccharide (EPS) production (Sellami *et al.*, 2015). Industrial wastewater from oil processing and fish industries demonstrated significant promise as alternative substrates, sustaining bacterial viability and metabolic activity. The composition of these waste streams influenced both bacterial proliferation and EPS yields, with fish processing wastewater (WWFP) facilitating growth comparable to standard yeast extract mannitol broth (YMB). Notably, combining wastewater from oil companies (WWOC2) and WWFP in a 50:50 ratio yielded the highest EPS production (42.4 g/l after 96 hours), surpassing conventional media (Sellami *et al.*, 2015). These findings underscore the feasibility of repurposing agro-industrial waste for microbial cultivation, enhancing sustainable biofertilizer production while reducing environmental waste disposal burdens. The ability of *Rhizobium leguminosarum* to thrive in such media opens new avenues for economically viable inoculant formulations with broad biotechnological applications. A different example from literature investigated the use of food industry wastewater, including mushroom and dill waste, as an alternative to conventional media for *Rhizobium* cultivation (Singh *et al.*, 2019). The results showed that a 90% food waste-based medium supported superior bacterial growth compared to standard media like yeast extract mannitol (YEM) and tryptone yeast (TY) agar. Similarly, a research study evaluated sugar waste (molasses) as a cost-effective medium for *Rhizobium trifolii* cultivation (Singh *et al.*, 2011). The study optimized pH, temperature, and incubation conditions, demonstrating that 10% sugar waste provided superior growth compared to traditional laboratory media. Recent findings by Maluk's group (Maluk *et al.* 2022) emphasize that effective rhizobial symbiosis can occur even in farmland without recent legume cropping, suggesting rhizobia persist in diverse soil environments. Furthermore, incorporating elite rhizobial strains as bioinoculants could enhance nitrogen fixation efficiency in specific soils, presenting opportunities for improving sustainable agriculture and reducing synthetic nitrogen fertilizer dependency. This study evaluates the BNF potential of faba bean (*Vicia faba L.*) in a crop rotation system without recent legume history. It explores how compost application influences nitrogen accumulation and assesses soil populations of *Rhizobium leguminosarum* sv. *viciae* (Rlv), identifying the genetic diversity of nodulating strains and comparing BNF across different sites in Britain. Using 15N natural abundance analysis, the study found that faba bean consistently acquired over 80% of its nitrogen from BNF, irrespective of variety or year. Compost application significantly improved nitrogen accumulation in specific cultivars, while soil rhizobial populations remained stable (~10^5–10^6 Rlv cells/g soil) across all fields. Genetic characterization revealed two major *nodAD* clades, with some strains demonstrating superior nodulation and nitrogen fixation compared to commercial inoculants.

Interestingly, new, highly effective rhizobial strains—absent in the initial inocula— may emerge during in vitro culture of inocula with mixed stocks, suggesting possible microbial evolution or recruitment from native populations. This peculiar example is from a very old study where five *Rhizobium trifolii* strains were used to inoculate subterranean clover (*Trifolium subterraneum* cv. Woogenellup) grown in soils containing native *R. trifolii* populations (Gibson *et al.*, 1976). In the first year, when strains were applied individually, all were well represented in the root nodules. However, when a mixed-strain inoculant was applied, one strain—WU95—became dominant across both field sites. Long-term monitoring over three years at one location revealed that WU95, CC2480a, and WU290 consistently maintained high occupancy in nodules, while the remaining two strains declined in frequency. This study underscores the importance of both strain selection and persistence when formulating rhizobial inoculants for legume crops. While some strains like WU95 and WU290 demonstrated strong competitiveness and long-term survival in nodules, others failed to persist, highlighting the need for ecological adaptability. The emergence of naturally occurring, highly effective strains over time also suggests that soil microbial dynamics and indigenous populations can influence inoculation outcomes. Furthermore, variability within a single strain like WU290 points to the need for genotypic and phenotypic monitoring to ensure inoculant reliability in field applications.

4.3.2 Mass multiplication techniques and formulation methods for rhizobial inoculants

Translating laboratory successes into commercially viable microbial inoculants requires scalable and robust mass multiplication techniques. Two primary approaches—liquid fermentation and solid-state fermentation with carrier-based formulations—are commonly employed to enhance microbial survival and efficacy in agricultural applications. In liquid fermentation, rhizobia are cultivated in controlled environments using stirred-tank bioreactors. These systems regulate key parameters such as pH, temperature, and dissolved oxygen to optimize bacterial growth. Fermentation media, often supplemented with agro-industrial substrates like molasses and whey, support high microbial proliferation and can be scaled up in batch or continuous culture systems. While liquid formulations allow rapid multiplication, maintaining culture purity and effective sterilization protocols during large-scale production is essential. Solid-state fermentation, on the other hand, involves the inoculation of rhizobia onto carriers such as peat, vermiculite, biochar, and lignite. These solid carriers facilitate microbial adhesion and survival, creating a stable matrix for long-term inoculant viability. Encapsulation techniques, including alginate or chitosan-based bead formulations, further enhance microbial stability, protecting bacteria from environmental stress and ensuring controlled release upon field application. Different formulation types, such as seed coatings, granules, and pelletized biofertilizers, improve delivery efficiency. Seed coatings foster early-stage symbiotic interactions, while granules and pellets provide prolonged microbial activity under diverse environmental conditions.

By integrating optimized fermentation strategies with advanced formulation techniques, commercially produced rhizobial inoculants can achieve higher viability and efficacy, bridging the gap between laboratory research and practical agricultural applications (Chandarana *et al.*, 2023). These innovations ensure enhanced microbial performance during storage and field application, contributing to sustainable and efficient biofertilizer development. For example, Begom and coworkers' study focuses on optimizing mass production of *Rhizobium leguminosarum* using a modified air-lift bioreactor, an efficient and cost-effective alternative to conventional techniques (Begom *et al.*, 2021). The research examines various nutrient media, environmental regulators, and aeration parameters to improve bacterial proliferation in submerged culture conditions. Key findings include the identification of yeast mannitol agar (YMA) as the most supportive medium for *Rhizobium* growth, with optimal cultivation conditions observed at 25–30°C, neutral pH (7.0), and low salinity (1%). Additionally, aeration and agitation were critical in maximizing biomass yield, with 0.1 VVM airflow yielding the highest bacterial density. The study highlights the significance of controlled oxygen supply in enhancing microbial viability and underscores the potential of using low-cost bioreactors and readily available substrates to scale up *Rhizobium* inoculant production efficiently. These findings offer valuable insights for sustainable biofertilizer development while addressing economic and environmental concerns in microbial culture systems. Hardy and group investigated biochar as an alternative carrier material for rhizobial inoculants, addressing the limitations of peat, which is vulnerable to environmental variability and availability constraints (Hardy *et al.*, 2021). The research evaluates the capacity of different biochars to support *Rhizobium leguminosarum* survival, assessing their physicochemical properties, phytotoxicity, and effectiveness in promoting nodulation in pea plants. Among the nine biochars tested, six maintained viable rhizobial populations for 84 days at 4°C, with two biochars sustaining $>1 \times 10^6$ cfu/g biochar. The study identifies C/N ratio and carbon content as key factors influencing microbial survival. Growth chamber experiments confirm that the selected biochars successfully delivered *R. leguminosarum*, enhancing nodulation, biomass accumulation, and nitrogen fixation beyond non-inoculated controls. These findings highlight the potential of biochar as a viable carrier for rhizobial inoculants, offering a sustainable and scalable alternative to peat in biofertilizer applications. Erdiyansyah and colleagues conducted studies at the Jember State Polytechnic Plant Protection Laboratory from May to August 2021, evaluating the effectiveness of household waste-derived media in supporting bacterial growth (Erdiansyah *et al.*, 2022). Four alternative media—rice washing water, rice bran water, soybean dregs water, and YMA as a control—were tested across 20 experimental units (five replicates per treatment). Data analysis included both quantitative assessments, using total plate count (TPC) and T-test techniques, and qualitative observations, such as bacterial purity and gram staining. Results indicated that rice bran water supported the highest bacterial population, with 6.80×10^{24} CFU, while rice washing water exhibited the lowest bacterial growth at 1.28×10^{24} CFU. Bacterial viability was monitored over four months, comparing aseptic and conventional media conditions, with T-test results (1.49×10^{-1} CFU) showing no statistically significant differences. These findings suggest that household waste media, particularly rice bran water, may serve as viable substrates for bacterial cultivation, providing a sustainable and cost-effective alternative to conventional media.

Rhizobial inoculants significantly enhance sustainable agriculture through their efficient nitrogen fixation in legume crops. The utilization of agro-industrial by-products such as molasses and whey not only reduces production costs but also supports environmental sustainability by turning waste into valuable resources. Moreover, advanced liquid and solid-state fermentation techniques combined with innovative formulation methods ensure that these biofertilizers are both effective and robust from the lab to the field.

An old study by Catroux and colleagues compellingly highlights the paradox at the heart of rhizobial inoculant use (Catroux *et al.* 2001). While these biological inputs have demonstrably improved nitrogen fixation and legume yield, the majority of products available globally remain suboptimal in quality and consistency. The authors bring necessary attention to this issue by unpacking the causes—ranging from poorly designed formulations to the absence of universal quality control protocols. A major strength of the study lies in its evidence-based emphasis on inoculation rates and their link to efficacy, offering a data-driven approach to improving inoculant design. It acknowledges the strides made in extending shelf life through new production technologies, yet rightly critiques the industry's slow uptake of available tools for assessing physiological viability—such as tests that go beyond cell counts and address microbial functionality. The discussion on quality assurance gaps is especially relevant: despite the availability of straightforward quality control methods, only a few large-scale producers implement them rigorously. As a result, many commercial inoculants deliver inconsistent rhizobia numbers or harbor unwanted contaminants, undermining farmer trust and agronomic potential. Finally, the study is forward-looking. By proposing an increase in standards—particularly in rhizobia delivery per seed and contamination control—it sets a practical roadmap for reform. Its mention of emerging technologies that enhance both efficacy and field reliability underlines a hopeful path forward, provided the industry embraces innovation and regulatory rigor (Catroux *et al.*, 2001). This work makes a clear and timely call for upgrading global biofertilizer quality through science-backed formulation, quality assurance, and technological integration. Appreciating the agronomic promise of rhizobial inoculants means holding them to higher standards—standards this study helps define with clarity and conviction.

4.4 *Azotobacter* Inoculants

4.4.1 Plant growth-promoting potential

Azotobacter species are free-living, nitrogen-fixing bacteria that play a multifaceted role in promoting plant growth and enhancing soil fertility, making them valuable bioinoculants in sustainable agriculture. Unlike symbiotic nitrogen fixers, *Azotobacter* operates independently in the rhizosphere, fixing atmospheric nitrogen into ammonia via nitrogenase activity, thereby enriching nitrogen-deficient soils and supporting non-leguminous crops such as wheat, maize, and chili (Aasfar *et al.*, 2021). Beyond nitrogen fixation, *Azotobacter* synthesizes a suite of phytohormones—including indole-3-acetic acid (IAA), gibberellins, and cytokinins—that stimulate root elongation, enhance cell division, and improve overall plant vigor (Ghatage *et al.*, 2024). These hormones have been directly linked to increased shoot and root biomass, chlorophyll content, and fruit yield in crops like *Capsicum annum* (Das and TK, 2025).

Another critical trait is the secretion of siderophores—low-molecular-weight iron-chelating compounds—that enhance iron uptake in iron-limited soils, a function essential for chlorophyll synthesis and enzymatic activity (Ahmed and Holmström, 2014). Studies have shown that *Azotobacter* strains maintain siderophore production even under saline and drought stress, contributing to their resilience and effectiveness in marginal environments. Additionally, *Azotobacter* produces EPS, which improve soil aggregation, water retention, and microbial colonization. EPS also protect nitrogenase enzymes from oxidative damage and facilitate biofilm formation, enhancing microbial survival and plant–microbe interactions (Gauri *et al.*, 2012).

Recent pot experiments and field trials have demonstrated that *Azotobacter* inoculation significantly improves plant growth parameters under abiotic stress conditions, including salinity, pH fluctuations, and drought. For instance, strains like AztRMD2 have shown superior EPS production and cyst formation, enabling better survival and performance in rice ecosystems under water-limited conditions (Sivapriya and Priya, 2017). These findings underscore the potential of *Azotobacter* as a robust PGPR capable of enhancing crop productivity while reducing dependency on chemical fertilizers.

4.4.2 Nutritional requirements and waste-derived media alternatives

Azotobacter requires nutrient-rich media to achieve high cell densities and maintain metabolic vigor during mass cultivation. Traditionally, synthetic media containing glucose or sucrose as carbon sources, along with mineral salts, have been used to support its growth. However, the economic and environmental costs associated with conventional media have prompted a shift toward waste-derived alternatives that are both cost-effective and sustainable. Agro-industrial by-products such as molasses—a sugar industry residue rich in sucrose—and whey—a dairy by-product abundant in lactose and proteins—have proven to be excellent substrates for *Azotobacter* cultivation. Studies have demonstrated that these substrates not only support robust bacterial growth but also enhance the production of valuable biopolymers like alginate and polyhydroxybutyrate (Khan *et al.*, 2015). Additionally, fruit and vegetable wastes, cereal bran, and spent brewery grains offer a complex nutrient matrix, including amino acids, vitamins, and trace elements, which further stimulate *Azotobacter* metabolism and nitrogenase activity (Jadhav *et al.*, 2018). These substrates have been successfully used to formulate low-cost culture media that rival or exceed the performance of standard formulations in terms of biomass yield and nitrogen fixation efficiency. Moreover, the valorization of such waste streams aligns with circular economy principles, reducing environmental burdens while enhancing the scalability of biofertilizer production. The integration of these waste-derived media not only addresses economic constraints but also supports sustainable agriculture by minimizing reliance on synthetic inputs and promoting microbial resource recovery.

4.4.3 Carrier materials from agro-waste

The successful commercialization and field application of *Azotobacter* inoculants depend heavily on the selection of suitable carrier materials that ensure microbial viability, ease of handling, and effective rhizosphere colonization. Agro-waste-derived carriers have emerged as promising alternatives due to their cost-effectiveness, local availability, and environmental sustainability (Table 4.3). Among these, composted rice or wheat husks and sawdust provide a nutrient-rich, organic matrix that supports long-term microbial survival while contributing organic matter to the soil ecosystem. Other carriers such as sugarcane bagasse, biochar from crop residues, and composted vegetable waste offer additional advantages, including high porosity and moisture retention, which create a favorable microenvironment for *Azotobacter* during storage and after soil application. Biochar, in particular, has been shown to enhance microbial colonization and nutrient retention, especially when derived from rice husk or sawdust at optimized pyrolysis temperatures. These carriers can be formulated into powders, granules, or pellets, each tailored for specific delivery methods such as seed coating, soil mixing, or furrow application. Granular formulations using immobilization matrices like press mud and molasses have demonstrated improved microbial survival and field efficacy under varying water regimes. By leveraging agro-waste carriers, producers not only reduce dependency on synthetic materials but also align with circular economy principles, enhancing the ecological footprint and agronomic performance of biofertilizer products.

Azotobacter inoculants thus represent a cornerstone of sustainable agriculture due to their ability to fix atmospheric nitrogen independently, produce phytohormones, and enhance soil structure and fertility. Their multifunctional role not only improves crop productivity but also reduces dependence on synthetic fertilizers, aligning with agroecological principles. The adoption of waste-derived media, such as molasses, whey, and fruit pulp, has emerged as a cost-effective and eco-friendly strategy for large-scale *Azotobacter* cultivation, significantly lowering production inputs while valorizing agro-industrial residues (Gatea *et al.*, 2019). Furthermore, the integration of agro-waste-based carrier materials—including composted husks, sawdust, and biochar—has proven effective in maintaining microbial viability, enhancing shelf life, and facilitating efficient delivery to the rhizosphere (Mankar *et al.*, 2021). These innovations not only improve the agronomic performance of *Azotobacter* formulations but also contribute to circular economy goals by transforming organic waste into high-value biofertilizer components. Collectively, these advancements bridge the gap between laboratory-scale research and real-world application, reinforcing the role of *Azotobacter* as a scalable, sustainable solution for modern agriculture.

Table 4.3. Comparison of carrier materials derived from biodegradable waste, highlighting their nutrient retention, microbial compatibility, cost-effectiveness, and other critical parameters for formulation. Table author's own.

Carrier material	Composition and properties	Moisture retention capacity	Microbial compatibility	Cost-effectiveness	Shelf life and stability	Potential limitations
Composted husk (rice/wheat)	High in organic carbon, retains nutrients well, porous texture for microbial adhesion	Moderate (~40–60%)	Supports rhizobia, PSB, *Azospirillum*	High (widely available agro-waste)	Good (long-term viability under dry conditions)	Requires sterilization to remove competing microbes
Sawdust and wood residues	Rich in carbon, provides physical protection	Low (~20–40%)	Suitable for solid-state fermentation, works well with fungi	High (common forestry waste)	Moderate (depends on processing)	Slow microbial degradation, may require nutrient enhancement
Sugarcane bagasse	Contains cellulose, lignin, provides structural support	High (~50–70%)	Good for AM fungi and cyanobacteria	Moderate (available in sugar industries)	Moderate (sensitive to moisture variations)	Needs pretreatment for microbial adhesion
Biochar (pyrolyzed agro-waste)	Highly porous, excellent carrier for long-term stability	Very high (~70–90%)	Effective for *Frankia*, rhizobia, and AM fungi	Moderate–low (processing required)	Excellent (high microbial survival rates)	Requires controlled pyrolysis, may alter pH of formulations
Spent mushroom substrate	Nutrient-dense, high microbial diversity, contains fungal residues	High (~60–80%)	Supports fungal inoculants, enhances root colonization	Moderate (availability varies)	Good (depends on formulation)	Competes with introduced microbes, requires proper strain selection
Composted food waste	Contains balanced nutrients, microbial-rich organic matter	High (~55–75%)	Works well for *Azotobacter*, PSB, and mycorrhizal fungi	High (diverse waste sources)	Moderate (depends on storage conditions)	Requires pH regulation and contamination control

4.5 *Azospirillum* Inoculants

4.5.1 Association with cereals and grasses

Azospirillum species are free-living PGPR that form beneficial associations with the roots of cereals and grasses, playing a pivotal role in sustainable crop production. These bacteria are adept at colonizing the rhizosphere of major cereals such as wheat, rice, maize, and sorghum, where they utilize root exudates to establish themselves and stimulate root development through biofilm formation and chemotactic responses (Nievas *et al.*, 2023). One of their hallmark traits is biological nitrogen fixation, which occurs under microaerophilic conditions and contributes significantly to nitrogen availability in low-input systems (Boddey and Dobereiner, 1995). In addition to nitrogen fixation, *Azospirillum* spp. produce a suite of phytohormones, including IAA, gibberellins, and cytokinins, which enhance root branching, increase nutrient uptake, and improve shoot biomass (Steenhoudt and Vanderleyden, 2000).

Moreover, *Azospirillum* inoculation has been shown to enhance stress tolerance in cereals by modulating antioxidant enzyme activity and upregulating stress-responsive genes. For instance, *Azospirillum brasilense* strains Ab-V5 and Ab-V6 have been reported to induce the expression of oxidative stress and defense-related genes in maize, contributing to improved drought resilience and photosynthetic efficiency (Fukami *et al.*, 2017). These effects are often attributed to both the bacterial cells and their secreted metabolites, which include IAA, salicylic acid, and indole derivatives. Field studies have consistently demonstrated yield improvements ranging from 10% to 30% in cereals inoculated with *Azospirillum*, with some reports noting nitrogen savings equivalent to 30–60 kg urea N/ha (Sahu *et al.*, 2017). Collectively, these findings underscore the value of *Azospirillum* as a bioinoculant for cereals and grasses, offering a biologically based strategy to enhance nutrient use efficiency, stress resilience, and overall crop productivity in both conventional and low-input farming systems.

4.5.2 Scalable production using agro-residues

The scalable production of *Azospirillum* inoculants can be significantly enhanced through the strategic use of agro-residues as alternative substrates, offering both economic and environmental benefits. Agro-industrial by-products such as wheat bran, rice husk, and sugarcane bagasse are rich in carbohydrates, lignocellulosic matter, and micronutrients, making them ideal low-cost substrates for microbial cultivation. These materials have been successfully employed in both liquid and solid-state fermentation (SSF) systems, supporting high cell densities and metabolic activity of *Azospirillum brasilense* strains (Martínez-Ramírez *et al.*, 2021). In SSF, substrates like composted bagasse and bran provide a porous matrix that enhances oxygen diffusion and microbial adhesion, while liquid fermentation in bioreactors allows for precise control of pH, aeration, and nutrient availability, ensuring reproducible yields at scale (Trujillo-Roldán *et al.*, 2013).

Moreover, the integration of agro-residues into fermentation media aligns with circular economy principles, transforming agricultural waste into value-added biofertilizer products. This not only reduces production costs but also mitigates environmental burdens associated with waste disposal (Akay *et al.*, 2012). Pilot-scale studies have demonstrated that *Azospirillum* cultures grown on such substrates maintain high viability and shelf life, with some formulations remaining effective for over 12 months under ambient storage conditions (Martínez-Ramírez *et al.*, 2021). These findings underscore the feasibility of scaling up *Azospirillum* production using agro-residues, offering a sustainable pathway for large-scale bioinoculant manufacturing that supports both soil health and resource efficiency.

4.5.3 Co-formulation with other waste-based biofertilizers

Co-formulating *Azospirillum* inoculants with other waste-based biofertilizers presents a promising strategy to enhance the functional diversity and agronomic impact of microbial inputs in sustainable agriculture. This approach leverages the complementary functions of multiple microbial groups—such as PSB, *Azotobacter*, and arbuscular mycorrhizal fungi (AMF)—to deliver synergistic benefits. While *Azospirillum* primarily contributes through nitrogen fixation and phytohormone production, PSB enhance phosphorus bioavailability, and AMF improve nutrient uptake and stress tolerance through extensive hyphal networks (Domínguez-Núñez *et al.*, 2015). These microbial consortia, when co-inoculated, have been shown to significantly improve root architecture, chlorophyll content, and yield in cereals and legumes (Di Barbaro *et al.*, 2023).

The use of waste-derived carrier systems—including composted crop residues, sawdust, and biochar—further enhances the viability and delivery efficiency of these mixed inoculants. These carriers provide a porous, nutrient-rich matrix that supports microbial survival during storage and facilitates gradual release into the rhizosphere (Sivasakthivelan *et al.*, 2023). Notably, biochar has been shown to improve microbial adhesion and moisture retention, making it particularly suitable for arid and semi-arid conditions. Field trials have demonstrated that co-formulated inoculants exhibit greater consistency across variable soil types and climatic conditions, often outperforming single-strain formulations in terms of nutrient uptake, plant biomass, and yield stability (Dohroo, 2024). Moreover, these integrated products reduce the reliance on chemical fertilizers, aligning with agroecological principles and circular economy goals by valorizing organic waste streams.

In essence, co-formulation strategies represent a scalable and ecologically sound innovation in biofertilizer development, offering farmers a robust, multifunctional tool to enhance soil health and crop productivity under diverse field conditions. *Azospirillum* inoculants have thus emerged as a cornerstone in the advancement of sustainable cereal and grass cultivation, owing to their multifaceted contributions to plant growth promotion. By enhancing root architecture, fixing atmospheric nitrogen, and bolstering plant resilience under abiotic stress conditions, these rhizobacteria significantly improve nutrient uptake and yield stability—particularly in low-input and climate-vulnerable systems (Sahu *et al.*, 2017). The transition toward scalable production using agro-residues—such as wheat bran, rice husk, and sugarcane bagasse—has not only reduced production costs but also aligned *Azospirillum*-based technologies with circular economy principles, transforming agricultural waste into high-value microbial inputs (Martínez-Ramírez *et al.*, 2021). Moreover, the co-formulation of *Azospirillum* with complementary biofertilizers—including phosphate-solubilizing bacteria, *Azotobacter*, and mycorrhizal fungi—has demonstrated synergistic effects on plant nutrition, stress tolerance, and soil health. These integrated formulations, often delivered through waste-derived carriers like biochar or composted residues, have shown superior field performance across diverse agroecological zones (Dohroo, 2024). As fermentation technologies and formulation science continue to evolve, the gap between laboratory-scale innovation and field-level adoption is narrowing, enabling the development of next-generation biofertilizer products that are both ecologically sound and agronomically effective.

4.6 Phosphate-solubilizing Bacteria (PSB)

4.6.1 Use of food waste derivatives in growth substrates

Integrating food waste derivatives—particularly fruit pulp residues—into the cultivation of PSB offers a compelling strategy for sustainable and cost-effective biofertilizer production. Fruit pulp waste from citrus, banana, mango, and guava processing industries is rich in fermentable sugars, organic acids, vitamins, and trace minerals, making it an ideal substrate for microbial growth and metabolic activity (Vieira *et al.*, 2025). These nutrient-dense residues support high-density PSB cultures and stimulate the secretion of key metabolites involved in phosphorus solubilization, such as gluconic and citric acids. Moreover, the use of such agro-industrial by-products aligns with circular economy principles, transforming organic waste into value-added fermentation media while reducing the environmental burden of waste disposal.

Studies have shown that PSB grown on fruit pulp-based media exhibit enhanced enzymatic activity, including increased phosphatase production, which accelerates the hydrolysis of organic phosphorus compounds during fermentation (Mohamed *et al.*, 2022). Additionally, the natural buffering capacity and carbon-to-nitrogen ratio of these substrates create favorable conditions for microbial proliferation and acidification of the growth medium—key factors in effective phosphate solubilization (Merrylin *et al.*, 2020). This approach not only reduces reliance on synthetic media components but also supports low-cost, decentralized production models that are particularly beneficial in resource-constrained agricultural settings.

4.6.2 Process optimization for enhanced solubilization activity

Optimizing the production and formulation of PSB is pivotal for enhancing their field efficacy and ensuring consistent phosphorus mobilization across diverse agroecosystems. A key component of this optimization lies in precise control of fermentation parameters—including pH, temperature, aeration, and nutrient concentration. Studies have shown that maintaining a slightly acidic pH (around 5.5–6.5) significantly boosts the secretion of organic acids like gluconic and citric acids, which are central to phosphate solubilization (Tao *et al.*, 2024). Similarly, optimal aeration and temperature (typically 28–32°C) enhance microbial growth and metabolic output, particularly in liquid and solid-state fermentation systems. To scale up production, batch and continuous cultivation strategies are employed using bioreactors equipped with real-time monitoring systems for pH, dissolved oxygen, and temperature. Continuous systems, though more complex, offer higher productivity and tighter control over microbial physiology, reducing variability in inoculant quality. These systems also allow for dynamic feeding strategies that prevent nutrient depletion and maintain high cell viability.

Advances in genetic and metabolic engineering have opened new avenues for enhancing PSB functionality. Genes such as gcd (glucose dehydrogenase) and pqq (pyrroloquinoline quinone biosynthesis) have been targeted to boost gluconic acid production, a key agent in phosphate solubilization (de Almeida Leite *et al.*, 2024). Engineered strains have demonstrated improved solubilization efficiency and resilience under stress conditions, making them more reliable for field deployment. Finally, integration with waste-derived carrier materials—such as composted fruit residues, sawdust, or biochar—enhances the shelf life and delivery efficiency of PSB formulations. These carriers provide a protective microenvironment that preserves microbial viability during storage and facilitates gradual release into the rhizosphere. Biochar, in particular, has been shown to improve microbial adhesion and moisture retention, making it ideal for arid soils (Bai *et al.*, 2024).

In sum, process optimization—spanning fermentation control, cultivation strategy, genetic enhancement, and carrier integration—forms the backbone of next-generation PSB biofertilizers, ensuring they are not only effective in solubilizing phosphorus but also scalable, stable, and environmentally aligned. This way, PSB represent a powerful biological tool for addressing the widespread challenge of phosphorus unavailability in agricultural soils. Through a combination of organic acid secretion, enzymatic hydrolysis, and proton extrusion, PSB are capable of mobilizing phosphorus locked in insoluble mineral and organic complexes, thereby enhancing its accessibility to plants (Pan and Cai, 2023). These mechanisms not only improve nutrient uptake but also contribute to root development and overall plant vigor, particularly in phosphorus-deficient or heavily fertilized soils where fixation limits efficiency.

The integration of food waste derivatives, such as citrus and banana pulp, into PSB cultivation media has emerged as a sustainable innovation that aligns with circular economy principles. These substrates are rich in fermentable sugars, vitamins, and trace elements, supporting high-density microbial growth while simultaneously valorizing agro-industrial waste (Vieira *et al.*, 2025). Moreover, process optimization strategies—including controlled fermentation parameters, advanced bioreactor systems, and metabolic engineering—have significantly enhanced PSB productivity and solubilization efficiency (Tao *et al.*, 2024). Equally important is the formulation of PSB inoculants with robust, waste-derived carrier systems such as biochar, composted fruit residues, or sawdust. These carriers not only preserve microbial viability during storage and transport but also contribute to soil organic matter and microbial colonization upon application (Bai *et al.*, 2024). Together, these innovations bridge the gap between laboratory-scale research and field-level implementation, enabling the development of next-generation PSB biofertilizers that are cost-effective, environmentally sound, and agronomically impactful.

4.7 Cyanobacterial Inoculants

4.7.1 Importance in paddy fields and degraded soils

Cyanobacterial inoculants, particularly those belonging to nitrogen-fixing genera such as *Nostoc*, *Anabaena*, and *Aulosira*, have long been recognized as vital contributors to sustainable rice cultivation and soil restoration. In paddy ecosystems, these photosynthetic prokaryotes form dense mats or biofilms that fix atmospheric nitrogen (N_2) into bioavailable ammonia, significantly reducing the need for synthetic nitrogen fertilizers. Studies conducted across rice-growing regions—including recent work in Mizoram, India—have demonstrated that diazotrophic cyanobacteria can contribute up to 30 kg of nitrogen per hectare per season, enhancing rice yields by 10–35%, depending on soil conditions and inoculant strains.

Beyond nitrogen fixation, cyanobacteria play a transformative role in degraded soils, including those affected by salinity, nutrient depletion, or erosion. Their ability to secrete EPS facilitates the formation of biological soil crusts (biocrusts), which stabilize soil aggregates, reduce erosion, and improve water retention. These biocrusts also enhance microbial diversity and organic carbon content, creating a more resilient and fertile soil environment (Garcia *et al.*, 2025). From an environmental sustainability perspective, cyanobacterial inoculants offer a nature-based solution that aligns with circular economy and regenerative agriculture principles. Their application reduces greenhouse gas emissions associated with synthetic fertilizer use and supports long-term soil health. Moreover, their adaptability to extreme environments—including high temperatures and salinity—makes them suitable for climate-resilient farming systems (Rocha *et al.*, 2020). In essence, cyanobacteria serve a dual function: they enhance productivity in flooded rice systems while simultaneously rehabilitating degraded soils, positioning them as a cornerstone of eco-friendly and climate-smart agriculture.

4.7.2 Outdoor cultivation using agro-effluents and organic leachates

Outdoor cultivation of cyanobacteria using agro-effluents and organic leachates presents a scalable, eco-efficient strategy for producing biofertilizer-grade biomass. Open systems such as raceway ponds and shallow tanks are particularly well-suited for this purpose, as they can be readily integrated with nutrient-rich agro-effluents—including wastewater from dairy, rice milling, and fruit processing industries. These effluents are naturally abundant in nitrogen, phosphorus, and trace elements, which support high cyanobacterial productivity without the need for synthetic nutrient supplementation (Attene *et al.*, 2023). In parallel, organic leachates derived from composted agricultural residues or fruit pulp waste provide a balanced matrix of macro- and micronutrients, along with organic carbon sources that enhance photosynthetic efficiency and biomass accumulation. These leachates have been shown to stimulate pigment production and carbohydrate synthesis in strains like *Anabaena* and *Nostoc*, improving both yield and biofertilizer quality (Tsolcha *et al.*, 2021).

From a sustainability standpoint, this approach exemplifies circular economy principles by transforming agro-industrial waste streams into valuable microbial inputs. It reduces the environmental burden of effluent disposal while minimizing the carbon and energy footprint of cyanobacterial cultivation. Moreover, the use of low-cost, decentralized infrastructure makes this method accessible to smallholder farmers and rural cooperatives, promoting inclusive biofertilizer production models. In essence, outdoor cultivation using agro-effluents and organic leachates not only enhances the economic feasibility of cyanobacterial inoculants but also reinforces their role as regenerative tools in climate-smart agriculture.

4.7.3 Harvesting and formulation techniques

The final stage in the development of cyanobacterial inoculants involves efficient harvesting and formulation, which are critical for preserving microbial viability, ensuring ease of application, and maintaining field efficacy. Among the most widely adopted harvesting techniques are centrifugation and flocculation, which enable rapid biomass recovery from open cultivation systems while minimizing cellular damage. Flocculation methods—such as chemical, bio-, or auto-flocculation—have been optimized using low-cost agents like calcium hydroxide or cationic polymers, achieving high recovery efficiencies with minimal energy input (Iasimone et al., 2021; Musteret *et al.*, 2021; Haddaji *et al.*, 2024). Additionally, membrane filtration and low-shear filtration systems are employed to concentrate biomass without compromising cell integrity, particularly for delicate filamentous strains (Schlesinger *et al.*, 2012).

Post-harvest, drying and stabilization are essential to extend shelf life and facilitate downstream formulation. Techniques such as spray drying, freeze drying, and controlled sun drying are commonly used, each with trade-offs in cost, energy use, and preservation of bioactive compounds. Spray drying, for instance, offers a lower energy footprint and continuous processing capability, making it suitable for large-scale production, though it requires careful optimization to avoid thermal degradation (Emami *et al.*, 2023). Freeze drying, while more protective of sensitive metabolites, is energy-intensive and better suited for high-value or small-batch applications.

Formulation strategies are tailored to application needs and storage conditions. Liquid concentrates are ideal for immediate use and seed treatments, offering high microbial counts and ease of handling. Powdered and granular formulations, often blended with carriers like biochar, composted residues, or sawdust, enhance field stability, facilitate uniform distribution, and support gradual microbial release.

Cyanobacterial inoculants thus stand at the forefront of regenerative agriculture, offering a biologically elegant solution to persistent challenges in soil fertility and sustainable crop production. Their ability to fix atmospheric nitrogen, enrich soil organic carbon, and foster microbial diversity makes them indispensable in both paddy ecosystems and degraded lands. For instance, inoculation with strains such as *Anabaena variabilis* and *Aulosira fertilissima* has been shown to enhance rice yields by 10–35% and reduce nitrogen fertilizer requirements by up to 30% (Garcia *et al.*, 2025).

The move toward outdoor cultivation using agro-effluents and organic leachates represents a sustainable production breakthrough. Nutrient-rich effluents from dairy and rice-processing industries, alongside leachates from composted residues, provide an abundant and eco-friendly nutrient base for cyanobacterial biomass production (Tsolcha *et al.*, 2021; Attene *et al.*, 2023). Equally transformative are the innovations in harvesting and formulation techniques. Methods such as low-shear filtration, bio-based flocculants, and spray or freeze drying help preserve the integrity of bioactive metabolites. When combined with advanced formulation strategies—like encapsulation using alginate or chitosan—cyanobacterial products exhibit enhanced viability and stability during storage and application (Dickson *et al.*, 2013; Emami *et al.*, 2023). In essence, when cultivated using waste-derived resources, harvested via optimized low-impact techniques, and formulated for resilience and precision delivery, cyanobacterial inoculants offer a holistic and scalable solution for enriching soils, improving rice productivity, and accelerating the shift toward climate-resilient agroecosystems.

4.8 *Frankia* Inoculants

4.8.1 Actinorhizal symbiosis and soil improvement

Frankia inoculants, composed of filamentous actinobacteria from the genus *Frankia*, form a unique and ecologically significant symbiosis with non-leguminous, nitrogen-fixing plants known as actinorhizal hosts—such as *Alnus*, *Casuarina*, *Myrica*, and *Elaeagnus*. This actinorhizal symbiosis mirrors the legume–rhizobia partnership in function but is phylogenetically and structurally distinct. Within specialized root nodules, *Frankia* fixes atmospheric nitrogen (N_2) into ammonia, supplying essential nutrients to the host plant and enriching the surrounding soil—particularly in nitrogen-deficient or degraded environments (Van Nguyen and Pawlowski, 2017). One of the most compelling features of this symbiosis is its resilience in challenging environments. *Frankia* strains have evolved vesicle structures with lipid envelopes that protect the oxygen-sensitive nitrogenase enzyme, enabling efficient nitrogen fixation even under low-oxygen or stress-prone conditions (Ghodhbane-Gtari *et al.*, 2014; Harriott *et al.*, 1991; Serazetdinova *et al.*, 2025). These nodules not only contribute to nitrogen input but also enhance soil structure by increasing organic matter, promoting microbial diversity, and improving water retention. As actinorhizal plants grow, they stabilize soil aggregates, reduce erosion, and facilitate nutrient cycling, making them ideal for revegetation and land reclamation efforts (Diagne *et al.*, 2013).

Over the long term, the cumulative effects of nitrogen enrichment and organic carbon deposition lead to the transformation of barren or chemically degraded soils into fertile, biologically active ecosystems. Actinorhizal plants, often considered pioneer species, initiate ecological succession by creating conditions favorable for the establishment of other plant species, thereby enhancing biodiversity and agroecosystem resilience (Bhattacharyya *et al.*, 2024). To summarize, *Frankia*-based inoculants offer a powerful, nature-based solution for soil fertility enhancement, ecological restoration, and sustainable land management, particularly in marginal or disturbed landscapes.

4.8.2 Cultivation challenges and innovative waste-supported strategies

Cultivating *Frankia* strains for inoculant production presents a unique set of challenges due to their slow growth rates, complex nutritional requirements, and sensitivity to artificial culture conditions. Unlike many other biofertilizer-producing microbes, *Frankia* often requires 4–8 weeks to form mature colonies, with growth highly dependent on medium composition and environmental parameters (Gtari *et al.*, 2024)). These actinobacteria are notoriously recalcitrant to axenic cultivation, particularly strains from cluster-2 lineages, which exhibit atypical auxotrophies and may require symbiotic-like cues for sustained growth (Gtari *et al.*, 2015). To address these limitations, researchers have begun exploring waste-supported culture strategies that mimic natural rhizosphere conditions. Nutrient-rich compost leachates and organic extracts from fruit residues—such as banana peel filtrates or citrus pulp infusions—have shown promise in supporting *Frankia* viability and metabolic activity. These substrates are rich in soluble carbon, amino acids, and micronutrients, and may contain signaling molecules analogous to plant root exudates, thereby enhancing vesicle formation and nitrogenase expression (Roy *et al.*, 2018). Moreover, the use of agro-industrial waste aligns with circular economy principles, offering a dual benefit: reducing production costs and valorizing organic waste streams. For instance, compost leachates not only provide a buffered pH and balanced nutrient profile but also reduce contamination risks when properly treated (Ogbu and Okey, 2023). These waste-derived media have been successfully integrated into *Frankia* cultivation protocols, particularly when combined with mild aeration and alkaline pH adjustments (28–29°C), which favor vesicle development and biomass accumulation (Thirugnanam and Dharumadurai, 2023).

Overall, innovative cultivation strategies using compost leachates and fruit-derived substrates offer a promising path forward for overcoming the intrinsic challenges of *Frankia* propagation. These approaches not only enhance microbial viability and metabolic output but also support the transition from lab-scale experimentation to commercial-scale inoculant production in an economically and environmentally sustainable manner.

4.8.3 Ecological restoration applications

Frankia inoculants hold exceptional promise in the realm of ecological restoration, particularly for rehabilitating degraded, nutrient-poor, or stress-impacted landscapes. When paired with actinorhizal host plants such as *Casuarina*, *Alnus*, and *Elaeagnus*, *Frankia* forms nitrogen-fixing root nodules that significantly enhance soil fertility and structure. These symbiotic systems have been successfully deployed in reforestation and land reclamation projects, especially in areas affected by industrial degradation, salinity, or drought stress. For example, inoculated *Casuarina equisetifolia* has demonstrated improved growth and nitrogen uptake in saline soils, accelerating the recovery of barren lands (Diagne *et al.*, 2013). Beyond nutrient enrichment,

Frankia-inoculated plants contribute to phytoremediation and land stabilization. Their deep root systems anchor soil, reduce erosion, and initiate the gradual buildup of organic matter and microbial diversity. This creates a self-sustaining fertility loop, enabling the colonization of other plant species and fostering natural succession in previously uninhabitable environments (Bhattacharyya *et al.*, 2024). Moreover, *Frankia* has shown potential in the biotransformation of allelochemicals and heavy metals, offering additional value in the remediation of chemically contaminated sites (Gtari *et al.*, 2024). Perhaps most importantly, these inoculants support biodiversity and long-term sustainability by facilitating the establishment of resilient plant communities. As pioneer species, actinorhizal plants inoculated with *Frankia* improve soil health and nutrient cycling, laying the groundwork for broader ecological recovery and climate-resilient land use (Bhattacharyya *et al.*, 2024).

In short, *Frankia*-based bioinoculants are not just soil enhancers—they are ecological engineers, catalyzing the transformation of degraded landscapes into thriving, biodiverse ecosystems. Through their unique actinorhizal symbiosis with non-leguminous host plants such as *Alnus*, *Casuarina*, and *Elaeagnus*, *Frankia* bacteria fix atmospheric nitrogen and enrich soil organic matter, thereby improving fertility, microbial diversity, and structural integrity in nutrient-depleted or degraded soils (Van Nguyen and Pawlowski, 2017). These symbiotic systems have been successfully deployed in reforestation, saline soil reclamation, and post-industrial land rehabilitation, where they initiate ecological succession and promote long-term soil resilience (Diagne *et al.*, 2013).

Despite the intrinsic cultivation challenges posed by *Frankia*—including slow growth and complex nutritional demands—emerging innovations in waste-supported culture media have opened new avenues for scalable production. The use of compost leachates and fruit-derived organic substrates has shown promise in mimicking natural rhizosphere conditions, enhancing microbial viability, and reducing production costs (Roy *et al.*, 2018). These strategies not only align with circular economy principles but also valorize agro-industrial waste streams, contributing to sustainable biomanufacturing. Moreover, *Frankia*-based inoculants are increasingly recognized as ecological engineers—capable of stabilizing soils, reducing erosion, and facilitating biodiversity recovery in marginal landscapes (Bhattacharyya *et al.*, 2024). Their integration into land management practices supports the restoration of ecosystem services, enhances carbon sequestration, and fosters agroecological resilience in the face of climate change. In conclusion, *Frankia* inoculants are not merely microbial inputs—they are strategic enablers of ecosystem regeneration and sustainable land stewardship. As cultivation technologies mature and field applications expand, these bioinoculants are poised to become indispensable tools in the global effort to rehabilitate degraded lands and build resilient agroecosystems.

4.9 Arbuscular Mycorrhizal (AM) Fungi

4.9.1 Role in enhancing nutrient uptake and stress resilience

AM fungi significantly enhance water absorption and drought tolerance. Their hyphal networks explore soil micropores inaccessible to roots, improving water uptake and maintaining plant turgor under moisture stress. This is mediated by increased expression of aquaporins and osmoprotective compounds in AM-colonized plants (Bahadur *et al.*, 2019). Moreover, AM symbiosis modulates hormonal signaling pathways—particularly abscisic acid and jasmonic acid—thereby enhancing plant resilience to abiotic stresses such as salinity, drought, and heavy metal toxicity (Cheng *et al.*, 2021).

AM fungi also play a pivotal role in biotic stress mitigation, reducing pathogen colonization through induced systemic resistance (ISR), competition for root niches, and secretion of antimicrobial compounds. These interactions are further supported by the upregulation of stress-responsive genes and reinforcement of cell wall structures, which collectively enhance plant immunity. A hallmark contribution of AM fungi to soil structure is the production of glomalin-related soil proteins (GRSPs)—notably glomalin, a recalcitrant glycoprotein that binds soil particles into stable aggregates. Glomalin improves soil aggregation, water retention, and carbon sequestration, making it a key determinant of soil fertility and resilience (Singh *et al.*, 2013). Its persistence in soil contributes to long-term organic matter stabilization and erosion control. In summary, AM fungi function as biological multipliers of plant performance, enhancing nutrient and water uptake, fortifying stress tolerance, and improving soil structure. Their integration into agroecosystems is essential for reducing chemical input dependency and achieving resilient, climate-smart agriculture.

4.9.2 Substrate formulation using composted agro-waste

Substrate formulation using composted agro-waste has emerged as a sustainable and effective strategy for the mass production of AM fungi, offering both agronomic and environmental benefits. These substrates—derived from crop residues, green manure, and fruit or vegetable waste—are rich in organic matter, micronutrients, and microbial stimulants that create a favorable microenvironment for AM fungal colonization and sporulation (Yang *et al.*, 2018). The porous structure and moisture-retentive properties of composted materials enhance aeration and nutrient diffusion, which are critical for spore germination and hyphal proliferation. Studies have shown that compost-amended substrates significantly increase spore density and root colonization rates, particularly when applied at moderate to high compost levels (22.5–45 Mg/ha), without disrupting AM fungal community composition (Yang *et al.*, 2018). From a sustainability perspective, the use of composted agro-waste aligns with circular economy principles by transforming agricultural residues into high-value microbial carriers. This reduces reliance on synthetic additives and mitigates the environmental burden of organic waste disposal (Chaiyasen *et al.*, 2017). Moreover, composted substrates have been shown to support enhanced fungal viability during storage, owing to their buffering capacity and microbial compatibility. The presence of humic substances and plant-like exudates in compost may also mimic natural rhizosphere conditions, further stimulating AM fungal activity.

Importantly, these substrates can be tailored to specific crop systems by adjusting parameters such as carbon-to-nitrogen ratio, pH, and moisture content. For example, composts derived from leguminous residues may enhance phosphorus solubilization, while those from fruit waste may boost carbohydrate availability for fungal metabolism (Nandikolmath, 2023). This adaptability makes composted agro-waste an ideal base for customized AM fungal formulations that meet the nutritional and ecological needs of diverse cropping systems. Therefore, at the outset, it can be concluded that composted agro-waste serves as a biologically active, cost-effective, and environmentally sound substrate for AM fungal inoculant production—bridging the goals of microbial biotechnology, sustainable agriculture, and waste valorization.

4.9.3 Mass multiplication and packaging technologies

Successful deployment of AM fungal inoculants in field settings hinges on the integration of robust mass multiplication techniques with advanced formulation and packaging technologies that preserve viability and ensure consistent field performance. Traditional bulk production methods—such as pot and trap cultures—remain widely used, wherein host plants like maize or sorghum are grown in sterilized substrates to propagate AM fungal spores and hyphae. These systems, often conducted in greenhouses or raised beds, allow for repeated inoculation cycles and yield high-density propagules suitable for downstream formulation (Raut *et al.*, 2019). Recent innovations have focused on in vitro cultivation using root organ culture (ROC) techniques, particularly with Ri T-DNA-transformed carrot roots, which enable the axenic propagation of AM fungi under sterile, controlled conditions. This method ensures the production of high-purity, genetically stable inoculants and facilitates the study of fungal development and host interactions (Cranenbrouck *et al.*, 2005). Once mass multiplication is achieved, formulation and packaging become critical for maintaining inoculum efficacy. Granular and pelletized formulations, often incorporating biochar or composted substrates, have demonstrated excellent moisture retention and UV protection, making them ideal for mechanical application and long-term storage. These carriers also enhance soil structure and microbial colonization upon application. Liquid suspensions offer a flexible alternative, particularly for seed coating or foliar application, where concentrated spores and hyphal fragments are suspended in nutrient-rich media. However, their shelf life is generally shorter and requires cold-chain logistics. To further enhance stability, encapsulation technologies using biopolymers like alginate and chitosan have gained traction. These encapsulated formulations provide controlled release, protect against desiccation and environmental stress, and extend shelf life under ambient conditions (da Silva Simões *et al.*, 2025).

Finally, quality control protocols—including spore viability assays, colonization efficiency tests, and field performance evaluations—are essential for standardizing commercial AM fungal products. Packaging innovations such as vacuum sealing and modified atmosphere packaging further preserve biological activity during storage and transport (Akhtar and Abdullah, 2014). The convergence of scalable propagation systems with precision formulation and packaging technologies ensures that AM fungal inoculants are not only biologically effective but also commercially viable—paving the way for their widespread adoption in regenerative and climate-smart agriculture.

AM fungi have emerged as cornerstone symbionts in sustainable agriculture, offering a biologically robust solution to nutrient limitation, water scarcity, and environmental stress. By forming mutualistic associations with plant roots, AM fungi significantly enhance the uptake of immobile nutrients—particularly phosphorus, potassium, and micronutrients—while simultaneously improving water absorption and drought resilience through their extensive hyphal networks (Nie *et al.*, 2024). These fungi also activate plant defense pathways, modulate stress-responsive gene expression, and contribute to systemic resistance against pathogens and abiotic stressors (Bahadur *et al.*, 2019).

The integration of composted agro-waste as a substrate has revolutionized AM fungal inoculant production by providing a nutrient-rich, porous, and biologically active medium that supports spore germination, hyphal proliferation, and long-term viability. Substrates derived from crop residues, green manure, and fruit waste not only reduce production costs but also align with circular economy principles by valorizing organic waste streams (Yang *et al.*, 2018). Advancements in mass multiplication and packaging technologies—including in vitro root organ cultures, granular and encapsulated formulations, and modified atmosphere packaging—have further ensured the scalability, stability, and field efficacy of AM fungal products (Cranenbrouck *et al.*, 2005; da Silva Simões *et al.*, 2025). These innovations enable the delivery of high-quality inoculants that retain biological activity during storage and perform consistently across diverse agroecological zones. In sum, the convergence of biological efficacy, eco-friendly substrate use, and technological innovation positions AM fungi as indispensable allies in the global transition toward regenerative, climate-resilient agriculture.

4.10 Waste-derived Substrates and Carrier Materials

4.10.1 Agricultural and food-processing wastes as culture media

Agricultural and food-processing wastes represent a largely untapped reservoir of nutrient-rich substrates that can be repurposed as culture media for microbial inoculant production. These materials—such as molasses, whey, fruit pulp, vegetable peels, and spent grains—are abundant in fermentable sugars, amino acids, vitamins, and trace minerals, making them ideal alternatives to synthetic or commercial media components. For instance, molasses, a by-product of sugar refining, is rich in sucrose and has been widely used to cultivate nitrogen-fixing bacteria and yeast strains, while whey provides lactose and proteins that support robust microbial metabolism (Hashempour-Baltork *et al.*, 2022).

Fruit and vegetable wastes, including banana peels, citrus pulp, and mango skins, have been successfully employed in the production of single-cell proteins and microbial biomass, owing to their high carbohydrate and micronutrient content (Thiviya *et al.*, 2022). These substrates not only support microbial proliferation but also enhance the production of bioactive metabolites such as organic acids, enzymes, and siderophores—key traits for plant growth-promoting inoculants. From a circular economy perspective, the valorization of agro-industrial residues as culture media aligns with sustainable waste management goals by reducing landfill burden, lowering greenhouse gas emissions, and creating value-added bioproducts (Gonçalves and Maximo, 2023). Moreover, these waste streams are often generated in large volumes by established industries, ensuring consistent availability and scalability for microbial fermentation processes. Hence, the strategic use of waste-derived substrates not only reduces production costs and environmental impact but also enhances the sustainability and accessibility of microbial inoculant technologies—particularly in resource-constrained agricultural systems.

4.10.2 Nutrient profiling of wastes for microbial support

Effective utilization of waste-derived substrates in microbial inoculant production hinges on precise nutrient profiling, which ensures that the selected materials can adequately support microbial growth and metabolite synthesis. This process involves a comprehensive analysis of the biochemical composition of agro-industrial residues, focusing on three primary nutrient categories:

- Carbohydrates and sugars: These serve as the principal carbon and energy sources for microbial metabolism. High levels of fermentable sugars—such as glucose, fructose, and sucrose—are particularly valuable for fast-growing strains like *Azotobacter* and *Azospirillum* (Hashempour-Baltork *et al.*, 2022).
- Proteins and amino acids: These are critical for biomass synthesis and enzymatic activity. Protein-rich wastes like whey or legume residues provide essential nitrogen and amino acid precursors that enhance microbial proliferation (Thiviya *et al.*, 2022).
- Minerals and trace elements: Elements such as phosphorus, potassium, calcium, magnesium, and iron are indispensable for cellular signaling, enzyme function, and structural integrity. Their presence in balanced proportions is vital for maintaining metabolic homeostasis during fermentation.

To accurately assess these parameters, advanced analytical tools—including gas chromatography (GC), high-performance liquid chromatography (HPLC), inductively coupled plasma mass spectrometry (ICP-MS), and Fourier-transform infrared spectroscopy (FTIR)—are employed. These techniques enable precise quantification of macro- and micronutrients, as well as detection of inhibitory compounds or imbalances that may hinder microbial activity (Palaniveloo *et al.*, 2020).

Armed with this data, researchers can further optimize substrate formulations by:

- supplementing deficient nutrients (e.g. adding phosphate salts or nitrogen sources);
- blending complementary waste streams (e.g. mixing sugar-rich fruit pulp with protein-rich whey);
- adjusting pH and moisture content to mimic ideal microbial growth conditions.

This tailored approach ensures that microbial cultures achieve high cell densities and metabolic efficiency, ultimately improving the yield and consistency of bioinoculant products.

4.10.3 Carrier materials from biodegradable waste streams

Biodegradable waste streams such as composted plant residues, sawdust, rice husks, and biochar have gained significant traction as carrier materials for microbial inoculants, offering a low-cost, eco-friendly alternative to synthetic carriers. These organic substrates create a protective microenvironment that sustains microbial viability during storage while facilitating gradual release and colonization when applied to soil. For instance, biochar's microporous structure enhances moisture retention and shields microbial cells from UV and desiccation, making it an ideal carrier in arid climates. Similarly, composted residues are rich in humic acids and secondary metabolites that improve inoculant compatibility and longevity.

Beyond microbial preservation, these carriers offer added agronomic benefits by contributing organic matter to the soil, improving aggregate stability, nutrient retention, and microbial diversity

(Table 4.3). When incorporated as part of a biofertilizer formulation, they do not merely serve as passive delivery vehicles but actively participate in soil regeneration and fertility enhancement (da Silva Simões *et al.*, 2025). Their biodegradability ensures they integrate seamlessly into soil ecosystems without leaving harmful residues. These carriers can be customized into granular, pelletized, or liquid forms, depending on the desired application method. Granular and pelletized formats—often blended with biochar or compost—are especially suited for broadcasting or seedling bed incorporation, whereas liquid suspensions cater to seed treatment and foliar sprays. The production workflow generally involves composting, drying, and microbial blending under controlled conditions, which are optimized to safeguard inoculum quality and consistency (Raut *et al.*, 2019). In essence, waste-derived biodegradable carriers are not only technically effective and agriculturally beneficial but also support circular economy goals by converting agricultural waste into high-value biofertilizer components. Their adoption enhances the sustainability, resilience, and field effectiveness of microbial inoculants, helping close the loop between waste management and regenerative farming.

The strategic use of agricultural and food-processing wastes as both culture media and carrier materials thus marks a transformative step toward sustainable and circular biofertilizer production. These organic residues—ranging from molasses and whey to fruit peels and spent grains—are rich in fermentable sugars, amino acids, and micronutrients, making them ideal substrates for microbial growth and metabolite production (Hashempour-Baltork *et al.*, 2022). When subjected to nutrient profiling using techniques like HPLC, ICP-MS, and FTIR, these wastes can be precisely tailored to meet the metabolic demands of specific microbial strains, ensuring high cell density and consistent inoculant quality (Palaniveloo *et al.*, 2020). Simultaneously, the conversion of biodegradable waste into carrier materials—such as composted residues, sawdust, and biochar—enhances microbial viability during storage and facilitates gradual release in the field. These carriers not only protect inoculants from desiccation and UV stress but also contribute to soil organic matter, aggregation, and microbial diversity, reinforcing long-term soil health (da Silva Simões *et al.*, 2025). This integrated approach—combining waste valorization, nutrient optimization, and biodegradable carrier development—reduces production costs, minimizes environmental impact, and aligns with circular economy principles. It exemplifies how microbial biotechnology can intersect with sustainable waste management to create scalable, eco-friendly solutions for modern agriculture (Fig. 4.3; Gonçalves and Maximo, 2023).

4.11 Technologies for Large-scale Production

4.11.1 Solid-state and liquid-state fermentation using food/agro-waste inputs

Large-scale production of microbial inoculants hinges on the deployment of efficient, scalable

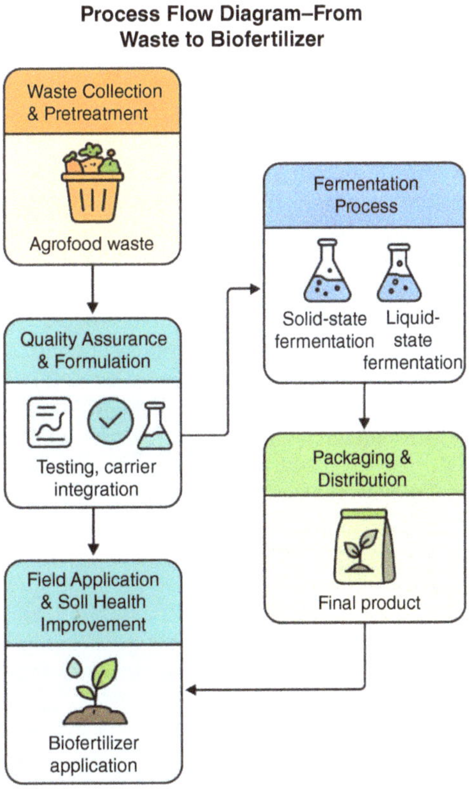

Fig. 4.3. Process flow from waste to biofertilizers. Figure author's own.

fermentation technologies, with solid-state fermentation (SSF) and liquid-state fermentation (LSF) emerging as the two dominant approaches (Table 4.4). Both methods can be effectively adapted to utilize food and agro-waste inputs, thereby reducing production costs and aligning with circular economy principles.

SSF involves microbial cultivation on solid substrates with minimal free water. Agro-industrial residues such as cereal bran, fruit pomace, rice husks, and spent grains serve as ideal substrates due to their high organic content, structural porosity, and phyto-stimulatory properties (Sadh *et al.* 2018). SSF systems are particularly suited for spore-forming microbes like *Trichoderma*, *Bacillus*, and *Aspergillus*, where the end product must be dry, shelf-stable, and resilient under ambient storage. The low moisture environment in SSF reduces contamination risks, minimizes energy inputs for drying, and often yields higher microbial densities with enhanced stress tolerance (Bautista-Hernández *et al.*, 2023).

In contrast, LSF—also known as submerged fermentation—employs nutrient-rich liquid media formulated with water-soluble agro-waste derivatives such as molasses, whey, and fruit pulp extracts. These substrates provide readily assimilable carbon and nitrogen sources, enabling rapid microbial proliferation and metabolite production (Suthar *et al.*, 2017). LSF offers precise control over fermentation parameters—including pH, temperature, aeration, and agitation—making it ideal for scaling up in bioreactors using batch, fed-batch, or continuous modes. This method is particularly advantageous for producing liquid inoculants used in seed coatings, foliar sprays, and drip irrigation systems, where rapid microbial colonization is essential.

Together, SSF and LSF provide complementary platforms for biofertilizer production, each suited to specific microbial traits, formulation goals, and field application strategies. Their integration with waste-derived substrates not only enhances economic feasibility but also reinforces the environmental sustainability of microbial inoculant technologies.

4.11.2 Bioreactor design and process control

Modern bioreactor design and process control are foundational to the success of large-scale microbial inoculant production, ensuring not only high yields and product consistency but also resource efficiency and environmental compliance. Among the most widely adopted configurations, stirred-tank bioreactors (STRs) remain the gold standard for liquid-state fermentation due to their robust mixing, efficient oxygen transfer, and uniform thermal distribution. These systems are highly adaptable to both aerobic and anaerobic cultures and allow precise control over agitation and aeration parameters (Uyar *et al.*, 2024).

In contrast, airlift and bubble column reactors offer energy-efficient alternatives by relying on gas circulation rather than mechanical agitation. These designs reduce shear stress and are particularly suited for shear-sensitive organisms or large-scale commodity production. For solid-state fermentation, tray and packed bed reactors are preferred, as they facilitate thin-layer microbial growth on substrates like bran or husks, optimizing surface area exposure and minimizing moisture gradients (Mahdinia *et al.*, 2019).

Equally critical is the implementation of advanced process control strategies. Real-time monitoring using integrated sensors for pH, dissolved oxygen, temperature, and substrate concentration enables dynamic adjustments to maintain optimal fermentation conditions. These systems are often coupled with automation platforms and data analytics, which not only streamline operations but also predict deviations and optimize fermentation cycles through feedback control loops and machine learning algorithms. When scaling up from lab to industrial scale, engineering equivalence in mass transfer, mixing, and energy efficiency becomes paramount. This is achieved through geometric similarity, impeller design optimization, and maintaining consistent oxygen transfer coefficients (kLa) across scales (Mahdinia *et al.*, 2019). Failure to replicate these parameters can lead to suboptimal microbial performance and reduced product quality. In essence, the convergence of intelligent bioreactor design, real-time process

Table 4.4. Fermentation techniques for microbial inoculant production. A detailed comparison of fermentation techniques tailored for microbial inoculant production using agro-waste inputs. Table author's own.

Fermentation type	Process description	Substrate options	Advantages	Challenges and limitations	Microbial applications
Solid-state fermentation (SSF)	Cultivation of microbes on solid substrates with low moisture content	Cereal bran, bagasse, composted husks, sawdust	Low contamination risk; energy efficient	Requires controlled moisture; slow microbial growth	Ideal for fungi (AM fungi, *Trichoderma*) and some bacteria (PSB, rhizobia)
Liquid-state fermentation (LSF)	Microbial growth in a liquid medium with nutrients dissolved in solution	Molasses, whey, fruit pulp extract	Fast growth; high yield; easier scalability	Contamination risk; requires sterilization	Widely used for *Azotobacter*, *Azospirillum*, *Frankia*, cyanobacteria
Batch fermentation	Cultivation occurs in a closed system with controlled conditions	Solid or liquid substrates	High product consistency; easy to monitor	Limited yield; requires precise control	Used for small-scale production or specialized inoculants
Continuous fermentation	Nutrient supply and microbial harvest occur in an uninterrupted cycle	Liquid waste-derived media (molasses, sugarcane extract)	Higher productivity; longer operational cycles	Complex control systems needed; risk of mutations	Used for large-scale microbial inoculants (PSB, *Azospirillum*)
Submerged fermentation	Microbes are grown in an aqueous environment with aeration	Liquid agro-waste derivatives	High biomass yield; suited for industrial scale	Requires aeration and pH control	Suitable for nitrogen-fixing bacteria and phosphate solubilizers

control, and scalable engineering ensures that microbial inoculants produced at industrial scale retain the efficacy, stability, and sustainability required for modern agricultural applications.

4.11.3 Contamination management and waste stream sterilization

Contamination management and sterilization of waste streams are pivotal to ensuring the quality, safety, and consistency of microbial inoculant production at industrial scales. Given the susceptibility of fermentation systems to microbial interference, especially when using nutrient-rich agro-waste substrates, a multi-tiered approach is essential. The first line of defense lies in the pretreatment of agro-waste and culture media, which must be rigorously sterilized to eliminate competing microorganisms and pathogens. Techniques such as autoclaving at 121°C for 20–60 min, pasteurization at 70–80°C, and chemical disinfection using agents like formalin or hydrogen peroxide are commonly employed depending on the substrate type and microbial sensitivity (Gadade and Rathod, 2022). In LSF, continuous sterilization systems—including steam injection loops, multi-stage heat exchangers, and UV irradiation units—enable inline sterilization of culture media, maintaining sterility without interrupting flow. During fermentation, aseptic processing protocols are critical. This includes the use of clean-in-place (CIP) and sterilize-in-place (SIP) systems integrated into bioreactors, which allow for automated, closed-loop sterilization of internal surfaces between batches. Additionally, cleanroom environments with high-efficiency particulate air (HEPA) filtration and positive pressure differentials help prevent airborne contamination, especially during sensitive operations like inoculation and sampling. To ensure ongoing sterility, real-time monitoring systems equipped with sensors for pH, dissolved oxygen, and microbial load are employed. These are supported by rapid-response protocols that include microbial plating, polymerase chain reaction (PCR)-based detection, and deviation-triggered corrective actions. Such systems enable early detection of contamination events and help maintain batch-to-batch consistency in inoculant quality.

In summary, contamination control in large-scale inoculant production is a multi-layered strategy—combining substrate sterilization, aseptic engineering, environmental safeguards, and real-time diagnostics—that ensures microbial purity, process reliability, and regulatory compliance. The integration of advanced fermentation technologies is central to the efficient and sustainable mass production of microbial inoculants. Both solid-state and liquid-state fermentation approaches, when optimized using food and agro-waste inputs, offer effective and eco-friendly production routes. Coupled with state-of-the-art bioreactor designs and rigorous process control systems, these technologies not only help maximize microbial yield and activity but also ensure product consistency and scalability. Furthermore, meticulous contamination management and waste stream sterilization are indispensable for maintaining high standards of quality in large-scale operations. Together, these technologies form the backbone of modern microbial inoculant production, paving the way for sustainable agricultural practices worldwide.

4.12 Quality Assurance and Environmental Safety

4.12.1 Viability and purity standards for waste-based inoculants

Ensuring the viability and purity of waste-based microbial inoculants is a cornerstone of quality assurance, particularly given the inherent variability and microbial load associated with organic waste substrates. Rigorous standards and analytical protocols are essential to guarantee that these bioinoculants are both effective and safe for agricultural deployment. Viability assessment begins with quantifying the number of live microbial cells, typically expressed as colony-forming units (CFU) per gram or milliliter. Traditional plate count assays remain the gold standard due to their simplicity and specificity, though they may underestimate viable but non-culturable (VBNC) cells. To address this, flow cytometry has emerged as a powerful complementary tool, offering rapid enumeration of both viable and VBNC populations using fluorescent staining technique. Additionally, respirometric methods—which measure oxygen uptake or CO_2 evolution—can provide indirect but real-time insights into microbial metabolic activity, especially in mixed cultures or spore-based formulations (Weitzel *et al.*, 2021). Purity testing is particularly critical for inoculants derived from waste, where the risk of contamination by opportunistic or pathogenic microbes is elevated. Techniques such as PCR and quantitative PCR (qPCR) enable precise detection of target strains and exclusion of non-target organisms. Selective media culturing helps isolate specific microbial groups, while high-throughput sequencing (HTS) provides a comprehensive profile of the microbial community, ensuring genetic fidelity and absence of undesirable taxa. To ensure reproducibility and regulatory compliance, manufacturers must implement standard operating procedures (SOPs) that span the entire production pipeline—from waste pretreatment and sterilization to fermentation, formulation, and packaging. These SOPs should include validated protocols for microbial enumeration, contamination control, and storage stability testing. Adherence to such protocols not only minimizes batch-to-batch variability but also supports traceability and product certification. In essence, a robust quality assurance framework—anchored in viability metrics, purity validation, and procedural standardization—is indispensable for the safe and effective use of waste-based microbial inoculants in sustainable agriculture (Table 4.5).

4.12.2 Waste-to-product traceability and eco-safety assessment

Ensuring traceability and eco-safety in the production of waste-derived microbial inoculants is essential for maintaining product integrity, regulatory compliance, and environmental stewardship. As these products originate from heterogeneous and potentially contaminated waste streams, robust systems must be in place to track their transformation and assess their ecological impact.

Traceability systems begin with meticulous documentation of the origin, handling, and pretreatment of food and agro-waste inputs. This includes recording the type of waste, its source (e.g. fruit processing, dairy effluent), and the sterilization or conditioning methods applied before fermentation. Increasingly, digital tracking platforms—including blockchain-based systems—are being adopted to create immutable, transparent records of each production step. These platforms enable decentralized data sharing among stakeholders and ensure that any deviation from standard operating procedures can be rapidly identified and corrected. Blockchain integration also supports smart contracts and real-time monitoring, enhancing accountability and enabling automated compliance verification across the supply chain.

On the environmental front, eco-safety assessments are conducted to evaluate the potential risks associated with both the production process and the final inoculant. A cornerstone of this evaluation is the life cycle assessment (LCA), which quantifies environmental impacts across all stages—from raw material acquisition and fermentation to field application and degradation. LCA frameworks increasingly incorporate freshwater ecotoxicity and soil health indicators as critical impact categories. Complementing LCA, toxicity screening ensures that any secondary metabolites or by-products generated during microbial fermentation do not pose risks to non-target organisms. This involves acute and chronic toxicity assays, often using model organisms such as *Daphnia magna* or *Eisenia fetida*, and may include USEtox® modeling to estimate potential ecosystem damage (Owsianiak *et al.*, 2023). Finally, soil and water impact studies assess the long-term ecological footprint of inoculant application. These studies monitor changes in soil microbial diversity, enzyme activity, and water quality parameters to ensure that the product supports sustainable agricultural practices without disrupting native ecosystems.

4.12.3 Certification and biosafety regulations

The successful commercialization of waste-derived microbial inoculants hinges on strict adherence to certification protocols and biosafety regulations that safeguard both environmental and human health. These frameworks ensure that products entering the market are not only effective but also compliant with national and international safety standards. Certification

Table 4.5. Quality assurance metrics and regulatory standards. A structured comparison of the critical parameters influencing microbial inoculant quality, regulatory standards, and biosafety compliance. Table author's own.

Quality metric	Parameters assessed	Testing methods	Industry standards and certification	Regulatory frameworks	Potential challenges
Viability assessment	Colony-forming units (CFU); metabolic activity	Plate count method, flow cytometry, respirometry	Minimum viable count per gram/ml per ISO and national standards	ISO 9001 for quality control, FAO guidelines	Variability in microbial survival during storage
Purity testing	Absence of contaminants; genetic integrity	PCR assays, selective culturing, sequencing	Defined purity levels in commercial inoculants	National biosafety protocols, USDA/EU microbial regulations	Risk of unwanted microbes in waste-based substrates
Shelf-life and stability	Viability over time; moisture control	Accelerated aging tests, desiccation studies	Minimum stability period per market standards	Labeling compliance, agro-product registration rules	Loss of microbial efficacy due to environmental factors
Eco-safety assessment	Soil and water impact; non-target effects	Life cycle assessment (LCA), toxicity screening	Environmental impact certification	FAO, UNEP sustainability guidelines	Long-term ecological effects need continuous monitoring
Waste-to-product traceability	Sourcing documentation; batch consistency	Blockchain-based tracking, digital traceability systems	Verified waste origin for organic certification	Circular economy policies, ISO environmental standards	Tracking accuracy in decentralized supply chains
Biosafety compliance	Genetic stability; risk analysis	Whole-genome sequencing, horizontal gene transfer studies	Biosafety clearance for commercial release	Cartagena Protocol on Biosafety, national regulations	Complexity in proving non-GMO status for certain strains

processes typically involve validation by recognized bodies such as the International Organization for Standardization (ISO). For instance, ISO 9001:2015 certification ensures that a manufacturer's quality management system meets global benchmarks for consistency, traceability, and continual improvement (ISO 9001 - URS India[1]). In the context of microbial inoculants, this certification confirms that production practices, microbial viability, and safety parameters are systematically monitored and optimized. In India, additional certifications may be required from agencies like the Fertilizer Control Order (FCO) or the Central Insecticide Board (CIB), depending on the product classification (Innovations – IPL Biological[2]). On the biosafety front, regulatory compliance is especially critical when dealing with genetically modified organisms (GMOs) or non-native microbial strains. The Cartagena Protocol on Biosafety, a legally binding international agreement under the Convention on Biological Diversity, provides a comprehensive framework for the safe handling, transport, and use of living modified organisms (LMOs) (Secretariat of the Convention on Biological Diversity, 2000; Cartagena Protocol Text[3]). National biosafety regulations often require risk assessment reports that evaluate potential impacts such as horizontal gene transfer, disruption of native microbial communities, and ecological imbalances. These assessments are typically supported by periodic audits, third-party testing, and compliance monitoring to ensure that biosafety standards are upheld throughout the product life cycle (Biosafety Clearing-House, 2025; Cartagena Protocol Portal[4]).

Equally important is labeling and transparency, which serve as the interface between producers, regulators, and end-users. Labels must specify the microbial strains, substrate origin, formulation type, storage conditions, and application guidelines. Transparent disclosure of safety metrics, such as CFU counts, shelf life, and biosafety clearance, builds trust and facilitates informed decision making by farmers and agricultural scientists. Thus, the transition of waste-derived inoculants from lab to market is governed by a multi-tiered regulatory ecosystem—anchored in certification, biosafety compliance, and transparent labeling—that ensures product efficacy, environmental safety, and public confidence. Ensuring the quality and environmental safety of waste-derived microbial inoculants is a multifaceted process that spans viability and purity testing, traceability, eco-safety analysis, and adherence to certification and regulatory protocols. By rigorously applying these quality assurance measures, producers not only guarantee the effectiveness of their biofertilizers but also support sustainable environmental practices. In turn, such stringent quality controls foster consumer confidence and facilitate the broader adoption of eco-friendly agricultural technologies.

4.13 Challenges and Future Prospects

4.13.1 Technical and socio-economic barriers to adoption

The transition of waste-derived microbial inoculants from experimental settings to widespread agricultural use is hindered by a combination of technical complexities and socio-economic constraints, each demanding targeted interventions to unlock their full potential. On the technical side, scale-up remains a formidable challenge. Laboratory protocols often rely on controlled, homogeneous conditions that are difficult to replicate at industrial scale. Variability in agro-waste composition—such as fluctuations in pH, carbon-to-nitrogen ratio, or microbial load—can disrupt fermentation kinetics and reduce inoculant efficacy (Kiruba N *et al.*, 2022). Moreover, maintaining microbial viability during storage and transport is particularly difficult for spore-forming and metabolically sensitive strains, necessitating innovations in formulation and cold-chain logistics.

Contamination control is another persistent issue, especially when using nutrient-rich waste substrates that can harbor opportunistic microbes. Ensuring sterility through pretreatment, aseptic processing, and real-time monitoring adds complexity and cost to the production pipeline. Compounding this is the lack of standardization across regions: agro-waste streams differ widely in composition, making it difficult to establish uniform production protocols. This calls for adaptive process control systems that can accommodate local variability while ensuring consistent product quality (Dubey *et al.*, 2023).

From a socio-economic perspective, capital investment remains a major barrier. The cost of bioreactors, sterilization units, and advanced packaging systems can be prohibitive for small-scale producers and cooperatives. Additionally, farmer adoption is often limited by low awareness, skepticism about efficacy, and lack of technical support. Many farmers remain reliant on chemical fertilizers due to their familiarity, immediate results, and subsidized pricing (Arora and Mishra, 2024). Finally, the policy and regulatory landscape for bio-based agricultural inputs is still evolving. In many regions, there is a lack of standardized certification systems, incentive structures, and market transparency mechanisms that would legitimize microbial inoculants and encourage broader adoption. While countries like India have begun implementing frameworks under the Fertilizer Control Order (FCO) and biostimulant regulations, gaps remain in enforcement, awareness, and cross-sector coordination. In essence, overcoming these barriers requires a multi-dimensional strategy: engineering innovations for scalable production, contamination-resistant systems, adaptive standardization protocols, targeted financial support for small producers, farmer education campaigns, and robust regulatory frameworks that legitimize and incentivize the use of waste-derived microbial technologies.

4.13.2 Integration with circular economy and sustainable waste management

Integrating waste-derived microbial inoculant production into a circular economy framework represents a transformative shift in how agriculture manages resources, waste, and sustainability. At its core, this approach emphasizes resource recovery, where agrifood and industrial by-products—such as fruit pomace, molasses, whey, and crop residues—are repurposed as nutrient-rich substrates for microbial cultivation. Rather than being incinerated or landfilled, these materials are biologically valorized, closing the loop between waste generation and agricultural input production (Kiruba N *et al.*, 2022).

This model also supports waste minimization by reducing dependency on synthetic culture media and chemical fertilizers. By incorporating waste streams into fermentation and formulation processes, producers not only lower input costs but also contribute to sustainable waste management practices that align with global environmental goals (Zainudin *et al.*, 2022). The microbial bioconversion of these residues into biofertilizers enhances nutrient cycling and reduces the environmental burden of agro-industrial waste disposal.

From an economic standpoint, this integration fosters localized synergies between waste generators and biofertilizer producers. Community-scale systems that link agricultural waste management with microbial inoculant production can generate employment opportunities, stimulate rural innovation, and create new revenue streams through the sale of value-added bio-products (Ahmad *et al.*, 2025). These systems are particularly impactful in regions with high volumes of organic waste and limited access to synthetic agricultural inputs.

Environmentally, the circular model delivers multiple co-benefits: it reduces landfill dependency, curbs greenhouse gas emissions, and enhances soil health through the application of organic carriers that improve microbial colonization and nutrient retention (Arora and Mishra, 2024). The use of biodegradable carriers derived from composted waste further reinforces the ecological integrity of this approach. In essence, embedding microbial inoculant production within a circular economy not only transforms waste into a resource but also redefines agriculture as a regenerative, low-impact system.

4.13.3 Biotechnological innovations in waste valorization for microbial production

Biotechnological innovations are rapidly transforming the landscape of waste valorization for microbial inoculant production, offering scalable, efficient, and environmentally sound solutions. At the forefront of this transformation is metabolic engineering, which enables the genetic modification of microbial strains to enhance their ability to metabolize complex organic substrates commonly found in agro-waste. Engineered microbes with improved organic acid production, stress tolerance, and substrate specificity are being developed to thrive in heterogeneous waste environments and deliver consistent field performance (Bose *et al.*, 2022). Complementing this, synthetic biology is being harnessed to design custom microbial consortia that work synergistically to exploit the full nutrient spectrum of waste streams (Abdi *et al.*, 2024a,b,c). These consortia are tailored to perform coordinated metabolic functions—such as nitrogen fixation, phosphate solubilization, and siderophore production—thereby enhancing the efficiency of waste-to-biofertilizer conversion (Jaiswal *et al.*, 2023).

In the realm of fermentation technology, innovations in continuous fermentation systems—integrated with real-time monitoring, automated feedback loops, and adaptive control algorithms—are enabling the production of high-density microbial cultures with minimal variability, even when using inconsistent waste inputs (Stikane *et al.*, 2023). Additionally, modular and scalable bioreactor designs are being developed to accommodate fluctuating substrate quality and streamline downstream processing, making decentralized and community-scale production more viable.

On the formulation front, biodegradable carrier materials derived from composted residues, lignocellulosic waste, and nanostructured polymers are being explored to improve inoculant shelf life, moisture retention, and microbial compatibility. These carriers not only protect microbial cells during storage and transport but also enhance colonization efficiency upon field application. Encapsulation technologies, including alginate beads, chitosan microcapsules, and spray-dried starch matrices, further shield microbes from desiccation, UV radiation, and temperature fluctuations, ensuring consistent performance under diverse agroecological conditions (Valdivia-Rivera *et al.*, 2021). In summary, the convergence of genetic engineering, precision fermentation, and advanced formulation science is redefining how agro-waste is transformed into high-performance microbial inoculants. These innovations not only enhance product efficacy and scalability but also reinforce the role of biotechnology in building a circular, climate-resilient agricultural future.

Hence, while the road to widespread adoption of waste-derived microbial inoculants is marked by significant technical and socio-economic challenges, the future appears promising. By integrating these innovative bio-based processes within a circular economy framework and leveraging cutting-edge biotechnological advancements, researchers and industry stakeholders can overcome these barriers. Sustainable waste valorization not only offers an efficient pathway to reduce production costs and enhance product performance but also contributes to environmental stewardship and agricultural resilience. Continued investment in research, supportive policy frameworks, and industry–academia collaborations are essential to catalyze the full potential of these technologies in global agricultural practices.

4.14 Conclusion

The production of microbial inoculants has evolved into a sustainable practice central to modern agriculture, directly supporting UN Sustainable Development Goal (SDG) 2: Zero Hunger by enhancing crop productivity and soil fertility (Fig. 4.4). This comprehensive exploration

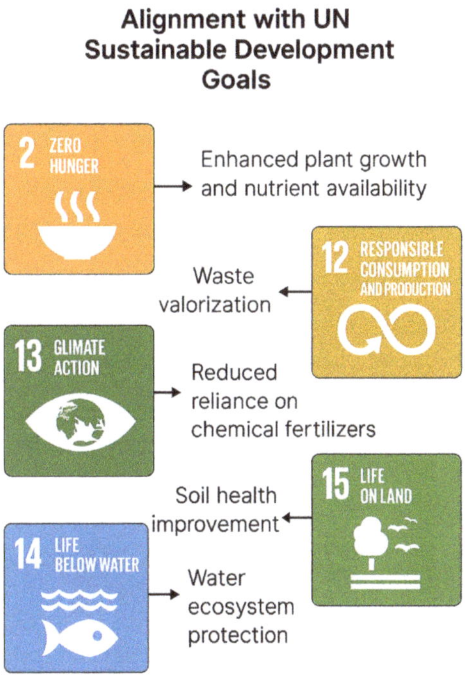

Fig. 4.4. Alignment of waste-derived bioinoculant production with UN SDGs. Figure author's own.

has demonstrated that integrating eco-friendly production methodologies—ranging from the use of agrifood waste substrates to innovative fermentation and formulation techniques—can significantly reduce dependency on synthetic inputs while restoring soil health. By prioritizing advanced bioprocessing technologies and rigorous quality assurance, producers are creating high-performance inoculants that not only enhance plant growth through improved nutrient uptake and stress resilience but also contribute to SDG 13: Climate Action by reducing greenhouse gas emissions associated with conventional fertilizers.

Agrifood waste has emerged as a particularly promising resource in this context. The repurposing of food and agricultural residues into nutrient-rich culture media and biodegradable carrier materials aligns perfectly with SDG 12: Responsible Consumption and Production, promoting circular economy principles. These waste-derived inputs not only lower production costs but also mitigate environmental impacts associated with waste disposal, thereby contributing to SDG 11: Sustainable Cities and Communities through improved waste management. This dual benefit reinforces the viability of integrating organic waste streams into microbial inoculant production, paving the way for the development of eco-friendly biofertilizers that support a range of agricultural ecosystems—from degraded soils to intensive cropping systems.

Looking ahead, the scalable production of microbial inoculants—bolstered by cutting-edge bioreactor designs, automated process controls, and robust contamination management protocols—holds great promise for SDG 9: Industry, Innovation, and Infrastructure by fostering sustainable biomanufacturing. These technological innovations, coupled with strategic policies and collaborative research efforts, are set to transform the landscape of sustainable farming. As these biofertilizers gain acceptance, they will not only boost crop productivity and soil health but also enhance adaptive capacities in the face of climate change, reinforcing SDG 15: Life on Land through improved ecosystem resilience. Such advancements mark a critical step toward establishing regenerative agricultural systems that meet global food security needs while preserving environmental integrity.

Notes

[1] https://www.ursindia.com/iso-certification/iso-9001-quality-management-system
[2] https://iplbiologicals.com/innovations/
[3] https://www.cbd.int/doc/legal/cartagena-protocol-en.pdf
[4] https://bch.cbd.int/protocol

References

A. Daanaa, H.-S., Abdou, M., Goda, H.A., Abbas, M.T., Hamza, M.A. *et al.* (2020) Plant pellets: a compatible vegan feedstock for preparation of plant-based culture media and production of value-added biomass of rhizobia. *Sustainability* 12(20), 8389. DOI: 10.3390/su12208389

Aasfar, A., Bargaz, A., Yaakoubi, K., Hilali, A., Bennis, I. *et al.* (2021) Nitrogen fixing *Azotobacter* species as potential soil biological enhancers for crop nutrition and yield stability. *Frontiers in Microbiology* 12, 628379. DOI: 10.3389/fmicb.2021.628379

Abdi, G., Firdous, S., Vasantharekha, R., Singh, G.B., Seetharaman, B. *et al.* (2024a) Progress on synthetic genomics. In: Singh, V. (ed.) *Advances in Genomics: Methods and Applications*. Springer Nature, Singapore, pp. 181–197.

Abdi, G., Patil, N., Mishra, P., Tariq, M., Dhariwal, R. *et al.* (2024b) Genomic precision: unveiling the transformative role of genome editing in advancing genomics research and applications. In: Singh, V. (ed.) *Advances in Genomics: Methods and Applications*. Springer Nature, Singapore, pp. 265–306.

Abdi, G., Patil, N., Tendulkar, R., Dhariwal, R., Mishra, P. *et al.* (2024c) Engineering genomic landscapes: synthetic biology approaches in genomic rearrangement. In: Singh, V. (ed.) *Advances in Genomics: Methods and Applications*. Springer Nature, Singapore, pp. 227–264.

Aguilar-Paredes, A., Valdés, G., and Nuti, M. (2020) Ecosystem functions of microbial consortia in sustainable agriculture. *Agronomy* 10(12), 1902. DOI: 10.3390/agronomy10121902

Aguilar-Rivera, N. and Olvera-Vargas, L.A. (2022) Management of food waste for sustainable economic development and circularity. In: Leal Filho, W., Azul, A.M., Doni, F., and Salvia, A.L. (eds) *Handbook of Sustainability Science in the Future*. Springer International Publishing, Cham, Switzerland, pp. 1895–1917.

Ahmad, A., Javed, M.H., Musharavati, F., Khan, M.I., Al-Muhtaseb, A.H. *et al.* (2025) Achieving circular economy through sustainable biofertilizer production from mixed municipal waste: a life cycle analysis approach. *Biomass Conversion and Biorefinery* Article 132403. DOI: 10.1007/s13399-025-06709-z

Ahmed, E. and Holmström, S.J.M. (2014) Siderophores in environmental research. *Microbial Biotechnology* 7, 196–208. DOI: 10.1111/1751-7915.12117

Akashdeep., Kumari, S., and Rani, N. (2024) Novel cereal bran based low-cost liquid medium for enhanced growth, multifunctional traits and shelf life of consortium biofertilizer containing *Azotobacter chroococcum*, *Bacillus subtilis* and *Pseudomonas* sp. *Journal of Microbiological Methods* 222, 106952. DOI: 10.1016/j.mimet.2024.106952

Akay, G. and Fleming, S. (2012) Agro-process intensification: soilborne micro-bioreactors with nitrogen fixing bacterium *Azospirillum brasilense* as self-sustaining biofertiliser source for enhanced nitrogen uptake by plants. *Green Processing and Synthesis* 1(5), 427–437. DOI: 10.1515/gps-2012-0041

Akhtar, M.S. and Abdullah, S.N.A. (2014) Mass production techniques of arbuscular mycorrhizal fungi: major advantages and disadvantages: a review. *Biosciences Biotechnology Research Asia* 11(3), 1199–1204. DOI: 10.13005/bbra/1506

Ali, Q., Shabaan, M., Ashraf, S., Kamran, M., Zulfiqar, U. *et al.* (2023) Comparative efficacy of different salt tolerant rhizobial inoculants in improving growth and productivity of *Vigna radiata* L. under salt stress. *Scientific Reports* 13(1), 17442. DOI: 10.1038/s41598-023-44433-8

Aloo, B.N., Mbega, E.R., Makumba, B.A., and Tumuhairwe, J.B. (2022) Effects of carrier materials and storage temperatures on the viability and stability of three biofertilizer inoculants obtained from potato (*Solanum tuberosum* L.) rhizosphere. *Agriculture* 12(2), 140. DOI: 10.3390/agriculture12020140

Ansori, A.N., Antonius, Y., Susilo, R.J., Hayaza, S., Kharisma, V.D. *et al.* (2023) Application of CRISPR-Cas9 genome editing technology in various fields: a review. *Narra J* 3(2), e184. DOI: 10.52225/narra.v3i2.184

Arora, N.K. and Mishra, J. (2024) Next generation microbe-based bioinoculants for sustainable agriculture and food security. *Environmental Sustainability* 7(1), 1–4. DOI: 10.1007/s42398-024-00308-w

Attene, L., Deiana, A., Carucci, A., Asunis, F., and Ledda, C. (2023) Efficient nitrogen recovery from agroenergy effluents for cyanobacteria cultivation (*Spirulina*). *Sustainability* 15(1), 675. DOI: 10.3390/su15010675

Awasthi, M.K., Harirchi, S., Sar, T., Vs, V., Rajendran, K. *et al.* (2022) Myco-biorefinery approaches for food waste valorization: present status and future prospects. *Bioresource Technology* 360, 127592. DOI: 10.1016/j.biortech.2022.127592

Bagga, D., Chauhan, S., Bhavanam, A., Nikhil, G.N., Meena, S.S. *et al.* (2024) Recent advancements in fermentation strategies for mass production and formulation of biofertilizers: towards waste valorization. *Journal of Soil Science and Plant Nutrition* 24(3), 5868–5897. DOI: 10.1007/s42729-024-01947-y

Bahadur, A., Batool, A., Nasir, F., Jiang, S., Mingsen, Q. *et al.* (2019) Mechanistic insights into arbuscular mycorrhizal fungi-mediated drought stress tolerance in plants. *International Journal of Molecular Sciences* 20(17), 4199. DOI: 10.3390/ijms20174199

Bai, K., Wang, W., Zhang, J., Yao, P., Cai, C. *et al.* (2024) Effects of phosphorus-solubilizing bacteria and biochar application on phosphorus availability and tomato growth under phosphorus stress. *BMC Biology* 22(1), 211. DOI: 10.1186/s12915-024-02011-y

Balla, A., Silini, A., Cherif-Silini, H., Chenari Bouket, A., Alenezi, F.N. *et al.* (2022) Recent advances in encapsulation techniques of plant growth-promoting microorganisms and their prospects in the sustainable agriculture. *Applied Sciences* 12(18), 9020. DOI: 10.3390/app12189020

Balu, U.R., Vasantharekha, R., Paromita, C., Ali, K., Mudgal, G. *et al.* (2024) Linking EDC-laden food consumption and modern lifestyle habits with preeclampsia: a non-animal approach to identifying early diagnostic biomarkers through biochemical alterations. *Food and Chemical Toxicology* 194, 115073. DOI: 10.1016/j.fct.2024.115073

Bautista-Hernández, I., Chávez-González, M.L., Sánchez, A.S., Guzmán, K.N.R., León, C.T. *et al.* (2023) Solid-state fermentation as strategy for food waste transformation. In: Gonzalez, C.N.A., Gómez-García, R., and Kuddus, M. (eds) *Food Waste Conversion*. Springer US, New York, pp. 147–160.

Begom, M., Ahmed, G.U., Sultana, R., and Akter, F. (2021) Culture optimization for mass production of *Rhizobium* using bioreactor made of readily available materials and agitated by air flow. *Agricultural Sciences* 12(06), 620–629. DOI: 10.4236/as.2021.126040

Ben Rebah, F., Prévost, D., Yezza, A., and Tyagi, R.D. (2007) Agro-industrial waste materials and wastewater sludge for rhizobial inoculant production: a review. *Bioresource Technology* 98(18), 3535–3546. DOI: 10.1016/j.biortech.2006.11.066

Berninger, T., González López, Ó., Bejarano, A., Preininger, C., and Sessitsch, A. (2018) Maintenance and assessment of cell viability in formulation of non-sporulating bacterial inoculants. *Microbial Biotechnology* 11(2), 277–301. DOI: 10.1111/1751-7915.12880

Bhatia, L., Jha, H., Sarkar, T., and Sarangi, P.K. (2023) Food waste utilization for reducing carbon footprints towards sustainable and cleaner environment: a review. *International Journal of Environmental Research and Public Health* 20(3), 2318. DOI: 10.3390/ijerph20032318

Bhattacharyya, P.N., Islam, N.F., Sarma, B., Nath, B.C., Al-Ani, L.K.T. *et al.* (2024) Frankia-actinorhizal symbiosis: a non-chemical biological assemblage for enhanced plant growth, nodulation and reclamation of degraded soils. *Symbiosis* 92(1), 1–26. DOI: 10.1007/s13199-023-00956-2

Bhowmick, M.K. (2018) Seed priming: a low-cost technology for resource-poor farmers in improving pulse productivity. In: Rakshit, A. and Singh, H.B. (eds) *Advances in Seed Priming*. Springer, Singapore, pp. 187–208.

Biosafety Clearing-House (2025) The Cartagena Protocol on Biosafety. Available at: https://bch.cbd.int/protocol (accessed September 10, 2025).

Boddey, R.M. and Dobereiner, J. (1995) Nitrogen fixation associated with grasses and cereals: recent progress and perspectives for the future. *Fertilizer Research* 42(1–3), 241–250. DOI: 10.1007/BF00750518

Bolan, S., Hou, D., Wang, L., Hale, L., Egamberdieva, D. *et al.* (2023) The potential of biochar as a microbial carrier for agricultural and environmental applications. *The Science of the Total Environment* 886, 163968. DOI: 10.1016/j.scitotenv.2023.163968

Bose, S.K., Upadhyay, S., and Srivastava, Y. (2022) Metabolic engineering of microbes for the production of plant-based compounds. In: Chowdhary, P., Mani, S., and Chaturvedi, P (eds) *Microbial Biotechnology: Role in Ecological Sustainability and Research*. Wiley, Oxford, UK, pp. 59–73.

Çakmakçı, R., Mosber, G., Milton, A.H., Alatürk, F., and Ali, B. (2020) The effect of auxin and auxin-producing bacteria on the growth, essential oil yield, and composition in medicinal and aromatic plants. *Current Microbiology* 77(4), 564–577. DOI: 10.1007/s00284-020-01917-4

Catroux, G., Hartmann, A., and Revellin, C. (2001) Trends in rhizobial inoculant production and use. *Plant and Soil* 230(1), 21–30. DOI: 10.1023/A:1004777115628

Chaiyasen, A., Chaiya, L., Douds, D.D., and Lumyong, S. (2017) Influence of host plants and soil diluents on arbuscular mycorrhizal fungus propagation for on-farm inoculum production using leaf litter compost and agrowastes. *Biological Agriculture & Horticulture* 33(1), 52–62. DOI: 10.1080/01448765.2016.1187670

Chandarana, K.A. and Amaresan, N. (2023) Rhizobium biofertilizers: mass production process and cost-benefit ratio analysis. In: Amaresan, N., Dharumadurai, D., and Babalola, O.O. (eds) *Agricultural Microbiology Based Entrepreneurship: Making Money from Microbes*. Springer Nature, Singapore, pp. 125–132.

Chaudhary, P., Xu, M., Ahamad, L., Chaudhary, A., Kumar, G. *et al.* (2023) Application of synthetic consortia for improvement of soil fertility, pollution remediation, and agricultural productivity: a review. *Agronomy* 13(3), 643. DOI: 10.3390/agronomy13030643

Chaudhary, T. and Shukla, P. (2020) Commercial bioinoculant development: techniques and challenges. In: Shukla, P. (ed.) *Microbial Enzymes and Biotechniques: Interdisciplinary Perspectives*. Springer, Singapore, pp. 57–70.

Chaudhary, T., Dixit, M., Gera, R., Shukla, A.K., Prakash, A. *et al.* (2020) Techniques for improving formulations of bioinoculants. *3 Biotech* 10(5), 199. DOI: 10.1007/s13205-020-02182-9

Checco, J., Azizan, F.A., Mitchell, J., and Aziz, A.A. (2023) Adoption of improved rice varieties in the global South: a review. *Rice Science* 30(3), 186–206. DOI: 10.1016/j.rsci.2023.03.004

Chemla, Y., Sweeney, C.J., Wozniak, C.A., and Voigt, C.A. (2025) Design and regulation of engineered bacteria for environmental release. *Nature Microbiology* 10(2), 281–300. DOI: 10.1038/s41564-024-01918-0

Chen, F., Chen, L., Yan, Z., Xu, J., Feng, L. *et al.* (2024) Recent advances of CRISPR-based genome editing for enhancing staple crops. *Frontiers in Plant Science* 15, 1478398. DOI: 10.3389/fpls.2024.1478398

Cheng, S., Zou, Y.-N., Kuča, K., Hashem, A., Abd_Allah, E.F. *et al.* (2021) Elucidating the mechanisms underlying enhanced drought tolerance in plants mediated by arbuscular mycorrhizal fungi. *Frontiers in Microbiology* 12, 809473. DOI: 10.3389/fmicb.2021.809473

Chiurazzi, M., Frugis, G., and Navazio, L. (2025) Symbiotic nitrogen fixation: a launchpad for investigating old and new challenges. *Journal of Experimental Botany* 76(6), 1473–1477. DOI: 10.1093/jxb/erae510

Chojnacka, K., Moustakas, K., and Mikulewicz, M. (2022) Valorisation of agri-food waste to fertilisers is a challenge in implementing the circular economy concept in practice. *Environmental Pollution* 312, 119906. DOI: 10.1016/j.envpol.2022.119906

Chopra, S., Sharma, S.G., Kaur, S., Kumar, V., and Guleria, P. (2025) Understanding the microRNA-mediated regulation of plant-microbe interaction and scope for regulation of abiotic and biotic stress tolerance in plants. *Physiological and Molecular Plant Pathology* 136, 102565. DOI: 10.1016/j.pmpp.2025.102565

Cranenbrouck, S., Voets, L., Bivort, C., Renard, L., Strullu, D.-G. *et al.* (2005) Methodologies for in vitro cultivation of arbuscular mycorrhizal fungi with root organs. In: Declerck, S., Fortin, J.A., and Strullu, D.-G. (eds) *In Vitro Culture of Mycorrhizas*. Springer, Berlin, Heidelberg, pp. 341–375.

da Silva, M.L.P., Moen, F.S., Liles, M.R., Feng, Y., and Sanz-Saez, A. (2023) Orange peel in combination with selected PGPR strains as seed treatment can improve soybean yield under field conditions. *Plant and Soil* 491(1–2), 401–420. DOI: 10.1007/s11104-023-06121-4

da Silva Simões, C.V., Stamford, T.C.M., Berger, L.R.R., Araújo, A.S., da Costa Medeiros, J.A. *et al.* (2025) Edible alginate-fungal Chitosan coatings as carriers for *Lacticaseibacillus casei* LC03 and their impact on quality parameters of strawberries during cold storage. *Foods* 14(2), 203. DOI: 10.3390/foods14020203

Das, S. and TK, D. (2025) Assessment of plant growth promoting parameters of azotobacter for their contribution to yield of chili (*Capsicum annum*) through path coefficient analysis. *International Journal of Current Microbiology and Applied Sciences* 14(6), 219–227. DOI: https://doi.org/10.20546/ijcmas.2025.1406.020

de Almeida Leite, R., Martins da Costa, E., Cabral Michel, D., do Amaral Leite, A., de Oliveira-Longatti, S.M. *et al.* (2024) Genomic insights into organic acid production and plant growth promotion by different species of phosphate-solubilizing bacteria. *World Journal of Microbiology & Biotechnology* 40(10), 311. DOI: 10.1007/s11274-024-04119-3

Dhar, S.K., Kaur, J., Singh, G.B., Chauhan, A., Tamang, J. *et al.* (2024) Novel *Bacillus* and *Prestia* isolates from dwarf century plant enhance crop yield and salinity tolerance. *Scientific Reports* 14(1), 14645. DOI: 10.1038/s41598-024-65632-x

Diagne, N., Arumugam, K., Ngom, M., Nambiar-Veetil, M., Franche, C. *et al.* (2013) Use of *Frankia* and actinorhizal plants for degraded lands reclamation. *BioMed Research International* 2013, 948258. DOI: 10.1155/2013/948258

Di Barbaro, M.G., Andrada, H., González Basso, V., Guzmán, P., Del Valle, E., and Brandán de Weth, C. (2023) Azospirillum sp. and mycorrhizal fungi: key microorganisms in sustainable agriculture. *Journal of Applied Biotechnology & Bioengineering* 10(6), 199–204. DOI: 10.15406/jabb.2023.10.00349

Díaz-Rodríguez, A.M., Cota, F.I.P., Chávez, L.A.C., Ortega, L.F.G., Alvarado, M.I.E. *et al.* (2025) Microbial inoculants in sustainable agriculture: advancements, challenges, and future directions. *Plants* 14(2), 191. DOI: 10.3390/plants14020191

Dickson, D.J. and Ely, R.L. (2013) Silica sol-gel encapsulation of cyanobacteria: lessons for academic and applied research. *Applied Microbiology and Biotechnology* 97(5), 1809–1819. DOI: 10.1007/s00253-012-4686-8

Dohroo, A. (2024) Azospirillum-a potent biofertilizer in agriculture. *Global Scientific and Academic Research Journal of Multidisciplinary Studies* 3, 8–11. Available at: https://gsarpublishers.com/wp-content/uploads/2024/02/GSARJMS192024-Gelary-script.pdf (accessed September 8, 2025).

Domínguez-Núñez, J.A., Berrocal-Lobo, M., and Albanesi, A.S. (2015) Interaction of *Azospirillum* and mycorrhiza. In: Cassán, F.D., Okon, Y., and Creus, C.M. (eds) *Handbook for Azospirillum: Technical Issues and Protocols*. Springer International Publishing, Cham, Switzerland, pp. 419–432.

dos Santos Sousa, W., Soratto, R.P., Peixoto, D.S., Campos, T.S., da Silva, M.B. *et al.* (2022) Effects of *Rhizobium* inoculum compared with mineral nitrogen fertilizer on nodulation and seed yield of common bean. A meta-analysis. *Agronomy for Sustainable Development* 42(3), 52. DOI: 10.1007/s13593-022-00784-6

Dubey, R., Rathore, D., and Dwivedi, A. (2023) Organic waste decomposition by microbial inoculants as an effective tool for environmental management. In: Sharma, V.K., Kumar, A., Passarini, M.R.Z., Parmar, S., and Singh, V.K. (eds) *Microbial Inoculants*. Academic Press, London, pp. 125–144.

Eaton, E., Hunt, A., Black, D., Frost, G., and Hargreaves, S. (2022) What are the environmental benefits and costs of reducing food waste? Bristol as a case study in the WASTE FEW Urban Living Lab Project. *Sustainability* 14, 5573. DOI: 10.3390/su14095573

Egamberdieva, D., Hua, M., Reckling, M., Wirth, S., and Bellingrath-Kimura, S.D. (2018) Potential effects of biochar-based microbial inoculants in agriculture. *Environmental Sustainability* 1(1), 19–24. DOI: 10.1007/s42398-018-0010-6

Ehis-Eriakha, C.B., Akemu, S.E., and Tiamiyu, A. (2023) Formulation and evaluation of sugarcane-bagasse-based biocontrol agents for sustainable phytopathogen management. In: Bellocchi, G. (ed.) *The 3rd International Electronic Conference on Agronomy* 27, p. 52. DOI: 10.3390/IECAG2023-15992

El Barnossi, A., Moussaid, F.Z., Saghrouchni, H., Zoubi, B., and Iraqi Housseini, A.I. (2022) Tangerine, pomegranate, and banana peels: a promising environmentally friendly bioorganic fertilizers for seed germination and cultivation of *Pisum sativum* L. *Waste and Biomass Valorization* 13(8), 3611–3627. DOI: 10.1007/s12649-022-01743-8

Emami, F., Keihan Shokooh, M., and Mostafavi Yazdi, S.J. (2023) Recent progress in drying technologies for improving the stability and delivery efficiency of biopharmaceuticals. *Journal of Pharmaceutical Investigation* 53(1), 35–57. DOI: 10.1007/s40005-022-00610-x

Erdiansyah, I., Taufika, R., Widodo, T., Damanhuri, Jannah, D.M. *et al.* (2022) Viability of biofertilizer bacteria *Rhizobium* spp. based on household waste. *IOP Conference Series* 980(1), 012009. DOI: 10.1088/1755-1315/980/1/012009

Ezeorba, T.P.C., Okeke, E.S., Mayel, M.H., Nwuche, C.O., and Ezike, T.C. (2024) Recent advances in biotechnological valorization of agro-food wastes (AFW): optimizing integrated approaches for sustainable biorefinery and circular bioeconomy. *Bioresource Technology Reports* 26, 101823. DOI: 10.1016/j.biteb.2024.101823

Fahad, M., Tariq, L., Li, W., and Wu, L. (2025) MicroRNA gatekeepers: orchestrating rhizospheric dynamics. *Journal of Integrative Plant Biology* 67(3), 845–876. DOI: 10.1111/jipb.13860

Faskhutdinova, E., Fotina, N.V., Neverova, O.A., Golubtsova, Y.V., Mudgal, G. *et al.* (2024) Extremophilic bacteria as biofertilizer for agricultural wheat. *Foods and Raw Materials* 12(2), 348–360. DOI: 10.21603/2308-4057-2024-2-613

Fukami, J., Ollero, F.J., Megías, M., and Hungria, M. (2017) Phytohormones and induction of plant-stress tolerance and defense genes by seed and foliar inoculation with *Azospirillum brasilense* cells and metabolites promote maize growth. *AMB Express* 7(1), 153. DOI: 10.1186/s13568-017-0453-7

Gabra, F.A., Abd-Alla, M.H., Danial, A.W., Abdel-Basset, R., and Abdel-Wahab, A.M. (2019) Production of biofuel from sugarcane molasses by diazotrophic *Bacillus* and recycle of spent bacterial biomass as biofertilizer inoculants for oil crops. *Biocatalysis and Agricultural Biotechnology* 19, 101112. DOI: 10.1016/j.bcab.2019.101112

Gadade, R. and Rathod, M. (2022) Sterilization and disinfection of agro-waste for mushroom cultivation. *Research Insights of Life Science Students* 3, 710–711.

Garcia, M., Bruna, P., Duran, P., and Abanto, M. (2025) Cyanobacteria and soil restoration: bridging molecular insights with practical solutions. *Microorganisms* 13(7), 1468. DOI: 10.3390/microorganisms13071468

Gatea, I.H., Sabr, A.B., Abdul Wahed, E.A., Abbas, A.H., Halob, A.A. *et al.* (2019) Isolation and characterization of local azotobacter isolate producing bio-plastics and consuming waste vegetable oils. *IOP Conference Series: Earth and Environmental Science* 388(1), 012082. DOI: 10.1088/1755-1315/388/1/012082

Gauri, S.S., Mandal, S.M., and Pati, B.R. (2012) Impact of azotobacter exopolysaccharides on sustainable agriculture. *Applied Microbiology and Biotechnology* 95(2), 331–338. DOI: 10.1007/s00253-012-4159-0

Ghatage, A., Patil, S.S., and Pathade, A. (2024) Isolation and screening of *Azotobacter* spp. for plant growth-promoting properties and their survival under various environmental stress conditions. *Bulletin of Pure & Applied Sciences-Zoology*, 43A(1S), 301–317.

Ghodhbane-Gtari, F., Hezbri, K., Ktari, A., Sbissi, I., Beauchemin, N. *et al.* (2014) Contrasted reactivity to oxygen tensions in *Frankia* sp. strain CcI3 throughout nitrogen fixation and assimilation. *BioMed Research International* 2014, 568549. DOI: 10.1155/2014/568549

Gibson, A.H., Date, R.A., Ireland, J.A., and Brockwell, J. (1976) A comparison of competitiveness and persistence amongst five strains of *Rhizobium tripolii*. *Soil Biology and Biochemistry* 8(5), 395–401. DOI: 10.1016/0038-0717(76)90040-7

Gnanaprakasam, P.D. and Vanisree, A.J. (2022) Recurring detrimental impact of agrochemicals on the ecosystem, and a glimpse of organic farming as a possible rescue. *Environmental Science and Pollution Research International* 29(50), 75103–75112. DOI: 10.1007/s11356-022-22750-1

Gonçalves, M.L.M.B.B. and Maximo, G.J. (2023) Circular economy in the food chain: production, processing and waste management. *Circular Economy and Sustainability* 3(3), 1405–1423. DOI: 10.1007/s43615-022-00243-0

Gopalakrishnan, S., Sathya, A., Vijayabharathi, R., Varshney, R.K., Gowda, C.L.L. *et al.* (2015) Plant growth promoting rhizobia: challenges and opportunities. *3 Biotech* 5(4), 355–377. DOI: 10.1007/s13205-014-0241-x

Goss, M.J., Carvalho, M., and Brito, I. (2017) Challenges to agriculture systems. In: Goss, M.J., Carvalho, M., and Brito, I. (eds) *Functional Diversity of Mycorrhiza and Sustainable Agriculture*. Academic Press, London, pp. 15–38.

Gtari, M., Ghodhbane-Gtari, F., Nouioui, I., Ktari, A., Hezbri, K. *et al.* (2015) Cultivating the uncultured: growing the recalcitrant cluster-2 *Frankia* strains. *Scientific Reports* 5, 13112. DOI: 10.1038/srep13112

Gtari, M., Maaoui, R., Ghodhbane-Gtari, F., Ben Slama, K., and Sbissi, I. (2024) MAGs-centric crack: how long will, spore-positive *Frankia* and most *Protofrankia*, microsymbionts remain recalcitrant to axenic growth? *Frontiers in Microbiology* 15, 1367490. DOI: 10.3389/fmicb.2024.1367490

Gunjal, A. and Kapadnis, B. (2020) Paddy husk, a lignocellulosic waste as potential carrier for *in-situ* production of plant growth promoting substances by *Bacillus circulans*. *Vigyan Varta* 1(1), 40–46.

Gupta, A., Bano, A., Rai, S., Sharma, S., and Pathak, N. (2022) Selection of carrier materials to formulate bioinoculant package for promoting seed germination. *Letters in Applied NanoBioScience* 12(3), 65. DOI: 10.33263/LIANBS123.065

Haddaji, C., Khattabi Rifi, S., Digua, K., Madinzi, A., Chatoui, M. *et al.* (2024) Optimization of the coagulation-flocculation process using response surface methodology for wastewater pretreatment generated by vegetable oil refineries: a path towards environmental sustainability. *Environmental Nanotechnology, Monitoring & Management* 22, 100973. DOI: 10.1016/j.enmm.2024.100973

Hardy, K. and Knight, J.D. (2021) Evaluation of biochars as carriers for *Rhizobium leguminosarum*. *Canadian Journal of Microbiology* 67(1), 53–63. DOI: 10.1139/cjm-2020-0416

Harika, R., Vasantharekha, R., Reddy, S.S., Chauhan, A., Achudhan, A.B. *et al.* (2025) Sweetener synapse: exploring non-nutritive sweeteners' role in zebrafish glucose homeostasis. *Environmental Technology & Innovation* 39, 104322. DOI: 10.1016/j.eti.2025.104322

Harriott, O.T., Khairallah, L., and Benson, D.R. (1991) Isolation and structure of the lipid envelopes from the nitrogen-fixing vesicles of *Frankia* sp. strain CpI1. *Journal of Bacteriology* 173(6), 2061–2067. DOI: 10.1128/jb.173.6.2061-2067.1991

Hashempour-Baltork, F., Farshi, P., and Khosravi-Darani, K. (2022) Vegetable and fruit wastes as substrate for production of single-cell protein and aquafeed meal. In: Ray, R.C. (ed.) *Fruits and Vegetable Wastes: Valorization to Bioproducts and Platform Chemicals*. Springer Nature, Singapore, pp. 169–187.

Huang, Y., Sheth, R.U., Zhao, S., Cohen, L.A., Dabaghi, K. *et al.* (2023) High-throughput microbial culturomics using automation and machine learning. *Nature Biotechnology* 41(10), 1424–1433. DOI: 10.1038/s41587-023-01674-2

Iasimone, F., Seira, J., Panico, A., De Felice, V., Pirozzi, F. *et al.* (2021) Insights into bioflocculation of filamentous cyanobacteria, microalgae and their mixture for a low-cost biomass harvesting system. *Environmental Research* 199, 111359. DOI: 10.1016/j.envres.2021.111359

Jabran, M., Ali, M.A., Acet, T., Zahoor, A., Abbas, A. *et al.* (2024) Growth regulation in bread wheat via novel bioinoculant formulation. *BMC Plant Biology* 24(1), 1039. DOI: 10.1186/s12870-024-05698-x

Jadhav, P., Sonne, M., Kadam, A., Patil, S., Dahigaonkar, K. *et al.* (2018) Formulation of cost effective alternative bacterial culture media using fruit and vegetables waste. *International Journal of Current Research Review* 10, 6–15.

Jaiswal, D.K., Verma, J.P., Belwal, T., Pereira, A.P.D.A., and Ade, A.B. (2023) Editorial: microbial co-cultures: a new era of synthetic biology and metabolic engineering. *Frontiers in Microbiology* 14, 2023. DOI: 10.3389/fmicb.2023.1235565

Kashyap, U., Garg, S., and Arora, P. (2024) Pesticide pollution in India: environmental and health risks, and policy challenges. *Toxicology Reports* 13, 101801. DOI: 10.1016/j.toxrep.2024.101801

Kaur, J., Mudgal, G., Chand, K., Singh, G.B., Perveen, K. *et al.* (2022) An exopolysaccharide-producing novel agrobacterium pusense strain JAS1 isolated from snake plant enhances plant growth and soil water retention. *Scientific Reports* 12(1), 21330. DOI: 10.1038/s41598-022-25225-y

Kaur, J., Mudgal, G., Negi, A., Tamang, J., Singh, S. *et al.* (2023) Reactive black-5, Congo red and methyl orange: chemical degradation of azo-dyes by agrobacterium. *Water* 15(9), 1664. DOI: 10.3390/w15091664

Kaur, J., Mudgal, G., and Dhar, S.K. (2025) Decoration plants and their role in remediation. In: Latef, A.A.H.A., Zayed, E.M., and Omar, A.A. (eds) *Sustainable Remediation for Pollution and Climate Resilience*. Springer Nature, Singapore, pp. 695–724.

Khan, A., Naseem, and Vardhini, B. (2015) Synthesis of extracellular and intracellular polymers in isolates of *Azotobacter* sp. *International Journal of Research in Engineering and Technology* 04(12), 231–233. DOI: 10.15623/ijret.2015.0412044

Kieliszek, M., Piwowarek, K., Kot, A.M., and Pobiega, K. (2020) The aspects of microbial biomass use in the utilization of selected waste from the agro-food industry. *Open Life Sciences* 15(1), 787–796. DOI: 10.1515/biol-2020-0099

Kiruba N., J.M. and Saeid, A. (2022) An insight into microbial inoculants for bioconversion of waste biomass into sustainable "bio-organic" fertilizers: a bibliometric analysis and systematic literature review. *International Journal of Molecular Sciences* 23(21), 13049. DOI: 10.3390/ijms232113049

Kopittke, P.M., Menzies, N.W., Wang, P., McKenna, B.A., and Lombi, E. (2019) Soil and the intensification of agriculture for global food security. *Environment International* 132, 105078. DOI: 10.1016/j.envint.2019.105078

Kumar, S., Diksha, Sindhu, S.S., and Kumar, R. (2022) Biofertilizers: An ecofriendly technology for nutrient recycling and environmental sustainability. *Current Research in Microbial Sciences* 3, 100094. DOI: 10.1016/j.crmicr.2021.100094

Kundu, M., Sarkar, M., Bisht, T.S., Chakraborty, B., Rakshit, A. *et al.* (2024) Bioinoculants: a sustainable tool for enhancement of productivity and nutritional quality in horticultural crops. In: Rakshit, A., Meena, V.S., Fraceto, L.F., Parihar, M., Mendonza, A.B. *et al* (eds) *Bio-Inoculants in Horticultural Crops*. Woodhead Publishing, Cambridge, UK, pp. 373–408.

Lata, C. and Jatan, R. (2019) Role of microRNAs in abiotic and biotic stress resistance in plants. *Proceedings of the Indian National Science Academy* 85(3), 553–567. DOI: 10.16943/ptinsa/2019/49586

Lin, L., Qin, J., Zhang, Y., Yin, J., Guo, G. *et al.* (2023) Assessing the suitability of municipal sewage sludge and coconut bran as breeding medium for *Oryza sativa* L. seedlings and developing a standardized substrate. *Journal of Environmental Management* 344, 118644. DOI: 10.1016/j.jenvman.2023.118644

Lindström, K. and Mousavi, S.A. (2020) Effectiveness of nitrogen fixation in rhizobia. *Microbial Biotechnology* 13(5), 1314–1335. DOI: 10.1111/1751-7915.13517

Liu, L., Yahaya, B.S., Li, J., and Wu, F. (2024) Enigmatic role of auxin response factors in plant growth and stress tolerance. *Frontiers in Plant Science* 15, 2024. DOI: 10.3389/fpls.2024.1398818

Mahdinia, E., Cekmecelioglu, D., and Demirci, A. (2019) Bioreactor scale-up. In: Berenjian, A. (ed.) *Essentials in Fermentation Technology*. Springer International Publishing, Cham, Switzerland, pp. 213–236.

Malos, I.G., Ghizdareanu, A.-I., Vidu, L., Matei, C.B., and Pasarin, D. (2025) The role of whey in functional microorganism growth and metabolite generation: a biotechnological perspective. *Foods* 14(9), 1488. DOI: 10.3390/foods14091488

Maluk, M., Ferrando-Molina, F., Lopez del Egido, L., Langarica-Fuentes, A., Yohannes, G.G. *et al.* (2022) Fields with no recent legume cultivation have sufficient nitrogen-fixing rhizobia for crops of faba bean (*Vicia faba* L.). *Plant and Soil* 472(1–2), 345–368. DOI: 10.1007/s11104-021-05246-8

Mankar, M.K., Sharma, U.S., and Sahay, S. (2021) Lantana charcoal as potent carrier material for *Azotobacter chroococcum*. *Die Bodenkultur: Journal of Land Management, Food and Environment* 72, 83–91. DOI: 10.2478/boku-2021-0008

Martin, N.H., Torres-Frenzel, P., and Wiedmann, M. (2021) Invited review: controlling dairy product spoilage to reduce food loss and waste. *Journal of Dairy Science* 104(2), 1251–1261. DOI: 10.3168/jds.2020-19130

Martínez-Ramírez, C., Esquivel-Cote, R., Ferrera-Cerrato, R., Martínez-Ruiz, J.A., Rodríguez-Serrano, G. *et al.* (2021) Solid-state culture of *Azospirillum brasilense*: a reliable technology for biofertilizer production from laboratory to pilot scale. *Bioprocess and Biosystems Engineering* 44(7), 1525–1538. DOI: 10.1007/s00449-021-02537-3

Masquelier, S., Sozzi, T., Bouvet, J.C., Bésiers, J., and Deogratias, J.-M. (2022) Conception and development of recycled raw materials (coconut fiber and bagasse)-based substrates enriched with soil microorganisms (arbuscular mycorrhizal fungi, *Trichoderma* spp. and *Pseudomonas* spp.) for the soilless cultivation of tomato (*S. lycopersicum*). *Agronomy* 12(4), 767. DOI: 10.3390/agronomy12040767

Merrylin, J., Kannah, R.Y., Banu, J.R., and Yeom, I.T. (2020) Production of organic acids and enzymes/biocatalysts from food waste. In: Banu, J.R., Kumar, G., Gunasekaran, M., and Kavitha, S. (eds) *Food Waste to Valuable Resources*. Academic Press, London, pp. 119–141.

Mohamed, T.A., Wu, J., Zhao, Y., Elgizawy, N., El Kholy, M. *et al.* (2022) Insights into enzyme activity and phosphorus conversion during kitchen waste composting utilizing phosphorus-solubilizing bacterial inoculation. *Bioresource Technology* 362, 127823. DOI: 10.1016/j.biortech.2022.127823

Mudgal, G., Kaur, J., Chand, K., Parashar, M., Dhar, S.K. *et al.* (2022) Mitigating the mistletoe menace: biotechnological and smart management approaches. *Biology* 11(11), 1645. DOI: 10.3390/biology11111645

Mudgal, G., Sivaji, S., Kaur, J., Dhar, S.K., Ramasamy, V. *et al.* (2025) Strategic conservation plan. In: Latef, A.A.H.A., Zayed, E.M., and Omar, A.A. (eds) *Sustainable Remediation for Pollution and Climate Resilience*. Springer Nature, Singapore, pp. 139–206

Munesue, Y. and Masui, T. (2019) The impacts of Japanese food losses and food waste on global natural resources and greenhouse gas emissions. *Journal of Industrial Ecology* 23(5), 1196–1210. DOI: 10.1111/jiec.12863

Mustafa, G., Arshad, M., Bano, I., and Abbas, M. (2023) Biotechnological applications of sugarcane bagasse and sugar beet molasses. *Biomass Conversion and Biorefinery* 13(2), 1489–1501. DOI: 10.1007/s13399-020-01141-x

Mustereț, C.P., Morosanu, I., Ciobanu, R., Plavan, O., Gherghel, A. *et al.* (2021) Assessment of coagulation–flocculation process efficiency for the natural organic matter removal in drinking water treatment. *Water* 13(21), 3073. DOI: 10.3390/w13213073

Muth, M.K., Birney, C., Cuéllar, A., Finn, S.M., Freeman, M. *et al.* (2019) A systems approach to assessing environmental and economic effects of food loss and waste interventions in the United States. *The Science of the Total Environment* 685, 1240–1254. DOI: 10.1016/j.scitotenv.2019.06.230

Naik, K., Mishra, S., Srichandan, H., Singh, P.K., and Choudhary, A. (2020) Microbial formulation and growth of cereals, pulses, oilseeds and vegetable crops. *Sustainable Environment Research* 30(1), 10. DOI: 10.1186/s42834-020-00051-x

Nandikolmath, V. (2023) Agro composite waste: a novel and economical substrate for the production of edible mushroom. *Kavaka* 59(2). DOI: 10.36460/Kavaka/59/2/2023/75-81

Nicastro, R. and Carillo, P. (2021) Food loss and waste prevention strategies from farm to fork. *Sustainability* 13(10), 5443. DOI: 10.3390/su13105443

Nie, W., He, Q., Guo, H., Zhang, W., Ma, L. *et al.* (2024) Arbuscular mycorrhizal fungi: boosting crop resilience to environmental stresses. *Microorganisms* 12(12), 2448. DOI: 10.3390/microorganisms12122448

Nievas, S., Coniglio, A., Takahashi, W.Y., López, G.A., Larama, G. *et al.* (2023) Unraveling *Azospirillum* 's colonization ability through microbiological and molecular evidence. *Journal of Applied Microbiology* 134(4). DOI: 10.1093/jambio/lxad071

O'Callaghan, M., Ballard, R.A., and Wright, D. (2022) Soil microbial inoculants for sustainable agriculture: limitations and opportunities. *Soil Use and Management* 38(3), 1340–1369. DOI: 10.1111/sum.12811

Ogbu, C.C. and Okey, S.N. (2023) Agro-industrial waste management: the circular and bioeconomic perspective. In: Ahmad, F. and Sultan, M. (eds) *Agricultural Waste - New Insights*. IntechOpen, Rijeka, Croatia. DOI: 10.5772/intechopen.109181

Owsianiak, M., Hauschild, M.Z., Posthuma, L., Saouter, E., Vijver, M.G. *et al.* (2023) Ecotoxicity characterization of chemicals: global recommendations and implementation in USEtox. *Chemosphere* 310, 136807. DOI: 10.1016/j.chemosphere.2022.136807

Palaniveloo, K., Amran, M.A., Norhashim, N.A., Mohamad-Fauzi, N., Peng-Hui, F. *et al.* (2020) Food waste composting and microbial community structure profiling. *Processes* 8(6), 723. DOI: 10.3390/pr8060723

Pan, L. and Cai, B. (2023) Phosphate-solubilizing bacteria: advances in their physiology. *Molecular Mechanisms and Microbial Community Effects* 11, 2904. DOI: 10.3390/microorganisms11122904

Papin, M. (2024) Efficiency and impact of recurrent microbial inoculation in soil, a lab to field assessment (*Efficacité et impact de l'inoculation microbienne récurrente, évaluation du laboratoire au champ*). Doctoral thesis, Université Bourgogne Franche-Comté, Besançon, France.

Parajuli, A., Khadka, A., Sapkota, L., and Ghimire, A. (2022) Effect of hydraulic retention time and organic-loading rate on two-staged, semi-continuous mesophilic anaerobic digestion of food waste during start-up. *Fermentation* 8(11), 620. DOI: 10.3390/fermentation8110620

Parashar, M., Dhar, S.K., Kaur, J., Chauhan, A., Tamang, J. *et al.* (2023) Two novel plant-growth-promoting *Lelliottia amnigena* isolates from *Euphorbia prostrata* aiton enhance the overall productivity of wheat and tomato. *Plants* 12(17), 3081. DOI: 10.3390/plants12173081

Pedersen, S.M. and Lind, K.M. (eds) (2017) *Precision Agriculture: Technology and Economic Perspectives*. Springer International Publishing AG, Cham, Switzerland.

Peralta, F.T., Shi, C., Widanagamage, G.W., Speight, R.E., O'Hara, I. *et al.* (2024) Pretreated sugarcane bagasse matches performance of synthetic media for lipid production with *Yarrowia lipolytica*. *Bioresource Technology* 413, 131558. DOI: 10.1016/j.biortech.2024.131558

Rafiq, S., Bhat, M.I., Sofi, S.A., Muzzafar, K., Majid, D. *et al.* (2023) Bioconversion of agri-food waste and by-products into microbial lipids: mechanism, cultivation strategies and potential in food applications. *Trends in Food Science & Technology* 139, 104118. DOI: 10.1016/j.tifs.2023.07.015

Raut, S.B., Deokar, C.D., Navale, A.M., and Dahatonde, J.A. (2019) On-farm production of arbuscular mycorrhizal (AM) fungi using trap crop cycles. *International Journal of Current Microbiology and Applied Sciences* 8(10), 1084–1101. DOI: 10.20546/ijcmas.2019.810.128

Read, Q.D., Brown, S., Cuéllar, A.D., Finn, S.M., Gephart, J.A. *et al.* (2020) Assessing the environmental impacts of halving food loss and waste along the food supply chain. *The Science of the Total Environment* 712, 136255. DOI: 10.1016/j.scitotenv.2019.136255

Rocha, F., Esteban Lucas-Borja, M., Pereira, P., and Muñoz-Rojas, M. (2020) Cyanobacteria as a nature-based biotechnological tool for restoring salt-affected soils. *Agronomy* 10(9), 1321. DOI: 10.3390/agronomy10091321

Rocha, I., Ma, Y., Souza-Alonso, P., Vosátka, M., Freitas, H. *et al.* (2019) Seed coating: a tool for delivering beneficial microbes to agricultural crops. *Frontiers in Plant Science* 10, 1357. DOI: 10.3389/fpls.2019.01357

Rojas-Padilla, J., Díaz-Rodríguez, A.M., and de los Santos Villalobos, S. (2024) Bioformulation of bacterial inoculants. In: de los Santos Villalobos, S. (ed.) *New Insights, Trends, and Challenges in the Development and Applications of Microbial Inoculants in Agriculture*. Academic Press, London, pp. 105–116.

Rojas-Sánchez, B., Guzmán-Guzmán, P., Morales-Cedeño, L.R., Orozco-Mosqueda, Ma. del C., Saucedo-Martínez, B.C. *et al.* (2022) Bioencapsulation of microbial inoculants: mechanisms, formulation types and application techniques. *Applied Biosciences* 1(2), 198–220. DOI: 10.3390/applbiosci1020013

Roosjen, M., Paque, S., and Weijers, D. (2018) Auxin response factors: output control in auxin biology. *Journal of Experimental Botany* 69(2), 179–188. DOI: 10.1093/jxb/erx237

Roy, D., Azaïs, A., Benkaraache, S., Drogui, P., and Tyagi, R.D. (2018) Composting leachate: characterization, treatment, and future perspectives. *Reviews in Environmental Science and Bio/Technology* 17(2), 323–349. DOI: 10.1007/s11157-018-9462-5

Sadh, P.K., Duhan, S., and Duhan, J.S. (2018) Agro-industrial wastes and their utilization using solid state fermentation: a review. *Bioresources and Bioprocessing* 5(1), 1. DOI: 10.1186/s40643-017-0187-z

Sahu, P.K., Gupta, A., Sharma, L., and Bakade, R. (2017) Mechanisms of *Azospirillum* in plant growth promotion. *Scholars Journal of Agriculture and Veterinary Sciences* 4(9), 338–343. DOI: 10.21276/sjavs.2017.4.9.3

Salomon, M.J., Demarmels, R., Watts-Williams, S.J., McLaughlin, M.J., Kafle, A. *et al.* (2022) Global evaluation of commercial arbuscular mycorrhizal inoculants under greenhouse and field conditions. *Applied Soil Ecology* 169, 104225. DOI: 10.1016/j.apsoil.2021.104225

Sammauria, R., Kumawat, S., Kumawat, P., Singh, J., and Jatwa, T.K. (2020) Microbial inoculants: potential tool for sustainability of agricultural production systems. *Archives of Microbiology* 202(4), 677–693. DOI: 10.1007/s00203-019-01795-w

Schlesinger, A., Eisenstadt, D., Einbinder, S., and Gressel, G. (2012) Method and system for efficient harvesting of microalgae and cyanobacteria. U.S. Patent Application No. 12/924,460.

Schütz, L., Gattinger, A., Meier, M., Müller, A., Boller, T. *et al.* (2018) Improving crop yield and nutrient use efficiency via biofertilization: a global meta-analysis. *Frontiers in Plant Science* 8, 2204. DOI: 10.3389/fpls.2017.02204

Secretariat of the Convention on Biological Diversity (2000) *Cartagena Protocol on Biosafety to the Convention on Biological Diversity: Text and Annexes*. Secretariat of the Convention of Biological Diversity, Montreal, Canada. Available at: https://www.cbd.int/doc/legal/cartagena-protocol-en.pdf (accessed September 10, 2025).

Seetharaman, B., Balu, U.R., Mudgal, G., Firdaus, S., Dash, S. *et al.* (2025) The piteous price of progress: the environmental and health costs of modern society. In: Prakash, C., Kesari, K.K., and Negi, A. (eds) *Sustainable Development Goals Towards Environmental Toxicity and Green Chemistry: Environment and Sustainability*. Springer Nature, Cham, Switzerland, pp. 291–313.

Sellami, M., Oszako, T., Miled, N., and Ben Rebah, F. (2015) Industrial wastewater as raw material for exopolysaccharide production by *Rhizobium leguminosarum*. *Brazilian Journal of Microbiology* 46(2), 407–413. DOI: 10.1590/S1517-838246220140153

Serazetdinova, Y., Borodina, E., Fotina, N., Naik, A., Mudgal, G. *et al.* (2025) Rhizobia as complex biofertilizers for wheat: biological nitrogen fixation and plant growth promotion. *Foods and Raw Materials* 14, 214–227. DOI: 10.21603/2308-4057-2026-1-669

Shahwar, D., Mushtaq, Z., Mushtaq, H., Alqarawi, A.A., Park, Y. *et al.* (2023) Role of microbial inoculants as bio fertilizers for improving crop productivity: a review. *Heliyon* 9(6), e16134. DOI: 10.1016/j.heliyon.2023.e16134

Sharma, A., Chand, K., Singh, G.B., and Mudgal, G. (2021) Toxicity with waste-generated ionizing radiations: blunders behind the scenes. In: Kesari, K.K. and Jha, N.K. (eds) *Free Radical Biology and Environmental Toxicity*. Springer International Publishing, Cham, Switzerland, pp. 305–325.

Sharma, V., Mishra, N., Thomas, S., Narasanna, R., Jambaladinni, K. *et al.* (2024) Advances in microbial study for crop improvement. In: Halami, P.M. and Sundararaman, A. (eds) *Genome Editing in Bacteria, Part 2*. Bentham Science Publishers, Sharjah, United Arab Emirates, pp. 1–42.
Sidana, A. and Farooq, U. (2014) Sugarcane bagasse: a potential medium for fungal cultures. *Chinese Journal of Biology* 2014(2), 1–5. DOI: 10.1155/2014/840505
Simon, S.A., Meyers, B.C., and Sherrier, D.J. (2009) MicroRNAs in the rhizobia legume symbiosis. *Plant Physiology* 151(3), 1002–1008. DOI: 10.1104/pp.109.144345
Singh, A.K., Gauri, S., Bhatt, R.P., and Pant, S. (2011) Optimization and comparative study of the sugar waste for the growth of rhizobium cells along with traditional laboratory media. *Research Journal of Microbiology* 6(9), 715–723. DOI: 10.3923/jm.2011.715.723
Singh, A.K., Singh, G., and Yadav, G. (2019) Food industry waste material as a growth medium for rhizobium. *International Journal of Pharmacy and Biological Sciences* 9, 144–155. Available at: https://www.academia.edu/74197259/Food_Industry_Waste_Material_as_A_Growth_Medium_for_Rhizobium (accessed September 8, 2025).
Singh, G.B., Mudgal, G., Vinayak, A., Kaur, J., Chand, K. *et al.* (2021) Molecular communications between plants and microbes. In: Tyagi, S., Kumar, R., Saharan, B.S., and Nadda, A.K. (eds) *Plant-Microbial Interactions and Smart Agricultural Biotechnology*. CRC Press, Boca Raton, Florida, pp. 147–184.
Singh, G.B., Dey, S., and Mudgal, G. (2025) Microbial contributions to a circular economy. In: Kaur, S., Dhiman, S., and Tripathi, M. (eds) *Microbial Metabolomics: Recent Developments, Challenges and Future Opportunities*. Springer Nature, Singapore, pp. 419–439.
Singh, P.K., Singh, M., and Tripathi, B.N. (2013) Glomalin: an arbuscular mycorrhizal fungal soil protein. *Protoplasma* 250(3), 663–669. DOI: 10.1007/s00709-012-0453-z
Singh, S., Bhople, N., Tomar, A., Savita, S., and Rana, M. (2018) Progress and potential of biofertilization in contemporary production practices of sustainable agriculture: review. *Annals of Biology* 34(2), 196–201.
Sivapriya, S.L. and Priya, P.R. (2017) Selection of hyper exopolysaccharide producing and cyst forming *Azotobacter* isolates for better survival under stress conditions. *International Journal of Current Microbiology and Applied Sciences* 6(6), 2310–2320. DOI: 10.20546/ijcmas.2017.606.274
Sivaram, A.K., Abinandan, S., Chen, C., Venkateswartlu, K., and Megharaj, M. (2023) Microbial inoculant carriers: soil health improvement and moisture retention in sustainable agriculture. In: Sparks, D.L. (ed.) *Advances in Agronomy*. Academic Press, London, pp. 35–91.
Sivasakthivelan, P., Saranraj, P., Sayyed, R.Z., Arivukkarasu, K., Kokila, M. *et al.* (2023) Inoculant production and formulation of *Azospirillum* species. In: Mawar, R., Sayyed, R.Z., Sharma, S.K., and Sattiraju, K.S. (eds) *Plant Growth Promoting Microorganisms of Arid Region*. Springer Nature Singapore, Singapore, pp. 423–455.
Steenhoudt, O. and Vanderleyden, J. (2000) Azospirillum, a free-living nitrogen-fixing bacterium closely associated with grasses: genetic, biochemical and ecological aspects. *FEMS Microbiology Reviews* 24(4), 487–506. DOI: 10.1111/j.1574-6976.2000.tb00552.x
Stephen, G.S., Shitindi, M.J., Bura, M.D., Kahangwa, C.A., and Nassary, E.K. (2024) Harnessing the potential of sugarcane-based liquid byproducts—molasses and spentwash (vinasse) for enhanced soil health and environmental quality. A systematic review. *Frontiers in Agronomy* 6, 1358076. DOI: 10.3389/fagro.2024.1358076
Stikane, A., Baumanis, M.R., Muiznieks, R., and Stalidzans, E. (2023) Impact of waste as a substrate on biomass formation, and optimization of spent microbial biomass re-use by sustainable metabolic engineering. *Fermentation* 9(6), 531. DOI: 10.3390/fermentation9060531
Suthar, H., Hingurao, K., Vaghashiya, J., and Parmar, J. (2017) Fermentation: a process for biofertilizer production. In: Panpatte, D.G., Jhala, Y.K., Vyas, R.V., and Shelat, H.N. (eds) *Microorganisms for Green Revolution*, Vol. 1. Springer, Singapore, pp. 229–252.
Tao, M., Huang, Y., Luo, J., Wang, Y., and Luo, X. (2024) The role of proton excreted by *Advenella kashmirensis* DF12 during ammonium assimilation in phosphate solubilization. *World Journal of Microbiology & Biotechnology* 40(11), 346. DOI: 10.1007/s11274-024-04087-8
Thirugnanam, T. and Dharumadurai, D. (2023) Mass multiplication and cost analysis of *Frankia* biofertilizer. In: Amaresan, N., Dharumadurai, D., and Babalola, O.O. (eds) *Agricultural Microbiology Based Entrepreneurship: Making Money from Microbes*. Springer Nature, Singapore, pp. 155–168.
Thiviya, P., Gamage, A., Kapilan, R., Merah, O., and Madhujith, T.J.S. (2022) Single cell protein production using different fruit waste: a review. *Separations* 9(7), 178. DOI: 10.3390/separations9070178
Tiwari, M., Pandey, V., Singh, B., and Bhatia, S. (2021) Dynamics of miRNA mediated regulation of legume symbiosis. *Plant, Cell & Environment* 44(5), 1279–1291. DOI: 10.1111/pce.13983
Trujillo-Roldán, M.A., Valdez-Cruz, N.A., Gonzalez-Monterrubio, C.F., Acevedo-Sánchez, E.V., Martínez-Salinas, C. *et al.* (2013) Scale-up from shake flasks to pilot-scale production of the plant growth-promoting

bacterium *Azospirillum brasilense* for preparing a liquid inoculant formulation. *Applied Microbiology and Biotechnology* 97(22), 9665–9674. DOI: 10.1007/s00253-013-5199-9

Tsolcha, O.N., Patrinou, V., Economou, C.N., Dourou, M., Aggelis, G. *et al.* (2021) Utilization of biomass derived from cyanobacteria-based agro-industrial wastewater treatment and raisin residue extract for bioethanol production. *Water* 13(4), 486. DOI: 10.3390/w13040486

Usmani, Z., Sharma, M., Gaffey, J., Sharma, M., Dewhurst, R.J. *et al.* (2022) Valorization of dairy waste and by-products through microbial bioprocesses. *Bioresource Technology* 346, 126444. DOI: 10.1016/j.biortech.2021.126444

Uwa, C., Oluwatosin, O., Ugah, U., Nwoba, S., and Ikeide, I. (2018) Microbial media formulation using rice husks, yam and cassava peels on some selected microorganisms. *Middle East Journal of Scientific Research* 26, 122–131. DOI: 10.5829/idosi.mejsr.2018.122.131

Uyar, B., Ali, M.D., and Uyar, G.E.O. (2024) Design parameters comparison of bubble column, airlift and stirred tank photobioreactors for microalgae production. *Bioprocess and Biosystems Engineering* 47(2), 195–209. DOI: 10.1007/s00449-023-02952-8

Valdivia-Rivera, S., Ayora-Talavera, T., Lizardi-Jiménez, M.A., García-Cruz, U., Cuevas-Bernardino, J.C. *et al.* (2021) Encapsulation of microorganisms for bioremediation: techniques and carriers. *Reviews in Environmental Science and Bio/Technology* 20(3), 815–838. DOI: 10.1007/s11157-021-09577-x

Van Nguyen, T. and Pawlowski, K. (2017) *Frankia* and actinorhizal plants: symbiotic nitrogen fixation. In: Mehnaz, S. (ed.) *Rhizotrophs: Plant Growth Promotion to Bioremediation*. Springer, Singapore, pp. 237–261. DOI: 10.1007/978-981-10-4862-3_12

Verma, S., Negi, N.P., Pareek, S., Mudgal, G., and Kumar, D. (2022) Auxin response factors in plant adaptation to drought and salinity stress. *Physiologia Plantarum* 174(3), e13714. DOI: 10.1111/ppl.13714

Vieira, R.M., de Freitas, C., Beluomini, M.A., Silva, R.D., Stradiotto, N.R. *et al.* (2025) Exploring fruit waste macromolecules and their derivatives to produce building blocks and materials. *Reviews in Environmental Science and Bio/Technology* 24(1), 167–189. DOI: 10.1007/s11157-024-09713-3

Vishwakarma, K., Kumar, N., Shandilya, C., Mohapatra, S., Bhayana, S. *et al.* (2020) Revisiting plant-microbe interactions and microbial consortia application for enhancing sustainable agriculture: a review. *Frontiers in Microbiology* 11, 560406. DOI: 10.3389/fmicb.2020.560406

Weis, C., Narang, A., Rickard, B., and Souza-Monteiro, D.M. (2021) Effects of date labels and freshness indicators on food waste patterns in the United States and the United Kingdom. *Sustainability* 13(14), 7897. DOI: 10.3390/su13147897

Weitzel, M.L.J., Vegge, C.S., Pane, M., Goldman, V.S., Koshy, B. *et al.* (2021) Improving and comparing probiotic plate count methods by analytical procedure lifecycle management. *Frontiers in Microbiology* 12, 693066. DOI: 10.3389/fmicb.2021.693066

Wunderlich, S.M. and Martinez, N.M. (2018) Conserving natural resources through food loss reduction: production and consumption stages of the food supply chain. *International Soil and Water Conservation Research* 6(4), 331–339. DOI: 10.1016/j.iswcr.2018.06.002

Yang, W., Gu, S., Xin, Y., Bello, A., Sun, W. *et al.* (2018) Compost addition enhanced hyphal growth and sporulation of arbuscular mycorrhizal fungi without affecting their community composition in the soil. *Frontiers in Microbiology* 9, 2018. DOI: 10.3389/fmicb.2018.00169

Yeremko, L., Czopek, K., Staniak, M., Marenych, M., and Hanhur, V. (2025) Role of environmental factors in legume-rhizobium symbiosis: a review. *Biomolecules* 15(1), 118. DOI: 10.3390/biom15010118

Zainudin, M.H.M., Zulkarnain, A., Azmi, A.S., Muniandy, S., Sakai, K. *et al.* (2022) Enhancement of agro-industrial waste composting process via the microbial inoculation: a brief review. *Agronomy* 12(1), 198. DOI: 10.3390/agronomy12010198

Zhang, F., Yang, J., Zhang, N., Wu, J., and Si, H. (2022) Roles of microRNAs in abiotic stress response and characteristics regulation of plant. *Frontiers in Plant Science* 13, 919243. DOI: 10.3389/fpls.2022.919243

Zhang, Y., Ku, Y.-S., Cheung, T.-Y., Cheng, S.-S., Xin, D. *et al.* (2024) Challenges to rhizobial adaptability in a changing climate: genetic engineering solutions for stress tolerance. *Microbiological Research* 288, 127886. DOI: 10.1016/j.micres.2024.127886

Zhu, S., He, Y., Dong, J., Dong, Y., Li, C. *et al.* (2023) Preparation of slow-release regulated Rs-198 bilayer microcapsules and application of its lyophilized bacterial inoculant on *Capsicum annuum* L. under salt stress. *Particuology* 79, 54–63. DOI: 10.1016/j.partic.2022.11.007

Zorn, M. and Komac, B. (2013) Land degradation. In: Bobrowsky, P.T. (ed.) *Encyclopedia of Natural Hazards*. Springer, Dordrecht, Netherlands, pp. 580–583.

5 Utilization of Agrifood Waste to Produce Solid Biofertilizer and Its Efficacy

Abha Kumari*

Centre for Biotechnology and Biochemical Engineering, Amity Institute of Biotechnology, Amity University, Noida, India

Abstract

The growing demand for sustainable agricultural practices has drawn attention to conversion of agrifood waste to value-added products, specifically solid biofertilizers. Solid biofertilizers are getting increased recognition as sustainable alternatives to chemical fertilizers, since they offer more environmental and agricultural benefits. This review evaluates the utilization of agrifood waste for the production of solid biofertilizers. It also explores their efficacy in soil fertility and crop performance. The importance of valorizing agrifood waste has also been detailed in the review so as to promote the concept of circular economy in a sustainable form.

5.1 Introduction

Rising global population levels—expected to reach 9.8 billion by 2050—pose serious challenges to maintaining a stable and sufficient food supply (Serraj and Pingali, 2018). These challenges are further intensified by global warming, climate change, and accelerating urbanization. At the same time, approximately 1.6 billion tonnes of uneaten food are generated annually as food waste (FW), representing a vast underutilized resource and a significant environmental burden. Effective and sustainable management of this waste is critical, not only to reduce ecological harm but also to recover valuable materials (Ma *et al.*, 2020). One promising solution lies in the valorization of food and agricultural waste to produce biofertilizers—an area that has not received sufficient attention. Recycling food waste into fertilizers offers a twofold advantage: mitigating the harmful environmental impacts of waste accumulation and reducing reliance on chemical fertilizers, which themselves contribute to soil degradation and greenhouse gas (GHG) emissions.

Food waste is a common concern in all the countries in the world, and the waste produced has been increasing exponentially. It has been calculated that around 1.05 billion tonnes of food waste is generated in all stages: food production, harvesting (Sravani *et al.*, 2024), retailing, and on the consumer level as well. Agrifood waste presents a major environmental and economic challenge due to limited infrastructure and management systems. A significant proportion of food produced around the world—close to

*Corresponding author: akumari@amity.edu

DOI: 10.1079/9781836991021.0005

33%—never gets consumed, despite 1 in every 11 people sleeping on an empty stomach (Guillou and Matheron, 2012). This not only causes financial loses but also causes environmental issues due to the release of greenhouse gases. The challenge here is to tackle the careless disposal of such waste when it can be valorized. Food waste originates from both domestic and commercial sources, ranging from kitchen scraps and fruit or vegetable peels to unsellable or spoiled goods from grocery outlets. Agro-waste, in contrast, includes agricultural residues such as crop straw, husks, and animal manure (Gustavsson *et al.*, 2011). Despite their differences, both waste streams can be transformed into biofertilizers through anaerobic digestion.

The Food and Agricultural Organization (FAO) makes an important distinction between two problems faced: food loss and food waste. Food loss is a phenomenon of early food chain processes such as transportation, harvesting, or processing (Koester and Galaktionova, 2021). It usually involves food becoming spoiled before reaching the consumer. Food waste, on the other hand, is a phenomenon that occurs later in the food chain processes. It happens during food distribution or due to over-purchasing or poor storage in houses and restaurants.

These problems have shifted the focus of the food industry toward more sustainable approaches to manage food waste. A growing area of interest is finding innovative ways to repurpose leftover food as part of a circular economy, where nothing goes to waste (Borello *et al.*, 2017). Food waste arises from both plant-based and animal-based sources. On the plant side, it commonly includes inedible or discarded parts such as peels, stems, seeds, shells, bran, pulp, and leftover residues. From animal sources, waste is generated during animal husbandry, dairy processing, seafood handling, and slaughtering activities. According to the FAO (Pérez *et al.*, 2023), approximately 17% of food produced globally becomes waste, while another 14% is lost throughout the production and supply chain. Industry-specific contributions to food waste are significant: the beverage industry accounts for 26%, followed by the dairy sector (21%), fruit and vegetable processing (14.5%), grain processing (12.5%), meat processing and preservation (8%), oil production (3.5%), and fish processing (0.5%).

5.2 Utilization of Agrifood Waste

Food waste holds significant potential as a source of value-added compounds. These include phytochemicals, antioxidants, natural colorants, and essential nutrients, all of which possess nutritional benefits and functional properties that could be harnessed.

Food waste treated inappropriately causes hazardous problems like human diseases and environmental issues (Vilas *et al.*, 2021). Food waste generally is not disposed of properly and ends up in landfill, causing water pollution, soil erosion, etc., while diseases due to the presence of bacteria and viruses may cause harm to humans and animals. This calls for a proper management protocol and innovative ways to treat the waste. While there are many sustainable ways to solve the problems, such as production of biofuel, fermentation feedstock, and eco-friendly packaging, production of biofertilizer can be a way to keep the nutrient flow in check and naturally enhance the quality of food and land without usage of any synthetic or inorganic fertilizer. Anaerobic digestion involves the microbial breakdown of organic matter in the absence of oxygen, facilitated by archaea-bacteria such as methanogens. These microorganisms act on nitrogenous substrates to produce methane gas. The activity and efficiency of methanogens are influenced by several factors, including substrate composition, temperature, pH, and the concentration of ammonia. The ammonia, produced during the biodegradation of nitrogen-rich material, is present in two forms: ammonium ion (NH_4^+) and free ammonia (NH_3). While nitrogen is an essential nutrient for microbial growth, excessive concentrations of ammonia can increase system alkalinity and inhibit methanogenesis. Thus, a careful balance must be maintained to ensure optimal microbial activity and system stability.

Equally important is the carbon-to-nitrogen (C/N) ratio of the waste. A high C/N ratio can lead to the excessive formation of organic acids during anaerobic digestion, reducing the pH and inhibiting methane production. On one hand, while nitrogen from ammonia is essential for microbial growth as it is a vital nutrient. However, in excess, it inhibits growth by increasing alkalinity. Reduced methane is beneficial for solid biofertilizers and the environment as methane is a greenhouse gas which causes global warming (Paul *et al.*, 2018). Moist mixtures for anaerobic digestion are preferred as they increase the degradation speed of the substrate.

Sustainable development based on the concept of biorefinery is an appropriate approach to utilizing this waste by producing commercially favorable products such as biofertilizers (Kathi *et al.*, 2023). In summary, the anaerobic treatment of agrifood waste provides a viable and environmentally sound approach to producing biofertilizers. This method offers a sustainable alternative to landfill and chemical fertilizer use, while simultaneously addressing the environmental and health concerns associated with urban waste accumulation.

A study conducted by Imran *et al.* (2022) focused on developing a tailored bioactive compost made from agricultural waste and evaluating its impact on the growth and yield of chili pepper and tomato plants. The compost was specifically formulated to enhance soil fertility and provide essential nutrients. The researchers found that applying this bioactive compost significantly improved plant growth parameters such as height, leaf area, and biomass. It also increased fruit yield and quality in both chili pepper and tomato plants compared to conventional fertilization methods. The study highlights the potential of using agro-waste-based compost as a sustainable and eco-friendly alternative to chemical fertilizers, promoting better crop productivity and soil health. Another study, carried out by Onyia *et al.* (2020), investigated the development of a biofertilizer using plant growth-promoting bacteria (GPB) isolated from organic waste materials. The waste used included cow dung and poultry manure, which served as substrates for cultivating beneficial microbial strains. The biofertilizer was applied to maize (Zea mays L.) to evaluate its effect on plant growth compared to untreated controls and chemical fertilizers. The key findings of this experiment were enhanced maize growth parameters, which included plant height, leaf number, and biomass. It improved the soil nutrient content by increasing nitrogen and phosphorus availability. Overall, it provides a sustainable alternative to chemical fertilizers.

Chakravarty (2024) objectivized the development of a low-cost, eco-friendly biofertilizer using solid-state fermentation (SSF) of agricultural wastes and tested its efficacy in promoting plant growth using indigenous strains of *Pseudomonas fluorescens* (Pf) and *Bacillus subtilis* (Bs). The co-substrate formulation (equal mix of all seven agro-wastes) with both Pf and Bs (named CSPfBs) showed the highest microbial load at 45 days after sowing (DAS), and the greatest stability by 90 DAS. The CSPfBs, vermicompost+Bs (VBs), and vermicompost+Pf (VPf) treatments significantly enhanced aubergine growth and yield parameters (e.g. plant height, leaf area, biomass, fruit number/weight) at 30 days after transplant compared to the control. Agricultural wastes can serve as effective carriers for beneficial microbes via SSF, creating stable biofertilizer formulations. The approach offers a cost-effective and sustainable alternative to chemical fertilizers, and supports a circular economy by valorizing organic waste.

Jiang *et al.* (2024) developed and evaluated a novel, sustainable method for converting food waste into biofertilizer using a biochar-augmented enzymatic hydrolysis process. The study aimed to address the growing challenges of food waste disposal and agricultural sustainability by transforming organic waste into high-quality biofertilizer products. The process involved treating 100 kg of mixed food waste with 10% biochar and 5% hydrolytic enzymes derived from a fungal fermentation mash of food waste. This treatment yielded approximately 22.3 kg of solid biofertilizer and 55.0 kg of liquid biofertilizer—both meeting national Chinese standards for biofertilizer safety and nutrient content. To assess the agricultural effectiveness of these products, the researchers conducted a planting trial using pak choi (*Brassica rapa var. chinensis*), a widely cultivated leafy vegetable. The results showed that the food waste-derived biofertilizers performed comparably to commercial biofertilizers, significantly enhancing plant growth and biomass yield. In addition to improving crop performance, the application of these biofertilizers did not negatively impact key soil chemical properties. Notably, they also contributed to an increase in soil microbial diversity, which can play a crucial role in long-term soil health and resilience. The presence of biochar in the formulation likely improved nutrient retention and microbial activity, further supporting plant growth. The study highlights the technical and economic viability of the method, pointing to a promising solution for waste valorization and circular agriculture. By efficiently integrating biochar, enzymatic treatment, and food waste management, this process offers a low-cost, eco-friendly strategy for producing both solid and liquid biofertilizers at scale. Overall, the research provides strong evidence for using biochar-enhanced enzymatic processes in food waste recycling and supports broader efforts to develop sustainable, closed-loop agricultural systems.

Wang *et al.* (2024) investigated how bio-organic fertilizer (Ft) alone, and in combination with biochar at two application rates (6.85 t/ha and 13.7 t/ha), compared to traditional compound chemical fertilizer (Fc), influenced the growth, yield, and quality of Chinese small cabbage (CSC) cultivated on newly reclaimed land. Researchers found that applying bio-organic fertilizer significantly enhanced plant height, stem diameter, leaf area index (LAI), biomass accumulation, and yield relative to compound chemical fertilizer. When biochar was added, its effect varied depending on the fertilizer type: compound fertilizer showed increased benefits with higher biochar doses, while bio-organic fertilizer achieved optimal results with the lower biochar rate (6.85 t/ha). The highest yields observed were 29.41 t/ha under high-rate biochar with compound fertilizer (B2Fc) and 37.93 t/ha with low-rate biochar and bio-organic fertilizer (B1Ft), with the latter being significantly greater. Water productivity also improved under both fertilizer types when paired with low biochar dosage. Interestingly, photosynthetic traits responded differently: they increased with compound fertilizer plus biochar, but decreased when bio-organic fertilizer was combined with biochar. Meanwhile, CSC quality—particularly soluble sugar and total phenolic levels—was higher under sole bio-organic fertilizer compared to compound fertilizer. However, increasing biochar rates led to higher nitrite content in the cabbage across both fertilizer regimes. The authors concluded that in newly reclaimed poor soils, replacing chemical fertilizers with bio-organic fertilizers, or partially substituting them with compound fertilizers plus low-dose biochar, is a highly effective, sustainable fertilization strategy. It not only boosts crop yield and quality but also supports fertilizer reduction goals.

5.3 Solid Biofertilizer

Fertilizers are essential for providing the nutrients plants need to grow and for enhancing agricultural productivity (Purnomo *et al.*, 2017). They are generally classified as chemical or biofertilizers, and can come in either liquid or solid form. Biofertilizers are organic formulations containing beneficial live microorganisms. When applied to seeds, plants, or soil, these microbes colonize the rhizosphere or plant tissues, improving nutrient availability and uptake. By promoting microbial processes that convert nutrients into forms more easily absorbed by plants, biofertilizers aid in nitrogen fixation, phosphate solubilization, and the production of plant growth-promoting substances (Patel *et al.*, 2022). This not only enhances soil fertility but also increases crop yields.

In recent years, biofertilizers have gained recognition as a natural and eco-friendly tool for sustainable nutrient management. Their benefits extend beyond agriculture, contributing to environmental remediation and ecological restoration (Fig. 5.1). As such, they play a key role in the shift toward more sustainable farming practices. A variety of commercial biofertilizer products are available, each claiming unique advantages (Kaur and Kaur, 2023). Among these, the most widely accepted formulations typically include either dormant or active bacterial cultures integrated with a carrier substance. A carrier substance here is a particular solid that the living organisms hold on to (Sipponen *et al.*, 2019). The choice of carrier material plays a vital role in determining the effectiveness and shelf life of biofertilizer by maintaining a sufficient count of viable microbes over time. When formulating biofertilizers for specific bacterial strains, carrier selection is made carefully, considering several key factors: the carrier should provide a basic carbon source to support microbial viability throughout storage and it must facilitate effective microbial delivery to plant roots while supporting their survival amidst native soil microorganisms. Carriers with microporous structures and fine particle size are particularly effective, as they harbor a denser microbial population. Carriers are usually classified into three main types (Kaur and Kaur, 2023): soil-derived, plant-derived, and inert substances.

Crops such as paddy, straw, rice bran/wheat bran, and bagasse are frequently used for the

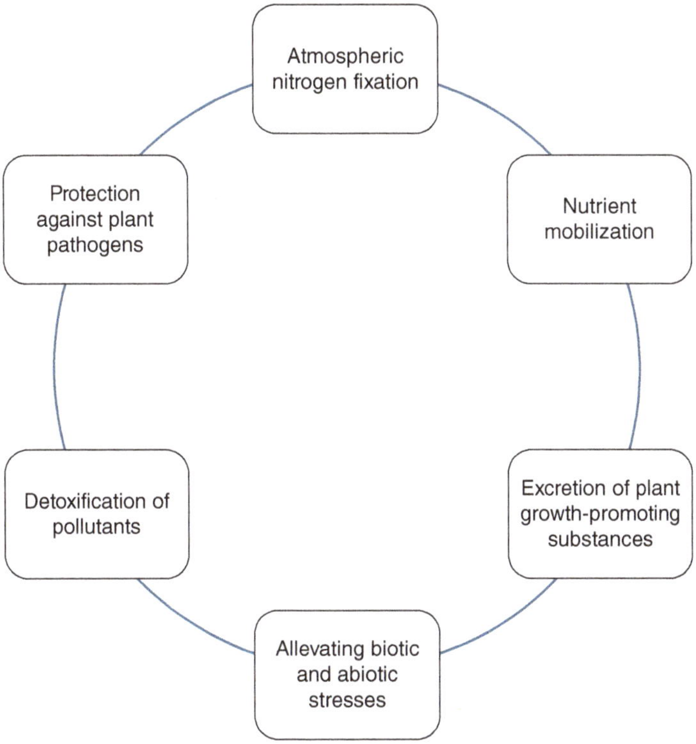

Fig. 5.1. The figure shows modes of action of a biofertilizer. Figure author's own.

formulation of *Rhizobium* and *Burkholderia* bacillus-based biofertilizers (Sharma *et al.*, 2023).

Several techniques are commonly employed to introduce beneficial bacteria into the soil ecosystem. These include seed coating, root dipping, bacterial suspensions, and the use of solid carriers. Despite their application, the number of viable bacterial cells often declines rapidly after inoculation due to unfavorable environmental and nutritional conditions. Fresh bacterial cultures generally have a limited shelf life, making it harder for them to get transported or stored and making field application impractical (Tripti *et al.*, 2022).

A wide variety of solid substrates have been explored as carriers in biofertilizers formulations (Tripti *et al.*, 2022). These include natural materials like soil, peat, and plant residues (e.g. ground maize cobs), as well as organic manures and composts. The use of industrial or agricultural waste, for instance, biochar, wood, or manure, as a carrier for bacterial inocula could be very beneficial to the food industry. While many other formulations exist for biofertilizers, agro-waste-based genesis has been the most promising (Table 5.1). It not only solves agricultural problems but also environmental and social issues. Agro-waste-based biofertilizers are now a cornerstone of sustainable agriculture as they decrease negative carbon footprints, greenhouse gases, and the cost of landfill.

5.3.1 Types of solid biofertilizers

5.3.1.1 Food processing waste

Maintaining the viability and optimal population of beneficial microbes in biofertilizer formulations requires the appropriate choice of carrier material (Das and Kim, 2024). An ideal carrier must meet several essential criteria: it should have good moisture retention, be non-toxic to both bacterial strain and host plant (Tabassam *et al.*, 2015), offer adequate pH buffering, and demonstrate good adhesion to seeds. It also

Table 5.1. Agro-waste biofertilizers: crop responses and yield effects. Table author's own.

Serial No.	Agro-waste source	Crop studied	Yield outcome	Reference
1.	Composted wheat and rice residues, enriched with microbes	Tomato	• Increased yield by ~75% • Improved plant height, chlorophyll, and nutrient content	Imran *et al.*, 2022
2.	Orange, banana, grape, wheat, and rice chaff; moringa leaves, brown sugar	Maize (*Zea mays* L.)	• Significant improvement in growth and yield compared to both NPK fertilizer and control • Treated plants also showed no insect damage, unlike other treatments	Onyia *et al.*, 2020
3.	Seven types of agricultural wastes equally, inoculated with *Pseudomonas fluorescens* and *Bacillus subtilis*	Aubergine (*Solanum melongena* L. cv. Pusa Kranti)	• The combined waste substrate (CSPfBs) produced the highest aubergine yield	Chakravarty, 2024
4.	Mixed food waste + 10% biochar, treated with food waste-derived enzymes (fungal mash)	Pak choi	• Yield comparable to commercial biofertilizer • Soil chemistry unchanged • Microbial diversity increased	Jiang *et al.*, 2024
5.	Bio-organic fertilizer combined with biochar	Chinese small cabbage (*Brassica rapa*)	• Improved growth and quality • Increased yield on newly reclaimed land	Wang *et al.*, 2024

needs to be easy to handle, sterilize (through autoclaving or gamma irradiation) and readily available at low cost, while also being easy to transport.

Agro-waste—such as crushed maize kernels, sawdust, crop residues (like husks, stalks, and leaves), animal dung, compost, biogas, slurry, biochar, and other organic by-products—has proven useful as sustainable carrier material in biofertilizer production (Das and Kim, 2024). However, to be suitable for use, these materials must first be processed and sterilized. This includes drying and grinding them to a particle size of 0.1–1.5 mm, as larger particles can cause lumps or poor seed adhesion when wet.

Agricultural waste plays a crucial role in advancing the principles of sustainable agriculture, as it helps to minimize reliance on chemical fertilizers and enhances the organic matter content in soil (Arfarita *et al.*, 2022). The addition of organic components promotes increased biological activity and supports improved biochemical nutrient cycling. An effective biofertilizer should maintain high microbial viability over extended storage periods without compromising its nutrition value.

Solid biofertilizers are mainly classified on the basis of the carrier materials used in their production. Carrier materials are substances that act as a medium to carry the microorganisms in sufficient quantities and keep them viable under various circumstances. Using an ideal carrier material is essential for the production of high quality biofertilizers. Furthermore, carrier materials can be organic or inorganic. Organic carriers include peat, compost, and, most importantly, agricultural waste. Inorganic carriers include clay, vermiculite, and rock phosphate (Kaur and Kaur, 2023). An ideal carrier should have high porosity, at least 50% absorptivity (water holding capacity), neutral pH to avoid harming the microorganisms, good adhesion to seeds for effective nourishment, and high organic matter content, as well as being low-cost and easily available.

Biofertilizers are added to the soil via two methods: seed inoculation and soil inoculation. In seed inoculation, a mixture of bacteria and the carrier, called inoculant, is prepared in the form of a slurry. This slurry is then mixed with seeds. For these purposes, the carrier must be a fine powder. To achieve good binding of seed with the inoculant, adhesives such as vegetable oils, sucrose, or methylcellulose are used. This method allows for large populations of bacteria to be introduced in a short span of time. Granular inoculant is preferred, which is placed into the furrows dug alongside the seeds. This allows the inoculated strain to be in contact with the plant roots. Soil inoculation involves introducing a larger quantity of microorganisms directly into the soil (Rocha *et al.*, 2019). The inoculant can be applied in the furrow with the seeds, broadcast over the soil, or mixed with it.

Several organic wastes serve as potential carriers for biofertilizers, including molasses waste, vermicompost, and seaweed solid residues.

5.3.1.2 Crop residues

Currently, biofertilizers are being manufactured on a large scale and are commonly applied in agriculture using various carrier materials, such as talc, lignite, peat, perlite, and vermiculite. Despite their widespread use, these carriers are associated with certain drawbacks. For instance, they often have a limited shelf life, are vulnerable to environmental fluctuations, and carry a high potential for contamination, especially during large-scale sterilization processes. Moreover, the global availability of such materials remains limited. Their extraction and processing may also lead to adverse impacts on human health, ecosystems, and the natural environment.

Utilization of agricultural waste, particularly waste with high carbon content, as potential carriers could solve the limitations of sustainability. Despite its current usage as a cooking fuel, which leads to air pollution, maize cob rind holds considerable potential for eco-friendly innovations, such as in the production of biofertilizers. Paygond *et al.* (2024) utilized cob rind-based fertilizer combined with NPK following a completely randomized design involving ten distinct treatments on capsicum plants. It recorded the highest populations of beneficial microorganisms, such as free-living nitrogen fixers (49.0×10^5 cfu/g), phosphate-solubilizing bacteria (31.0×10^5 cfu/g), and potassium-solubilizing bacteria (30.66×10^5 cfu/g). Additionally, the treatment (Paygond *et al.*, 2024) significantly improved the soil nutrient profile, particularly in terms of nitrogen, phosphorus, and potassium. Enhanced plant growth parameters were observed.

Another study was done where press mud from the sugar industry was used with plant growth-promoting rhizobacteria (PGPR) as a carrier on okra plants (Mushtaq *et al.*, 2021). This resulted in increased NPK and also enhanced the fruit structure, i.e. the diameter of fruit (1.55 cm), and the yield (1.73 t/ha).

Another option could be guar gum: banana peels dried at 50°C for 8 h before being ground into powder or cassava starch gelatinized in dispersion at 95°C for 15 min with 2% of alginate added to each starch dispersion and stirred to produce a bead matrix solution (Ariani and Simarmata, 2023).

In a similar study, it was discovered that maltodextrin inoculated with *B. subtilis* can be used as a carrier biofertilizer (Ariani and Simarmata, 2023). The mixture is subjected to spray drying with airflow rate and atomization pressure of 600 l/h and 0.55 bar. Other potential carrier materials are sago dregs, rice straw, and pea fiber. Different crop residues are beneficial for different microbial strains and plants (Yuliani and Pratiwi, 2024). The waste material used also determines the pathway followed for their production and the capital invested.

5.3.1.3 Manure

Manure is formed after the decomposition of organic waste such as dead and decaying plant and animal parts. The environmental consequences of manure production and its disposal now go beyond the traditional role of manure as a simple organic fertilizer (Fig. 5.2). Intensive confined animal feeding operations produce far more manure than can be economically transported to crop lands situated away from the farms. The increasing accumulation of manure

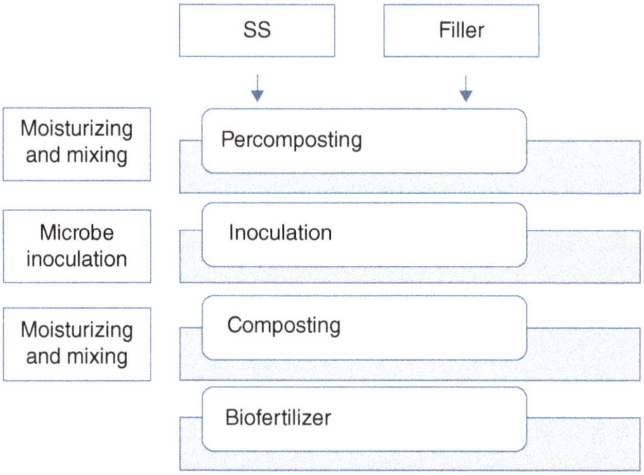

Fig. 5.2. A flow chart showing the process of manure formation. Figure author's own.

waste poses a serious threat to both the environment and agricultural ecosystems. To address this growing issue, scientific and sustainable approaches are essential for the safe reuse of manure. However, a significant barrier to its reutilization is the risk of pathogen contamination. Techniques like vermicomposting and composting have proven effective for recycling crop nutrients, especially nitrogen.

Recent advancements also focus on managing contamination from various organic wastes, including human waste, industrial effluents, animal manure, and agro-industrial by-products.

In the case of poultry farming, the substantial amounts of manure generated are often applied directly to agricultural soils without undergoing any form of pretreatment. This unprocessed application can contribute to an unpleasant smell and also raises concerns regarding soil contamination by harmful pathogens. Therefore, manure alone is not suitable for direct application; it needs to undergo processing for it to be better for the soil and plants.

By inoculating manure with bacterial or microbial strains, we can form a carrier biofertilizer that is good for soil health and provides a suitable environment for plants to grow in.

Mortola *et al.* (2019) conducted an experiment using poultry digestate as a biofertilizer for *Lactuca sativa*. Despite the absence of immediate increases in soil inorganic nitrogen and available pores following the application of digestate, a notable improvement in crop fresh weight was observed with higher dosage treatments. This was due to the gradual release of nutrients in the soil. Additionally, shifts in the soil microbial community were observed, but these changes were temporary and occurred mainly the early stages of the experiment.

Manure can also be generated by sheltered and domesticated animals such as cattle (Khan *et al.*, 2023). Animal manure is composed of a mixture of feces and urine, along with additional materials such as bedding, spilled feed, wasted drinking water, and wash water from animal housing areas. Due to this, the composition of manure can vary widely. Furthermore, the method of manure removal and handling significantly influences its characteristics. Manure is often classified based on the types of livestock it originates from; each species contributes differently to its physical and chemical makeup.

Important attributes are particle size and the technology used for making the manure (Khan *et al.*, 2023). There are various methods that can be used, such as anaerobic digestion, composting, and vermicomposting (Zhou *et al.*, 2022). In these methods, breakdown of lignin, cellulose, and hemicellulose occurs.

5.3.1.4 Sludge

Urban wastewater sludge contains a significant amount of organic matter, often reaching up to 75% (Zhang *et al.*, 2016). It serves as a potential source of essential nutrients, particularly nitrogen and phosphorus, for soil enrichment. However, a large proportion of the potassium in sludge exists in the liquid phase, making it susceptible to leaching losses. The concentration of key nutrients in sludge varies widely: nitrogen ranges from 0.8 to 6.0%, phosphorus from 0.6 to 5.6%, and potassium from 0.1 to 0.5% (Krutiakova *et al.*, 2021).

Studies have shown that compost produced by combining sewage sludge with agricultural waste meets established quality standards for agricultural use. This approach serves a dual purpose; it not only addresses the issue of waste disposal—specifically waste that poses significant risks to the environment—but also supports the expansion of the production of organic fertilizer, which is in growing demand. To increase the carbon concentration in compost derived from sewage, organic fillers such as sawdust, shredded bark, straw, leaves, peat, and other biodegradable substances are commonly added during the composting process.

A key factor in achieving high agricultural yields is maintaining adequate phosphorus availability. Soil microorganisms play a vital role in converting insoluble phosphates into forms accessible to plants, thereby enhancing plant growth, nutrient uptake, and productivity. These phosphate solubilizing microbes can be integrated as functional components in bio-organic fertilizers. Krutiakova *et al.* (2021) investigated a novel production technology for bio-organic fertilizers, focusing on the use of sewage sludge sourced from a biological treatment facility in southern Ukraine. The process involved incorporating different organic fillers—winter wheat straw and sunflower seed husks—in combination with a functional microbial strain, *Microbacterium barkeri* to enhance compost quality.

With the development of this biofertilizer technology, the final products demonstrated significantly enhanced nutrient content, including nitrogen levels ranging from 21.7 to 30.7 g/kg (compared to the standard 18–20 g/kg), phosphorus levels between 14.4 and 33.2 g/kg (compared to the usual 10–12 g/kg), and potassium concentrations ranging from 5.7 to 11.5 g/kg (above the norm of 2–5 g/kg). These values confirm the superior agronomic characteristics of the newly formulated biofertilizers. Further validation from a 2017–2019 study on maize cultivation confirmed that the application of these biofertilizers resulted in a notable increase in yield. Specifically, maize yield rose by 31.6% and 33% when using fertilizers derived from biologically treated sewage sludge.

5.3.1.5 *Biochar*

Biochar-based soil enhancers are generated through the pyrolysis of carbon-rich biological wastes, including woody materials, agricultural residues, animal remains, and biosolids. Pyrolysis thermally decompresses organic matter in the absence of oxygen, yielding three main products: syngas, bio-oil, and biochar. Among these, biochar functions like a long-term carbon sink (Bolan *et al.*, 2023). Over time, biochar use has expanded beyond carbon sequestration (Spokas *et al.*, 2012). It now finds diverse applications in agriculture and environmental management. When used as a soil amendment, biochar enhances soil structure, nutrient cycling, and overall soil health, which collectively contribute to improved crop productivity.

Numerous studies (Bolan *et al.*, 2023) have demonstrated that biochar serves as an effective carrier for microbial inoculants, thanks to its favorable structural and chemical attributes that support microbial survival and functionality, even under harsh environment conditions (Saravanan *et al.*, 2023). The unique features of biochar, such as its large surface area and porosity, facilitate the immobilization of beneficial microbes, thereby playing a significant role in the remediation of organic contaminants. Furthermore, its porous matrix offers a protective and nutrient-rich habitat, making it an ideal medium for microbial colonization and persistence.

Biochar's high organic carbon content, along with essential nutrients like nitrogen, phosphorus, and potassium, combined with its ability to retain moisture and support aeration, makes it particularly conducive to microbial habitation. When biochar is co-applied with microbial inoculants, it enhances their survivability and establishment in the soil and rhizosphere.

In a recent study, Kumar *et al.* (2017) describe the use of biochar with fly ash as a carrier material for developing biofertilizers, utilizing two plant growth-promoting rhizobacteria: *Bacillus* sp. strain A30 and *Burkholderia* sp. strain L2. The viability of the bacterial strains in various formulations was monitored over a period of 240 days.

Seeds of tomato plants were coated with these carrier-based bioformulations and the impact on their growth was assessed through pot experiments. Among all the treatments, the biochar-based formulation inoculated with strain L2 demonstrated the highest microbial viability, maintaining a CFU count of 10^7 g^{-1} at the end of the 240-day storage period. The treatment also achieved the highest seed germination rates.

The pot trials revealed a statistically significant ($p<0.05$) improvement in plant growth parameters, such as plant height, fresh and dry biomass, flower number, and overall yield, with the biochar + strain L2 formulation consistently outperforming other combinations (Kumar *et al.*, 2017). The pot trials turned out to be successful in aiding plant growth. Moreover, after the crop had been harvested, the physiological and chemical properties of the soil as well as its nutrient content and the activity of enzymes in the soil remained enhanced. In a sense, there is no loss of nutrients from the soil.

In another study (Araujo *et al.*, 2020), inoculants formulated with strains of *Bradyrhizobia*, isolated from the root nodules of pigeon pea (*Cajanus cajan*), using pine bark biochar as carrier material, have shown great potential as an effective strategy to enhance the productivity of pigeon pea in tropical agro-ecosystems. These carrier materials provide a favorable microenvironment for the survival, proliferation, and sustained activity of the rhizobia, thereby improving modulation and biological fixation

5.4 Importance of Using Agrifood Waste

Agrifood waste-based fertilizer has advanced greatly and can gradually be transformed into a circular economy concept as it favors reusing and recycling. It initiates a cycle of close-looped agriculture and diminishes the involvement of external resources, particularly the dependence on chemical fertilizers.

Chemical fertilizers are not cost-effective and pose environmental threats. Exploitation of chemical fertilizers causes eutrophication in water bodies due to groundwater and air pollution (Youssef and Eissa, 2014). Traditionally processed chemical fertilizers destabilize the soil biome and have adverse effects on human health (Ayala and Rao, 2002). In the long run, the practice of chemical-based farming is a risk to the agricultural and farming community. In light of this, biofertilizers have emerged as a crucial link in the nutrient supply chain since they have the potential to revamp crop yields by supplying better nutrients.

Biofertilizers can be defined as fertilizers that contain live soil microorganisms, which aid in the uptake of nutrients for crops (Vessey, 2003). Biofertilizers also include organic substances like manure and organic waste (kitchen waste, agrifood waste, etc.) (Table 5.2). They need to undergo the action of microorganisms or have some sort of a symbiosis with the plant before they are functionally active and help in nutrient assimilation. They are developed with the aim of providing micro- and macronutrients to the soil by nitrogen fixation, production of antibiotics, or degrading the organic matter in soil (Sinha *et al.*, 2010). Ultimately, they facilitate better uptake of nutrients and stronger resistance to climatic stressors. The major difference that can be observed between chemical fertilizers and biofertilizers is that the latter do not supply the nutrients directly but are the cultures of specific bacteria and fungi which aid in nutrient supply. Thus, they are cost-effective and have lower installation costs (Alam and Seth, 2014; Suhag, 2016).

The replacement of chemically synthesized fertilizers by biofertilizers would directly benefit food production and decrease the accumulation of food waste, thereby decreasing its environmental impact. The production of biofertilizers has been studied worldwide.

5.5 Efficacy of Solid Biofertilizer

Biofertilizer efficacy is the ability of microbial inoculants to improve plant nutrition, development, and yield by increasing nutrient availability, absorption, and accessibility. According to Malusà *et al.* (2016), solid biofertilizers outperform liquid biofertilizers in terms of storage stability, microbiological shelf life, and compatibility with farm equipment. They can cut the use of chemical fertilizers by 20–50% while maintaining crop output. Beneficial microbes, including *Rhizobium*,

Table 5.2. Production of Biofertilizers. Table author's own.

Serial No.	Material	Crop	Process	Microorganisms used	Applications	Reference
1.	Organic municipal solid waste (food, garden and kitchen)	–	Anaerobic digestion in reactor	Consortium of anaerobic bacteria and archaea	Increased plant lifespan	Aghbashlo *et al.*, 2019
2.	*Sargassum*, *K. alvarezii*, molasses	–	Treatment by addition of microorganisms	Marine fungus (strain RS6A)	Better growth due to increased growth hormone levels	Basmal and Kusumawati, 2023
3.	Microbial inoculants	Maize, *Brassica napus, Triticum aestivum*	Treatment by addition of microorganisms	*Paenibacillus polymyxa*, *Azospirillum*, and *Enterobacter* consortium, Mycorrhizal fungi	Increased nitrogen level in the soil promotes better growth	Meena *et al.*, 2017
4.	Agricultural waste	Winter wheat	Aerobic digestion	–	Improved water holding capacity and air permeability	Wei *et al.*, 2015; Wang *et al.*, 2017; Du *et al.*, 2018
5.	Fruit waste (watermelon, papaya, orange, banana, and pineapple)	*Brassica juncea* var. *Rugosa*	Solid state fermentation	–	High acidic values hinder nutrient uptake of the crop whilst low pH shows strong growth	Lim and Matu, 2015
6.	Food waste	–	Chemical hydrolysis of fermented bio-waste	*Bacillus proteolyticus*, *Bacillus sanguinis*, *Bacillus spizizenii*, *Bacillus paramycoides*, *Bacillus paranthracis*, and *Neobacillus fumarioli*	Increased amounts of humic acid (HA) are beneficial for plant development and soil health	Rosso *et al.*, 2015; Chukwuma *et al.*, 2023
7.	Agro-waste (rice straw)	Rice	In situ straw degradation, addition of soil microorganisms	*Bacillus* sp., *Azospirillum* sp.	Degrades the straw waste added to the soil as a biofertilizer	Borah *et al.*, 2016; Du *et al.*, 2018
8.	Crop residue	Sugarcane	Incineration	–	Decreased levels of carbon and nitrogen, increased levels of oxygen in the soil	Du *et al.*, 2018; Rossi *et al.*, 2016

Continued

Table 5.2. Continued.

Serial No.	Material	Crop	Process	Microorganisms used	Applications	Reference
9.	*Carica papaya* (pawpaw) peels	Maize	Anaerobic digestion	Microbes from bovine rumen	Increased plant height, leaf area, biomass, better than chemical fertilizer	Dahunsi *et al*., 2021
10.	Kitchen waste oil	Cabbage	Microbial digestion in anaerobic conditions	*Pseudomonas aeruginosa* PA-3	Improved plant height, leaf area, stem diameter	Li *et al*., 2018
11.	Vegetable and fruit waste	Cucumber	Composting	*Mesophilic bacteria*	>80% germination index, enhanced shoot/root growth	Sangamithirai *et al*., 2015
12.	Fruit peels (banana, papaya, watermelon), crop residue	Vegetables (e.g. leafy greens)	Solid-state fermentation (SSF)	Native thermophilic and decomposer microbial communities	Nutrient-rich organic fertilizer, enhanced growth	Lim and Matu, 2015; Bala *et al*., 2023

phosphate-solubilizing bacteria (PSB), potassium-solubilizing bacteria (KSB), and arbuscular mycorrhizal fungus (AMF), help to improve nutrient utilization efficiency. Multi-strain consortia improve plant stress tolerance and nutrient absorption. However, the performance of biofertilizers is influenced by a variety of factors such as soil type, native microbial competition, application methods, and agricultural practices, as well as by factors such as poor formulation, contamination, and insufficient storage limit effectiveness. As a result, standard quality control, appropriate carriers, and well-designed application tactics are critical for achieving dependable and consistent field results.

Solid biofertilizers derived from agricultural waste such as banana, papaya, and watermelon dramatically increased mustard plant growth. Lim and Matu (2015) found that watermelon-based fertilizer improved root length (70.8 mm), weight (0.208 g), and leaf count the most, while banana and papaya also had a significant impact. Similarly, in a radish study, mixing organic manure and inorganic fertilizers with a solid microbial consortia (IIHR—Indian Institute of Horticultural Research) resulted in the highest biomass and nutrient absorption. Pathak *et al.* (2018) found that the most effective treatment boosted yield to 60,947 kg/ha and improved NPK levels in roots and shoots. These findings demonstrate that solid biofertilizers not only increase crop production and quality, but also provide an environmentally acceptable alternative to chemical fertilizers when employed alone or in integrated nutrient management systems.

5.6 Combined Effects of Chemical Fertilizers and Biofertilizers

Mazumdar *et al.* (2022) studied the comparative effectiveness of a novel bio-organic fertilizer mix (containing beneficial microbes and organic matter) and traditional chemical fertilizers. Multiple test crops, such as *Brassica juncea* (brown mustard), *Basella alba* (climbing spinach), and *Amaranthus dubius* (red spinach), were studied. They concluded that the bio-organic combination improved soil microbial activity, plant nutrient uptake, and growth performance. Yields under the bio-organic fertilizer were comparable or superior to those under chemical fertilizer. Most importantly, the bio-organic system proved more sustainable and environmentally friendly (Mazumdar *et al.*, 2022).

Umesha *et al.* (2014) conducted a field experiment to evaluate the comparative effects of organic inputs and biofertilizers on the growth and yield of maize (*Zea mays* L.) during the 2011 Kharif season in Tumkur district, Karnataka, India (Table 5.3). The study used the maize variety Nithyashree (NAH 2049) and employed a randomized complete block design with 14 treatments replicated three times. Treatments included various combinations of chemical fertilizers, enriched compost, and a biofertilizer consortium containing *Azotobacter chroococcum*, *Bacillus megaterium*, and *Pseudomonas fluorescens*. Among all treatments, the combination of recommended NPK fertilizer + enriched compost + biofertilizer consortium (T_{13}) resulted in the most favorable outcomes for plant growth, yield, and soil health. Plants in the T_{13} treatment recorded the highest plant height at all stages of growth (up to 188.13 cm at 120 days after sowing), the greatest total dry matter (375.80 g/plant), and the highest cob weight (207.63 g) and grain yield (158.93 g/plant). This translated to a significantly higher grain yield per hectare (5453 kg/ha), along with improved seed weight and quality. In addition to crop performance, this integrated treatment also led to the most favorable post-harvest soil nutrient levels, particularly in nitrogen (185.40 kg/ha), phosphorus (38.83 kg/ha), and potassium (181.47 kg/ha). The results suggest that the integrated use of biofertilizers and organic compost, in combination with a reduced chemical fertilizer dose, can enhance both crop productivity and soil fertility. The study highlights the potential for a 25% reduction in nitrogen and phosphorus chemical fertilizer usage without compromising yield, promoting a more sustainable and cost-effective maize cultivation strategy. Overall, the research supports integrated nutrient management as a viable alternative to purely chemical-based farming systems.

Patel *et al.* (2024) evaluated the effects of biofertilizers and chemical fertilizers on the physical and biological properties of soil, as well as on chickpea (*Cicer arietinum* L.) yield, during the 2020–2021 Rabi season at the Instructional Farm of Acharya Narendra Dev University,

Table 5.3. Comparative effects of chemical fertilizers, biofertilizers, and their combination on growth and yield of various crops. Table author's own.

Serial No.	Crop	Effects of chemical fertilizer	Effects of biofertilizers	Effects when combined (chemical fertilizers + biofertilizers)	References
1.	*Zea mays* L. (maize)	Provided baseline growth parameters	Increased yield with improved soil health indicators	Improved biomass, cob weight, plant height, seed quality, and post-harvest soil fertility	Umesha *et al*., 2014
2.	*Cicer arietinum* L. (chickpea)	Basic yield benefits but no long-term soil improvement	Improved soil carbon, root development, and water retention	Optimized soil health and overall productivity	Patel *et al*., 2024
3.	Rainfed wheat (Azar-2 cultivar)	Increased grain yield	Improved grain yield and protein content	Further enhancement in grain protein concentration	Sedri *et al*., 2022
4.	*Vigna unguiculata* L. (cowpea)	Moderate improvements in yield	Improved plant height and grain yield	Highest yield, with increased grain mass and harvest index	Gautam *et al*., 2024
5.	*Oryza sativa* (paddy crop)	Immediate yield response, but deteriorated soil health	Conferred disease resistance and enhanced growth and productivity	Maximized yield combining synthetic inputs with ecological resilience	Basotra and Mohan, 2023

Ayodhya, India. The study used a randomized block design with 11 treatments, including control, 100% and 50% recommended dose of chemical fertilizer (RDF), farmyard manure (FYM) and Jeevamrit (a nutrient-rich microbial solution), agro-residue mulch + FYM and Jeevamrit, *Rhizobium* + phosphate-solubilizing bacteria (PSB), and various combinations of these with chemical fertilizers and organic amendments. The chickpea variety KPG-59 was cultivated under these treatments to assess their impact. Results indicated that treatments combining FYM, Jeevamrit, and *Rhizobium* + PSB (specifically T_8) produced the most pronounced improvements in soil biological properties and chickpea yield. These treatments enhanced soil organic carbon (OC) and organic matter (OM) content, while electrical conductivity (EC) also increased, though changes in physical soil properties were not significant. Chickpea yield was significantly higher under these integrated organic and microbial treatments compared to conventional chemical fertilizer application. The inclusion of *Rhizobium* and PSB stimulated beneficial soil microbial activity, indicating improved biological health. Interestingly, a combination of 50% RDF with organic amendments such as FYM + Jeevamrit (T_9) also yielded significant benefits, suggesting that partial substitution of chemical fertilizers with organic and biofertilizer inputs can be effective. Overall, the study underscores the potential of integrating solid biofertilizers and organic soil amendments with reduced chemical fertilizer usage to enhance soil health and chickpea productivity. This integrated nutrient management approach offers promising benefits in soil fertility, crop yield, and environmental sustainability under chickpea cultivation.

Sedri *et al.* (2022) conducted a comprehensive 3-year field study (2016–2019) at a rainfed research station in Iran to compare the effects of five PGPR products—Flawheat (F), Barvar2 (B), Nitroxin (N1), Nitrokara (N2), and SWRI (Soil and Water Research Institute)—against conventional chemical fertilizer regimes on rainfed wheat (*Triticum aestivum*) productivity and quality. Utilizing a randomized complete block design with four replicates each season, treatments combined PGPR inoculation with either 50% or 100% of the recommended nitrogen (N), phosphorus (P), and zinc (Zn) chemical fertilizers. The study's core goal was to assess whether PGPR inoculation could reduce chemical fertilizer dependency while maintaining or enhancing yields and grain quality. Results demonstrated substantial benefits from PGPR + 50% chemical fertilizer compared to uninoculated control plots. Grain yield increased by 28% (F), 28% (B), 37% (N1), and 33% (N2), and grain protein content improved by 0.54%, 0.88%, and 0.34% respectively with F, N1, and N2 treatments. The most remarkable result came from the combination of Nitroxin (N1) with 100% chemical fertilizer, which boosted grain yield by 56% and protein content by 1% relative to control. All PGPR-treated plots, especially those with chemical fertilizer supplementation, exhibited significantly higher plant height, grain weight, and quality markers. The authors concluded that integrated nutrient management—combining PGPR with reduced chemical fertilizer levels—offers a viable, eco-friendly strategy for enhancing rainfed wheat production and quality in semi-arid regions. This approach not only cuts down chemical fertilizer use but also supports environmental sustainability by mitigating potential pollution and soil degradation.

Gautam *et al.* (2024) evaluated the effects of biofertilizers and chemical fertilizers on two cowpea (*Vigna unguiculata*) varieties—Malepatan1 and Stickless—in a field trial conducted from February to June 2022 in Nawalparasi West, Nepal. A double-factor randomized complete block design was used, comparing six fertilizer treatments: control (no fertilizer), mycorrhizal inoculation, rhizobial inoculation, recommended dose of chemical fertilizer (RDF), a combination of mycorrhiza + rhizobia, and simultaneous application of mycorrhiza + rhizobia + RDF. The results demonstrated that Malepatan1 outperformed Stickless in all measured parameters, achieving the greatest plant height (125.7 cm), seed weight per plant (72.3 g), 1000-grain weight (151.6 g), and yield (3536.8 kg ha^{-1}). Regarding fertilizer treatment, the combined mycorrhiza + rhizobia treatment produced yields and grain quality nearly matching those of the full chemical fertilizer treatment. Specifically, combined biofertilizers alone achieved 4321 kg ha^{-1} yield and 167.2 g 1000-grain weight, compared to 4714 kg ha^{-1} and 176.8 g when RDF was included. Crucially, biofertilizer treatments yielded a benefit–cost ratio of 3.76, surpassing the mixed bio+chemical treatment, making it economically and environmentally superior. The authors concluded that the mycorrhiza + rhizobia inoculation, especially when paired with the high-performing Malepatan1 variety, offers a sustainable strategy for cowpea cultivation. This approach delivers high yields, quality grain, and excellent economic returns while reducing reliance on chemical inputs. Overall, the study supports integrating microbial biofertilizers as an eco-friendly alternative or complement to chemical fertilizers in cowpea production systems.

Basotra and Mohan (2023) investigated the comparative impacts of biofertilizers vs. chemical fertilizers on the growth, development, and yield of paddy (*Oryza sativa*), with a broader focus on nutrient content in grains and soil health. The study, published in the *Journal of Chemical Health Risks*, emphasizes that while chemical fertilizers boost crop productivity, their long-term use can degrade soil fertility, alter pH, suppress microbial populations, and cause environmental pollution through runoff and leaching. Biofertilizers—comprising beneficial microbes such as nitrogen-fixing bacteria, phosphate-solubilizing organisms, and phytohormone-producers—are presented as a sustainable alternative. These organisms enhance nutrient availability, promote plant growth, and improve resistance to abiotic and biotic stress without the harmful side effects linked to chemical inputs. The study explains mechanisms through which biofertilizers operate, including nitrogen fixation, potassium solubilization, siderophore production, and phytohormone secretion. Although full experimental details (specific treatments, crop stages, or quantitative yield data) were not disclosed in the public abstract, the key message is clear: biofertilizers can support comparable plant growth and yields while preserving or enhancing soil biological activity and nutrient balance. The study suggests that combining biofertilizers with reduced doses of chemical fertilizers—or replacing them entirely—could help mitigate the environmental costs of conventional agriculture. In conclusion, Basotra and Mohan advocate using biofertilizers as an eco-friendly, effective alternative to chemical fertilizers in rice cultivation. Their analysis contributes to the growing body of evidence supporting integrated nutrient management systems that optimize crop performance, promote environmental stewardship, and foster soil resilience in paddy production systems.

5.7 Factors Affecting Efficacy of Biofertilizers

Biofertilizer efficacy is determined by a number of interrelated factors, including microbial strain selection, soil conditions, carrier materials, delivery methods, and crop compatibility. According to Malusà *et al.* (2016), one of the most important factors is microbial viability and formulation type; solid biofertilizers provide longer shelf life, field stability, and compatibility with agricultural equipment, boosting efficacy in real-world situations. According to Gulshan *et al.* (2022), matching biofertilizers to crop nutrient demands and applying them at suitable growth phases improves nutrient use efficiency. Microbial features like nitrogen fixation, phosphate solubilization, and phytohormone synthesis must be compatible with plant physiology and soil conditions.

In a meta-analysis, Pei *et al.* (2025) discovered that biofertilizer efficiency varies depending on microbial species, crop type, soil pH, and application timing. For example, the highest production improvements occurred when biofertilizers were coupled with organic matter and administered prior to planting. Adequate organic matter, neutral pH, and minimal pathogen loads promote microbial colonization and nutrient uptake. Chakraborty and Akhtar (2021) emphasize that mass-cultivation characteristics such as temperature, pH, and nutrient balance during fermentation, as well as the choice of suitable carriers, influence microbial stability and performance.

Schütz *et al.* (2018) shows that climate has a substantial impact on efficacy, with biofertilizers performing better in dry and tropical zones than in continental climates. Increasing soil phosphorus levels led to stronger reactions, particularly among phosphate-solubilizing organisms and AMF, according to meta-regression analysis. Thus, biofertilizer success is context-dependent and must be optimized holistically.

Several variables can reduce the effectiveness of biofertilizers. Their nutrient content is often lower than that of inorganic fertilizers, which may result in deficiency symptoms if not supplemented. Storage conditions are critical; exposure to high temperatures or direct sunshine might diminish their efficiency, necessitating room temperature or cold storage. Other constraints include the scarcity of suitable microbial strains, a lack of appropriate carrier materials for long shelf life, insufficient farmer awareness of their benefits, insufficient human resources for proper application, and environmental factors such as soil salinity, acidity, drought, or waterlogging. (Itelima *et al.*, 2018)

5.8 Future Perspectives and Conclusion

Solid biofertilizers hold immense promise as sustainable alternatives to chemical fertilizers, offering environmentally friendly solutions to enhance crop productivity, restore soil fertility, and manage agricultural and industrial waste. With growing concerns about soil degradation, climate change, and pollution from synthetic inputs, solid biofertilizers are gaining momentum as viable tools for regenerative agriculture. Future biofertilizer development is expected to capitalize on diverse organic wastes such as maize cob rind, molasses, seaweed residues, vermicomposting, poultry manure, sewage sludge, and fly ash. These materials serve as cost-effective carriers and also solve waste disposal problems. High-efficiency carrier materials like biochar, composted residues, and processed agro-waste with enhanced microbial viability, moisture retention, and nutrient content will be central to the next generation of biofertilizer. Biochar in particular have shown superior performance due to their porosity, water holding capacity, and support for microbial colonization.

Progress in microbial biotechnology and strain selection, including the use of multifunctional plant growth-promoting bacteria and phosphate-solubilizing or nitrogen-fixing microbes, will improve the effectiveness of solid biofertilizers. Genetic and metabolic engineering may also enable the development of stress-resilient strains for adverse environments. The combination of solid biofertilizers with precision agriculture, soil health monitoring, and smart delivery systems (seed coating or controlled release pellets) will enhance field application efficiency and reduce nutrient losses. With rising global focus on climate-smart agriculture, biofertilizers are likely to receive stronger regulatory and policy backing. Government incentives, certification systems, and quality standards will support their large-scale adoption, especially in developing countries.

Solid biofertilizers hold the key to unlocking sustainable agricultural practices in the future. It is an economically viable approach to tackling the problems circling waste management. This review has demonstrated the valorization of agrifood residues, which, when properly treated, can be transformed to nutrient-packed fertilizers that are capable of improving plant health and soil fertility and bringing about an overall increase in crop production. Additionally, dependency on chemically synthesized fertilizers can be reduced, thereby promoting environmental friendly agricultural practices. By turning waste into a valuable resource, this method supports circular economy principles and contributes to long-term agricultural sustainability. Future efforts should focus on optimizing production methods, ensuring nutrient consistency, and promoting widespread adoption among farming communities.

This chapter has demonstrated the potential for producing solid biofertilizers from agrifood waste, emphasizing the sustainable utilization and valorization of such waste. It has explored how food waste can be transformed into eco-friendly and cost-effective alternatives to chemical fertilizers. The chapter also reviewed the types of food wastes suitable for this purpose, the mechanisms by which biofertilizers function, and their specific impacts on crop performance. Furthermore, a comparative analysis of chemical fertilizers, biofertilizers, and their combined applications on crops has been presented, highlighting their respective benefits and synergies.

References

Aghbashlo, M., Tabatabaei, M., Soltanian, S., and Ghanavati, H. (2019) Biopower and biofertilizer production from organic municipal solid waste: an exergoenvironmental analysis. *Renewable Energy* 143, 64–76. DOI: 10.1016/j.renene.2019.04.109

Alam, S. and Seth, R.K. (2014) Comparative study on effect of chemical and bio-fertilizer on growth, development and yield production of paddy crop (*Oryza sativa*). *International Journal of Science and Research* 3(9), 411–414.

Araujo, J., Díaz-Alcántara, C.A., Urbano, B., and González-Andrés, F. (2020) Inoculation with native *Bradyrhizobium strains* formulated with biochar as carrier improves the performance of pigeonpea (*Cajanus cajan* L.). *European Journal of Agronomy* 113, 125985. DOI: 10.1016/j.eja.2019.125985

Arfarita, N., Imai, T., and Prayogo, C. (2022) Utilization of various organic wastes as liquid biofertilizer carrier agents towards viability of bacteria and green bean growth. *Journal of Tropical Life Science* 12(1), 1–10. DOI: 10.11594/jtls.12.01.01

Ariani, N.S. and Simarmata, T. (2023) A systematic review: current technology of solid carrier formulation to improve viability and effectiveness of nitrogen-fixing inoculant. *Agrikultura* 34(1), 48. DOI: 10.24198/agrikultura.v34i1.43138

Ayala, S. and Rao, E.P. (2002) Perspectives of soil fertility management with a focus on fertilizer use for crop productivity. *Current Science* 82(7), 797–807.

Bala, S., Garg, D., Sridhar, K., Inbaraj, B.S., Singh, R. *et al.* (2023) Transformation of agro-waste into value-added bioproducts and bioactive compounds: micro/nano formulations and application in the agri-food-pharma sector. *Bioengineering* 10(2), 152. DOI: 10.3390/bioengineering10020152

Basmal, J. and Kusumawati, R. (2023) The role of microorganisms in solid biofertilizer production. *IOP Conference Series: Earth and Environmental Science* 1137(1), 012037.

Basotra, R. and Mohan, A. (2023) Comparative study on effect of bio-fertilizers and chemical fertilizers on growth, development and yield of paddy crop (*Oryza sativa*). *Journal of Chemical Health Risks* 13(4).

Bolan, S., Hou, D., Wang, L., Hale, L., Egamberdieva, D. *et al.* (2023) The potential of biochar as a microbial carrier for agricultural and environmental applications. *The Science of the Total Environment* 886, 163968. DOI: 10.1016/j.scitotenv.2023.163968

Borah, N., Barua, R., Nath, D., Hazarika, K., Phukon, A. *et al.* (2016) Low energy rice stubble management through in situ decomposition. *Procedia Environmental Sciences* 35, 771–780. DOI: 10.1016/j.proenv.2016.07.092

Borrello, M., Caracciolo, F., Lombardi, A., Pascucci, S., and Cembalo, L. (2017) Consumers' perspective on circular economy strategy for reducing food waste. *Sustainability* 9(1), 141. DOI: 10.3390/su9010141

Chakraborty, T. and Akhtar, N. (2021) Biofertilizers: prospects and challenges for future. In: Inamuddin, Ahamed, M.I., Boddula, R., and Rezakazemi, M. (eds) *Biofertilizers: Study and Impact*. Wiley, Oxford, UK, pp. 575–590.

Chakravarty, G. (2024) Preparation of plant growth enhancing bioformulation from agricultural wastes by solid state fermentation. *Current Agriculture Research Journal* 12(1), 316–325. DOI: 10.12944/CARJ.12.1.25

Chukwuma, O.B., Rafatullah, M., Kapoor, R.T., Tajarudin, H.A., Ismail, N. *et al.* (2023) Isolation and characterization of lignocellulolytic bacteria from municipal solid waste landfill for identification of potential hydrolytic enzyme. *Fermentation* 9(3), 298. DOI: 10.3390/fermentation9030298

Dahunsi, S.O., Oranusi, S., Efeovbokhan, V.E., Adesulu-Dahunsi, A.T., and Ogunwole, J.O. (2021) Crop performance and soil fertility improvement using organic fertilizer produced from valorization of *Carica* papaya fruit peel. *Scientific Reports* 11(1), 4696. DOI: 10.1038/s41598-021-84206-9

Das, S. and Kim, P.J. (2024) Biofertilizers from agro-wastes: a path towards sustainable agriculture. DOI: 10.56669/NCBE8246

Du, C., Abdullah, J.J., Greetham, D., Fu, D., Yu, M. *et al.* (2018) Valorization of food waste into biofertiliser and its field application. *Journal of Cleaner Production* 187, 273–284. DOI: 10.1016/j.jclepro.2018.03.211

Gautam, N., Ghimire, S., Kafle, S., and Dawadi, B. (2024) Efficacy of bio-fertilizers and chemical fertilizers on growth and yield of cowpea varieties. *Technology in Agronomy* 4(1), 1–10. DOI: 10.48130/tia-0024-0004

Guillou, M. and Matheron, G. (2012) *The World's Challenge: Feeding 9 Billion People*. Editions Quae, Versailles, France.

Gulshan, T., Verma, A., Ayoub, L., Sharma, J., Sharma, T. *et al.* (2022) Increasing nutrient use efficiency in crops through biofertilizers. *The Pharma Innovation Journal* 11(6), 2003–2010.

Gustavsson, J., Cederberg, C., Sonesson, U., Otterdijk, R., and Meybeck, A. (2011) *Global Food Losses and Food Waste – Extent, Causes and Prevention*. FAO, Rural Infrastructure and Agro-Industries Division, Rome.
Imran, A., Sardar, F., Khaliq, Z., Nawaz, M.S., Shehzad, A. *et al.* (2022) Tailored bioactive compost from agri-waste improves the growth and yield of chili pepper and tomato. *Frontiers in Bioengineering and Biotechnology* 9, 787764. DOI: 10.3389/fbioe.2021.787764
Itelima, J.U., Bang, W.J., Onyimba, I.A., and Oj, E. (2018) A review: biofertilizer; a key player in enhancing soil fertility and crop productivity. *Journal of Microbiology and Biotechnology Reports* 2(1), 22–28. DOI: 10.26765/DRJAFS.2018.4815
Jiang, Y., Zhang, X., An, L., and Liu, Y. (2024) A novel biochar-augmented enzymatic process for conversion of food waste to biofertilizers: planting trial with leafy vegetable. *Bioresource Technology* 399, 130554. DOI: 10.1016/j.biortech.2024.130554
Kathi, S., Singh, S., Yadav, R., Singh, A.N., and Mahmoud, A.E.D. (2023) Wastewater and sludge valorisation: a novel approach for treatment and resource recovery to achieve circular economy concept. *Frontiers in Chemical Engineering* 5, 1129783. DOI: 10.3389/fceng.2023.1129783
Kaur, R. and Kaur, S. (2023) Carrier-based biofertilizers. In: Kaur, S., Dwibedi, V., Sahu, P.K., and Kocher, G.S. (eds) *Metabolomics, Proteomes and Gene Editing Approaches in Biofertilizer Industry*. Springer Nature, Singapore, pp. 57–75.
Khan, R., Jan, H.A., Shaheen, B., Abidin, S.Z.U., Ullah, A. *et al.* (2023) Biofertiliser from animal wastes. In: Arshad, M. (ed.) *Climate Changes Mitigation and Sustainable Bioenergy Harvest Through Animal Waste: Sustainable Environmental Implications of Animal Waste*. Springer Nature, Cham, Switzerland, pp. 413–429.
Koester, U. and Galaktionova, E. (2021) FAO food loss index methodology and policy implications. *Studies in Agricultural Economics* 123(1), 1–7. DOI: 10.7896/j.2093
Krutiakova, V., Pyliak, N., Nikipelova, O., Bulgakov, V., Rucins, A. *et al.* (2021) Investigation of technology for obtaining biofertilisers based on sewage sludge. *Engineering for Rural Development* 20, 1080–1087. DOI: 10.22616/ERDev.2021.20.TF234
Kumar, A., Usmani, Z., and Kumar, V. (2017) Biochar and flyash inoculated with plant growth promoting rhizobacteria act as potential biofertilizer for luxuriant growth and yield of tomato plant. *Journal of Environmental Management* 190, 20–27. DOI: 10.1016/j.jenvman.2016.11.060
Li, Y., Cui, T., Wang, Y., and Ge, X. (2018) Isolation and characterization of a novel bacterium *Pseudomonas aeruginosa* for biofertilizer production from kitchen waste oil. *RSC Advances* 8(73), 41966–41975. DOI: 10.1039/C8RA09779H
Lim, S.F. and Matu, S.U. (2015) Utilization of agro-wastes to produce biofertilizer. *International Journal of Energy and Environmental Engineering* 6(1), 31–35. DOI: 10.1007/s40095-014-0147-8
Ma, Y., Shen, Y., and Liu, Y. (2020) Food waste to biofertilizer: a potential game changer of global circular agricultural economy. *Journal of Agricultural and Food Chemistry* 68(18), 5021–5023. DOI: 10.1021/acs.jafc.0c02210
Malusà, E., Pinzari, F., and Canfora, L. (2016) Efficacy of biofertilizers: challenges to improve crop production. In: Singh, D., Singh, H., and Prabha, R. (eds) *Microbial Inoculants in Sustainable Agricultural Productivity: Vol. 2: Functional Applications*. Springer, New Delhi, pp. 17–40.
Mazumdar, M., Sultana, T., Sinha, S., Sadhukhan, S., Datta, A. *et al.* (2022) Comparison of a novel combination of bio-organic fertilizers vis-à-vis a chemical fertilizer. *Indian Journal of Biochemistry and Biophysics (IJBB)* 59(2), 197–204. DOI: 10.56042/ijbb.v59i2.46743
Meena, V.S., Mishra, P.K., Bisht, J.K., and Pattanayak, A. (eds) (2017) *Agriculturally Important Microbes for Sustainable Agriculture*, Vol. 2. Springer, Singapore.
Mortola, N., Romaniuk, R., Cosentino, V., Eiza, M., Carfagno, P. *et al.* (2019) Potential use of a poultry manure digestate as a biofertiliser: evaluation of soil properties and *Lactuca sativa* growth. *Pedosphere* 29(1), 60–69. DOI: 10.1016/S1002-0160(18)60057-8
Mushtaq, Z., Asghar, H.N., and Zahir, Z.A. (2021) Comparative growth analysis of okra (*Abelmoschus esculentus*) in the presence of PGPR and press mud in chromium contaminated soil. *Chemosphere* 262, 127865. DOI: 10.1016/j.chemosphere.2020.127865
Onyia, C.O., Okoh, A.M., and Irene, O. (2020) Production of plant growth-promoting bacteria biofertilizer from organic waste material and evaluation of its performance on the growth of corn (*Zea mays*). *American Journal of Plant Sciences* 11(02), 189–200. DOI: 10.4236/ajps.2020.112015
Patel, H.K., Thumar, N.K., Patel, P.D., and Gohil, A.V. (2022) Bioprospecting of microbial diversity for sustainable agriculture and environment. In: Kumar, V., Garg, V.K., Kumar, S., and Biswas, J.K. (eds)

Omics for Environmental Engineering and Microbiology Systems. CRC Press, Boca Raton, Florida, pp. 283–314.

Patel, K.K., Pandey, A.K., Baheliya, A.K., Singh, V., Yadav, R. *et al.* (2024) A comparative study of the effects of biofertilizers and chemical fertilizers on soil physical and biological properties under chickpea crop (*Cicer arietinum* L.). *International Journal of Environment and Climate Change* 14(3), 72–80. DOI: 10.9734/ijecc/2024/v14i34020

Pathak, M., Tripathy, P., Dash, S.K., Sahu, G.S., and Pattanayak, S.K. (2018) Efficacy of biofertilizer, organic and inorganic fertilizer on yield and quality of radish (*Raphanus sativus* L.). *International Journal of Chemical Studies* 6(4), 1671–1673.

Paul, S., Dutta, A., Defersha, F., and Dubey, B. (2018) Municipal food waste to biomethane and biofertilizer: a circular economy concept. *Waste and Biomass Valorization* 9(4), 601–611. DOI: 10.1007/s12649-017-0014-y

Paygond, S.Y., Umashankar, N., Raghu, H., Naik, L.K., Benherlal, P. *et al.* (2024) Assessing the impact of corn-rind-carrier based biofertilizer on the growth and yield of capsicum (*Capsicum annuum* L.). *Mysore Journal of Agricultural Sciences* 58(1), 295–305.

Pei, B., Liu, T., Xue, Z., Cao, J., Zhang, Y. *et al.* (2025) Effects of biofertilizer on yield and quality of crops and properties of soil under field conditions in China: a meta-analysis. *Agriculture* 15(10), 1066. DOI: 10.3390/agriculture15101066

Pérez-Marroquín, X.A., Estrada-Fernández, A.G., García-Ceja, A., Aguirre-Álvarez, G., and León-López, A. (2023) Agro-food waste as an ingredient in functional beverage processing: sources, functionality, market and regulation. *Foods* 12(8), 1583. DOI: 10.3390/foods12081583

Purnomo, C.W., Indarti, S., Wulandari, C., Hinode, H., and Nakasaki, K. (2017) Slow release fertiliser production from poultry manure. *Chemical Engineering Transactions* 56, 1531–1536. DOI: 10.3303/CET1756256

Rocha, L., Leite, J.T., and Ramos, R. (2019) Comparison of different inoculation methods for improved colonization of PGPR and plant growth promotion in *Phaseolus vulgaris*. *Frontiers in Microbiology* 10, 2341. DOI: 10.3389/fmicb.2019.02341

Rossi, C.Q., Pereira, M.G., García, A.C., Berbara, R.L.L., Gazolla, P.R. *et al.* (2016) Effects on the composition and structural properties of the humified organic matter of soil in sugarcane strawburning: a chronosequence study in the Brazilian Cerrado of Goiás state. *Agriculture, Ecosystems & Environment* 216, 34–43. DOI: 10.1016/j.agee.2015.09.022

Rosso, D., Fan, J., Montoneri, E., Negre, M., Clark, J. *et al.* (2015) Conventional and microwave assisted hydrolysis of urban biowastes to added value lignin-like products. *Green Chemistry* 17(6), 3424–3435. DOI: 10.1039/C5GC00357A

Sangamithirai, K.M., Jayapriya, J., Hema, J., and Manoj, R. (2015) Evaluation of in-vessel co-composting of yard waste and development of kinetic models for co-composting. *International Journal of Recycling of Organic Waste in Agriculture* 4(3), 157–165. DOI: 10.1007/s40093-015-0095-1

Saravanan, A., Swaminaathan, P., Kumar, P.S., Yaashikaa, P.R., Kamalesh, R. *et al.* (2023) A comprehensive review on immobilized microbes - biochar and their environmental remediation: mechanism, challenges and future perspectives. *Environmental Research* 236(Pt 1), 116723. DOI: 10.1016/j.envres.2023.116723

Schütz, L., Gattinger, A., Meier, M., Müller, A., Boller, T., Mäder, P., and Mathimaran, N. (2018) Improving crop yield and nutrient use efficiency via biofertilization—a global meta-analysis. *Frontiers in Plant Science*, 8, 2204. DOI: 10.3389/fpls.2017.02204

Sedri, M.H., Niedbała, G., Roohi, E., Niazian, M., Szulc, P. *et al.* (2022) Comparative analysis of plant growth-promoting rhizobacteria (PGPR) and chemical fertilizers on quantitative and qualitative characteristics of rainfed wheat. *Agronomy* 12(7), 1524. DOI: 10.3390/agronomy12071524

Serraj, R., and Pingali, P. (eds) (2018) *Agriculture & Food Systems to 2050: Global Trends, Challenges and Opportunities*. World Scientific, Singapore.

Sharma, V.K., Kumar, A., Passarini, M.R.Z., Parmar, S., and Singh, V.K. (eds) (2023) *Microbial Inoculants: Recent Progress and Applications*. Academic Press, London.

Sinha, R.K., Valani, D., Chauhan, K., and Agarwal, S. (2010) Embarking on a second green revolution for sustainable agriculture by vermiculture biotechnology using earthworms: reviving the dreams of Sir Charles Darwin. *Journal of Agricultural Biotechnology and Sustainable Development* 2(7), 113.

Sipponen, M.H., Lange, H., Crestini, C., Henn, A., and Österberg, M. (2019) Lignin for nano- and microscaled carrier systems: applications, trends, and challenges. *ChemSusChem* 12(10), 2039–2054. DOI: 10.1002/cssc.201900480

Spokas, K.A., Cantrell, K.B., Novak, J.M., Archer, D.W., Ippolito, J.A. *et al.* (2012) Biochar: a synthesis of its agronomic impact beyond carbon sequestration. *Journal of Environmental Quality* 41(4), 973–989. DOI: 10.2134/jeq2011.0069

Sravani, A., Patil, C.R., and Sharma, S. (2024) Value-added product development utilising the food wastes. In: Srivastav, A.L., Kumar, A., and Kumar, M. (eds) *Valorization of Biomass Wastes for Environmental Sustainability: Green Practices for the Rural Circular Economy*. Springer Nature, Cham, Switzerland, pp. 287–301.

Suhag, M. (2016) Potential of biofertilizers to replace chemical fertilizers. *International Advanced Research Journal in Science, Engineering and Technology* 3(5), 163–167. DOI: 10.17148/IARJSET.2016.3534

Tabassam, T., Sultan, T., Akhtar, M.E., Mahmood-ul-Hassan, M., and Arshad, A. (2015) Suitability of different formulated carriers for sustaining microbial shelf life. *Pakistan Journal of Agricultural Research* 28(2).

Tripti, Kumar, A., Kumar, V., Anshumali, Bruno, L.B. *et al.* (2022) Synergism of industrial and agricultural waste as a suitable carrier material for developing potential biofertilizer for sustainable agricultural production of eggplant. *Horticulturae* 8(5), 444. DOI: 10.3390/horticulturae8050444

Umesha, S., Divya, M., Prasanna, K., Lakshmipathi, R., and Sreeramulu, K. (2014) Comparative effect of organics and biofertilizers on growth and yield of maize (*Zea mays*. L). *Current Agriculture Research Journal* 2(1), 55–62. DOI: 10.12944/CARJ.2.1.08

Vessey, J.K. (2003) Plant growth promoting rhizobacteria as biofertilizers. *Plant and Soil* 255(2), 571–586. DOI: 10.1023/A:1026037216893

Vilas-Boas, A.A., Pintado, M., and Oliveira, A.L.S. (2021) Natural bioactive compounds from food waste: toxicity and safety concerns. *Foods* 10(7), 1564. DOI: 10.3390/foods10071564

Wang, J., Zhai, B., Shi, D., Chen, A., and Liu, C. (2024) How does bio-organic fertilizer combined with biochar affect Chinese small cabbage's growth and quality on newly reclaimed land? *Plants* 13(5), 598. DOI: 10.3390/plants13050598

Wang, X., Pan, S., Zhang, Z., Lin, X., Zhang, Y. *et al.* (2017) Effects of the feeding ratio of food waste on fed-batch aerobic composting and its microbial community. *Bioresource Technology* 224, 397–404. DOI: 10.1016/j.biortech.2016.11.076

Wei, T., Zhang, P., Wang, K., Ding, R., Yang, B. *et al.* (2015) Effects of wheat straw incorporation on the availability of soil nutrients and enzyme activities in semiarid areas. *PLOS One* 10(4), e0120994. DOI: 10.1371/journal.pone.0120994

Youssef, M.M.A. and Eissa, M.F.M. (2014) Biofertilizers and their role in management of plant parasitic nematodes: a review. *Journal of Biotechnology and Pharmaceutical Research* 5(1), 1–6.

Yuliani, S. and Pratiwi, E. (2024) Potential carriers for biofertilizers: microstructural and entrapment properties. *IOP Conference Series: Earth and Environmental Science* 1377(1), 012004). DOI: 10.1088/1755-1315/1377/1/012004

Zhang, Q.H., Yang, W.N., Ngo, H.H., Guo, W.S., Jin, P.K. *et al.* (2016) Current status of urban wastewater treatment plants in China. *Environment International* 92–93, 11–22. DOI: 10.1016/j.envint.2016.03.024

Zhou, Y., Xiao, R., Klammsteiner, T., Kong, X., Yan, B. *et al.* (2022) Recent trends and advances in composting and vermicomposting technologies: a review. *Bioresource Technology* 360, 127591. DOI: 10.1016/j.biortech.2022.127591

6 Utilization of Agrifood Waste to Produce Liquid Biofertilizer and Its Efficacy

Abha Kumari*
Centre for Biotechnology and Biochemical Engineering, Amity Institute of Biotechnology, Amity University, Noida, India

Abstract
The increasing demand for sustainable agriculture and environmental conservation has driven interest in organic fertilizers derived from agricultural waste. The creation of liquid fertilizer from locally sourced agricultural waste, including fruit peels, vegetable scraps, and animal manure, is investigated in this study. The waste materials were converted into nutrient-rich biofertilizer that could be applied topically and in soil by using basic fermentation processes. The effectiveness of the liquid fertilizer was evaluated by analyzing its pH, microbial activity, and key macro- and micronutrients (N, P, K, Ca, and Mg). The finished product, according to the results, enhanced soil fertility and decreased reliance on chemical inputs, encouraging a circular economy in farming communities. This method solves agricultural waste management issues while offering a more affordable and environmentally responsible substitute for petrochemical fertilizers.

6.1 Introduction

There are several ways to fertilize a plant through fertilizers, which can be categorized into granular and liquid fertilizers. While plants cannot necessarily show the differences between the two methods, there are certain advantages and disadvantages that come to light. In recent years, liquid fertilizers have become popular in the agricultural world. Liquid fertilizer is a concentrated homogeneous solution rich in nutrients; its purpose is to supply nutrients to plants in a form that is easily absorbed through their roots (Malhotra, 2016). Unlike solid fertilizers, liquid fertilizer is already in a dissolved state, making it easy to use. This kind of fertilization offers several benefits, comprising rapid nutrient uptake, application flexibility, and the ability to address specific nutrient deficiencies. Liquid fertilizers contain balanced proportions of essential nutrients like nitrogen, phosphorus, and potassium along with micronutrients. While these fertilizers can provide a complete balance of nutrients, they can also focus on a singular nutrient. Liquid fertilizers have many characteristics that help make them a better choice than granular or solid fertilizers (Timilsena *et al.*, 2015). Unlike inorganic fertilizers, organic liquid fertilizers can overcome certain drawbacks such as by supporting a circular economy (Fernández-Delgado *et al.*, 2022). Inorganic fertilizers have monopolized agriculture for a long time, causing many disadvantages, such as greenhouse gas emissions and surface water eutrophication. Apart from

*Corresponding author: akumari@amity.edu

DOI: 10.1079/9781836991021.0006

the environmental benefits, liquid fertilizers also provide slow release of nutrients in the soil, ensuring greater stability of nutrients in plants (Sanadi *et al.*, 2019), and are user friendly—with the right tools, it is easier to apply to soil or leaves, saving on labor expense and time. Systems like drip irrigation can be used, which requires much less liquid fertilizer as it involves a high concentration of nutrients.

Formation of liquid organic fertilizers from agricultural residues and industrial waste has become a resourceful way to manage waste (Sharma *et al.*, 2019). Usually these wastes have been left to decompose, contributing to environmental issues such as greenhouse gas emissions and soil degradation.

This chapter will cover the process, advantages, and difficulties of turning agricultural waste into liquid fertilizer, the types of agricultural waste that can be utilized, and the nutrient content of these fertilizers.

6.2 Liquid Fertilizers

Liquid biofertilizers, known as liquid microbial consortia (LMCs), are formulations that contain beneficial bacteria, fungi, or algae that are specifically designed for fertigation, hydroponics, and soil drenching (Richa, 2023). They have a higher contamination resistance when compared to solid fertilizers. According to Allouzi *et al.* (2022), liquid fertilizer is "a concentrated liquid formulation that contains live beneficial microorganisms—such as nitrogen-fixing, phosphate-solubilizing, or plant growth-promoting bacteria—which are applied to crops to enhance plant growth, nutrient uptake, and soil fertility." These formulations of liquid fertilizers are made so that they can be readily put to use and easily applicable to methods such as seed treatment, soil drenching, or foliar spraying. In comparison with solid biofertilizers, liquid biofertilizers have a longer shelf life. They deliver microbes in a viable, stable, and effective state for plant growth and cultivation. Microbial strains like *Rhizobium*, *Azotobacter*, *Bacillus*, and *Pseudomonas* can be used to produce liquid biofertilizers using organic waste, especially agrifood waste streams, in an eco-friendly manner.

LBFs present a compelling, sustainable alternative to traditional fertilizers, offering environmental and agronomic advantages. However, to fully realize their potential, it is essential to address barriers related to awareness, technological readiness, and market-scale deployment.

6.3 Types of Liquid Fertilizers

Liquid fertilizers are an invaluable tool for gardening and agriculture because they deliver important nutrients in a form that plants can easily absorb (El-Ramady *et al.*, 2014). There are several varieties of these fertilizers, each designed to satisfy certain plant requirements. Knowing these different kinds of fertilizers makes it easier to choose the best product for promoting healthy soil and plant growth. While liquid fertilizers can be tailored to different needs of the crop or soil, there are mainly five types of liquid fertilizers, as outlined below.

6.3.1 NPK fertilizers

Theses are broad-spectrum fertilizers containing the three major nutrients necessary for plant growth: nitrogen (N), phosphorus (P), and potassium (K). These nutrients are foundational; they support the overall wellbeing of plants (Bednarek *et al.*, 2012). NPK fertilizers are the most common type of mineral fertilizers, containing the three essential macronutrients. Each plays a unique role in plant physiology—nitrogen promotes vegetative growth and chlorophyll production; phosphorus supports energy transfer and root development; and potassium aids in water regulation and stress resistance (Savci, 2012). These fertilizers are produced in various N:P:K ratios tailored to specific crop requirements and soil conditions. Their fast-acting nature ensures immediate nutrient availability, which is particularly beneficial in high-demand agricultural settings. However, the overuse or mismanagement of NPK fertilizers can lead to negative environmental consequences, such as soil acidification, nutrient runoff, and eutrophication of aquatic systems (Zhang *et al.*, 2015). Therefore, integrated nutrient management strategies—such as combining NPK with organic amendments or precision farming practices—are increasingly encouraged to improve nutrient-use efficiency and sustainability. NPK fertilizers remain foundational in modern agriculture due to their consistent yield-improving results, especially in staple crops like rice, wheat, and maize.

6.3.2 Micronutrient fertilizers

Micronutrient fertilizers supply trace elements such as iron (Fe), zinc (Zn), manganese (Mn), copper (Cu), boron (B), molybdenum (Mo), and chlorine (Cl). Though required in trace amounts, these are still vital for plant growth (Hettiarachchi *et al.*, 2010), and their deficiency can lead to severe physiological and developmental problems in crops. For instance, iron is vital for chlorophyll synthesis and respiration, zinc is involved in enzyme activation and growth regulation, and manganese aids in nitrogen metabolism and photosynthesis (Hettiarachchi *et al.*, 2010). These micronutrients often become unavailable in high-pH or calcareous soils, making supplementation critical. Chelated forms of micronutrients, such as Fe-EDTA (ethylenediaminetetraacetic acid) or Zn-EDTA, enhance solubility and uptake efficiency, especially in alkaline soils (Alloway, 2008). With increasing attention to food quality and soil health, micronutrient fertilizers have gained importance not only in correcting visible deficiency symptoms but also in improving nutrient density in food crops (White and Broadley, 2009). As global agriculture intensifies, the inclusion of micronutrients in fertilizer regimes is essential to sustain productivity and reduce hidden hunger caused by micronutrient deficiencies in human diets.

6.3.3 Organic liquid fertilizers

Organic liquid fertilizers are derived from natural sources such as animal manure, compost extracts, vermiwash, and crop residues. Unlike synthetic fertilizers, they offer a dual advantage: supplying essential nutrients and enhancing soil structure, microbial activity, and biodiversity. These fertilizers typically contain both macro- and micronutrients in bioavailable forms, along with beneficial microorganisms that improve nutrient cycling and plant resilience (Pant *et al.*, 2012). Their use also supports sustainable agriculture by recycling agricultural waste, thereby contributing to a circular economy and reducing dependency on chemical inputs. Furthermore, organic liquid fertilizers help retain soil moisture, reduce erosion, and promote the formation of humus, all of which are crucial for long-term soil fertility (Gopinath *et al.*, 2008). They are particularly beneficial in organic farming systems, horticulture, and smallholder farming where chemical fertilizers are either restricted or unaffordable. While nutrient concentrations in organic liquid fertilizers may be lower than in synthetic alternatives, their long-term benefits on soil health and environmental sustainability are well documented.

6.3.4 Foliar fertilizers

Foliar fertilizers are specially formulated to be applied directly to plant leaves, allowing nutrients to be absorbed through the stomata or epidermal cells. This method provides rapid nutrient uptake, making it particularly effective during critical growth stages or when soil conditions limit nutrient availability (Niu *et al.*, 2021b). Foliar application is especially useful for micronutrients like iron, zinc, and manganese, which may be immobilized in the soil under certain pH conditions. This technique is widely used to correct mid-season deficiencies, support reproductive development, and enhance crop quality and yield. In stress conditions such as drought or compacted soils, foliar feeding can bypass root uptake limitations and ensure continued nutrient supply (Fernández and Ebert, 2005). However, the success of foliar fertilization depends on factors such as leaf age, environmental conditions, and the formulation of the fertilizer. Proper timing, concentration, and use of surfactants are essential to prevent phytotoxicity. Overall, foliar fertilizers offer a fast, targeted, and resource-efficient approach to nutrient management in both conventional and precision agriculture.

6.3.5 Custom tailored fertilizers

These kinds of fertilizers are made according to the specific needs of the plant and soil. As the requirements of plants and soil vary, so a custom blended fertilizer can be used to address their particular deficiencies accordingly. Custom tailored fertilizers are formulations specifically designed to address the unique nutrient needs of particular soils and crops. Unlike standard fertilizers, these blends are created based on comprehensive soil tests, crop requirements, and local agro-climatic conditions. This approach is rooted in the understanding that plant nutritional needs vary significantly with growth stage, soil type, and environmental factors. Custom fertilizers often include precise ratios of NPK, secondary nutrients like calcium and magnesium, and micronutrients such as iron, zinc, or boron to address identified deficiencies (Fageria, 2009). By aligning fertilizer inputs with crop demands, custom formulations enhance nutrient-use efficiency, reduce environmental loss, and promote sustainable yield increases. This is particularly important in degraded or nutrient-imbalanced soils where standard fertilizers may be ineffective or even harmful. Technological advancements in precision agriculture, including GPS-based application systems and data analytics, have made custom fertilization more accessible and scalable. As global agriculture shifts toward sustainability and resource efficiency, custom tailored fertilizers are becoming essential tools for site-specific nutrient management.

6.4 Utilization of Different Wastes

Waste materials that are often considered liabilities can actually be valuable for the formation of liquid fertilizers (Fig. 6.1). These waste streams can be transformed into fertilizers, thereby aligning with the principle of circular economy where recycling or repurposing helps environmental management.

6.4.1 Food waste

As the human population elevates, so does the waste produced by them. Humans produce food waste like vegetable scraps and fruit peels, spoiled and unfit food, and coffee grounds. Restaurants can produce large amounts of wasted food. A study by Unnisa (2015), highlighting the usage of restaurant food waste repurposed as fertilizer using anaerobic digestion, showed parameters of 1.15% nitrogen, 0.308% phosphorus, and 0.77% potassium.

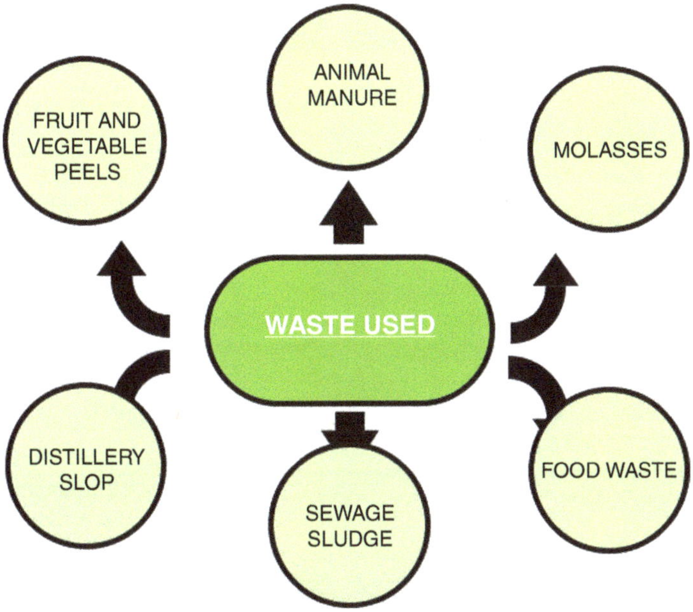

Fig. 6.1. Different types of wastes that can be utilized. Figure author's own.

A seed germination test was done to determine that there was no toxicity.

6.4.2 Animal manure

The fecal matter and urine of animals such as cows, chickens, and sheep goes to form animal manure. The nutrient content of the fertilizer depends on which animal the manure comes from.

In an experiment, seven manures were considered (Prado *et al.*, 2022): two contrasting cattle slurries, one caprine, two swine slurries, and two contrasting swine slurries. Slurries were further treated by solid–liquid separation and acidification. It was shown that pig slurry can not be used as a fertilizer due to its low nutrient content. The rest of the slurries can be combined and blended accordingly to have a stronger potential.

6.4.3 Molasses

Molasses is a by-product of sugar extraction from sugarcane or sugarbeet and also from other sources like sorghum, pomegranates, and dates. It is rich in nutrients, including micronutrients like iron, magnesium, and calcium. Depending on the maturity of the sugarcane, its source, the extraction method, and the extent of sugar extraction, molasses can be classified into light, dark, and blackstrap varieties. Light molasses is the lightest in color and sweetest in taste, while blackstrap molasses is the darkest, thickest, and has a bitter flavor with the least sugar content. In Indonesia (Indrawati *et al.*, 2020), when the production of banana pseudo-stem rose from 6.28 million tonnes to 7.04 million tonnes, scientists and researchers started looking for a sustainable solution for the better use of increasing banana pseudo-stem waste. The study concluded with a liquid fertilizer with the composition of 100 ml rice water : 100 g banana pseudo-stem : EM4 3.0 ml : 30 ml molasses; this provided the largest amount of N, P, and K (potassium 0.47%, phosphate 0.32%, nitrogen 0.076%).

6.4.4 Distillery slop

Distillery slop is derived during fermentation and distillation of alcohol. It is high in content for protein, iron, magnesium, and calcium. An experiment (Shao *et al.*, 2023) was conducted to verify the effect of liquid fertilizer made from alcoholic waste (distillery slop); the trial was conducted on cherry tomatoes using five treatments (conventional fertilizers T1, alcohol waste liquid fertilizer T2, commercial fertilizer A T3, commercial fertilizer B T4, blank control T0). The treatment was applied at a rate of 100 ml/pot. The results showed that compared to T1, T2 to T4 showed significant growth, and yield per plant increased by a range of 8.3–16.7%, with T2 being the highest, with root length increased by 25.7%, root surface area by 23.3%, and average root diameter by 18.8%. Therefore, liquid fertilizer from alcoholic waste improves yield and quality by promoting the root system of cherry tomatoes.

6.4.5 Sewage sludge

Wastewater, often considered a problem for disposal, is, however, a source of many valuable nutrients and hence can be used as a precursor in the formation of liquid fertilizer. This targets both agricultural advantage and waste management. In addition, it can be highly cost-effective and its use very feasible. (Reig *et al.*, 2021) conducted an experiment where wastewater was used to recover nutrients in the formation of liquid fertilizer. Recovery of nitrogen was achieved through procedures such as air stripping and stream stripping, while ammonia was recovered using membrane contractors. It was possible to extract 96% of ammonia and 4.6% of nitrogen.

Another study (Mekki *et al.*, 2006) used water waste from an olive mill on tomatoes (*Lycopersicon esculentum*), chickpeas (*Cicer airetinum*), beans (*Vicia faba*), wheat (*Triticum durum*), and barley (*Hordeum vulgar*) to check the effects on seed germination, plant growth, and soil fertility. The results showed positive effects. Treated plants showed increased biomass and plant growth.

6.4.6 Agriwood in liquid fertilizer production: a circular bioeconomy perspective

The incorporation of agriwood—woody biomass derived from agricultural residues such as pruned branches, sawdust, coconut husks, or forestry by-products—into liquid fertilizer production presents a sustainable and resource-efficient strategy aligned with modern circular economy principles. Traditionally considered a waste product or used for low-value applications like fuelwood, agriwood is now gaining recognition for its potential to be biochemically processed into nutrient-rich liquid inputs for agriculture.

Agriwood materials are high in lignocellulosic content, making them suitable for composting, fermentation, and bio-digestion. These biological processes break down complex carbon structures into humic substances, fulvic acids, and bioavailable micronutrients, which are vital for soil health and plant growth. When used in the production of organic liquid fertilizers, agriwood not only serves as a source of nutrients but also improves the soil-conditioning properties of the final product by enhancing microbial diversity and organic matter content (Bernal *et al.*, 2009). Moreover, compost extracts made from sawdust and woody residues have shown beneficial effects on soil aggregation and microbial activity, key indicators of long-term soil fertility.

From an industrial standpoint, using agriwood as a raw material for liquid fertilizers contributes to waste valorization, lowers production costs, and enhances product sustainability. By converting what would otherwise be discarded or openly burned into a valuable agricultural input, industries support climate-smart agriculture and reduce carbon emissions (Bhattacharyya et al., 2016). In countries with high agricultural biomass output, such as India, Indonesia, and Brazil, the use of woody biomass for fertilizer production offers a scalable and regionally adaptable solution.

Furthermore, agriwood-based inputs align well with the EU Bioeconomy Strategy and UN Sustainable Development Goals (SDGs), particularly SDG 12 (responsible consumption and production) and SDG 13 (climate action). The European Commission (2018) highlights biomass utilization as a key driver of the bioeconomy, urging innovation in converting forestry and agricultural residues into value-added products. Agriwood, in this context, becomes a bridge between sustainable waste management and nutrient recycling.

Challenges remain, particularly in terms of processing complexity. The high lignin content in woody biomass can slow decomposition, requiring pretreatment technologies such as shredding, microbial inoculation, or alkaline hydrolysis to speed up the breakdown process and enhance extractability (Bernal *et al.*, 2009). However, with advancements in bioconversion technologies, industries are increasingly able to overcome these limitations.

In conclusion, agriwood represents an untapped yet promising resource in the liquid fertilizer industry. By integrating agriwood into production systems, industries can advance soil regeneration, promote circularity in agriculture, and contribute to a more sustainable and low-emission food system.

6.4.7 Fruit and vegetable peels

Fruit and vegetable peels that are often disposed of as waste can be a valuable source of nutrients. They are rich in vitamins, organic compounds, and minerals. Production of liquid fertilizer from these can be an easy and helpful process that farmers can adopt (Table 6.1), as it generally involves fermentation of the peels which requires no extra cost for machinery.

An experiment was conducted where liquid fertilizer was produced from orange peels using gas-permeable membrane to recover nitrogen from anaerobic digestate. Of the total ammonia, about 62% was recovered (Horta *et al.*, 2022). Another attempt was made to transform potato peel and chromium-free wet blue leather into liquid fertilizer, where the nitrogen, phosphorus, and potassium content was found to be 13.10%, 2.41%, and 20.20%, respectively. Its effect on okra (*Abelmoschos esculentus*) was monitored, which showed higher plant growth and overall productivity (Majee *et al.*, 2021).

6.5 Advantages of Liquid Fertilizers

Liquid fertilizers hold a lot advantages in the modern world. These benefits are consistent with enhanced plant health, economic efficiency, and environmental sustainability. Below are a few benefits of liquid fertilizers from agricultural waste.

6.5.1 Sustainability and waste reduction

The circular economy's tenets are supported by liquid fertilizers derived from agricultural waste

Table 6.1. Overview of liquid biofertilizer formulations, application methods, and crop responses. Table author's own.

Serial No.	Liquid biofertilizer source and formulation	Crop studied	Yield outcome	Reference
1	Consortium of *Rhizobium* + phosphate-solubilizing bacteria (liquid seed inoculant + soil drench with vermicompost)	Field pea (*Pisum sativum*)	Seed yield ~1,624 kg/ha (75% RDF + LBF); stover ~3,551 kg/ha – on par with 100% RDF + LBF	Kumari *et al.*, 2024a
2	Liquid biofertilizer (seed + soil application)	Finger millet (*Eleusine coracana*)	Grain yield up to 23.58 q/ha when using 100% RDF + seed + soil LBF	Yadav *et al.*, 2025b
3	*Rhizobium* + liquid organic nutrients (vermiwash, fish amino acids, seaweed extract)	Black gram (*Vigna mungo*)	Highest seed yield 1.57 t/ha, improved growth traits	Molla *et al.*, 2024
4	Liquid formulations of *Rhizobium* + PSB + KMB applied as seed treatments	Black gram	Best yield characters and highest net return, B:C ratio ~2.02	Yadav *et al.*, 2024
5	Liquid biofertilizer drench + NPK	Rice (*Oryza sativa*)	With 0.75 × NPK + LBF: agronomic and economic performance matched full NPK; R/C = 1.83	Hazra and Santosa, 2024
6	LBF consortium (lucerne inoculation with *Sinorhizobium meliloti*)	Lucerne (*Medicago sativa*)	Improved growth, yield, N uptake (specific stats in paper)	Gatabazi *et al.*, 2023
7	Daily LBF drench (organic-waste-derived) vs. NPK	Maize (*Zea mays*)	Height ~225 cm by week 8; yield higher than control and NPK	SCRiP Research Team, 2022
8	Liquid biofertilizer + charcoal + additives (CMC, PVP)	Chickpea (*Cicer arietinum*)	Increased P solubilization, emergence, height, chlorophyll vs. control	Anjali *et al.*, 2019

(Chojnacka *et al.*, 2022), which lessen waste and the environmental effect of fertilizer manufacturing. A good example is fertilizer made from sugarcane waste in India.

Usually the waste produced by sugarcane production, such as molasses, distillery slop, and sugarcane leaves, is either thrown away or burned, which pollutes the environment (Singh, 2021). But now, in states like Maharashtra, sugarcane waste is being turned into liquid fertilizer, which has environmental benefits as well as contributing to cutting agricultural waste (Dotaniya *et al.*, 2016).

6.5.2 Environmentally friendly

Compared to synthetic fertilizers, liquid fertilisers made from agricultural waste are more environmentally friendly since they are organic (Chew *et al.*, 2019). They have a smaller carbon footprint and cause less pollution of the soil and water. For example, farmers use anaerobic digestate as a liquid fertilizer to nourish their fields and improve soil structure and fertility without causing water pollution (Kovačić *et al.*, 2022).

6.5.3 Cost-effectiveness

Liquid fertilizers made from agricultural waste are frequently less expensive than synthetic ones, especially for small-scale farmers (Case *et al.*, 2017). Farmers may save money on fertilizers while nourishing their fields and increasing crop yields by recycling waste materials.

By using these fertilizers, there is reduction in input costs and also the purchasing cost, giving farmers relief. One example would be coffee waste in Ethiopia. Coffee is a significant crop in Ethiopia. The growing and processing of the crop creates a lot of waste, such as coffee husks and pulp (Ademe *et al.*, 2019). Now farmers have adapted the situation in their favor by producing fertilizers from the waste. Coffee plants now get a consistent supply of nitrogen and potassium thanks to this waste.

6.5.4 Slow release of nutrients

Liquid fertilizers made from agricultural waste distribute nutrients gradually and steadily (Osman, 2013), unlike synthetic fertilizers, which may release nutrients too rapidly and cause leaching. The slow release guarantees that plants absorbs nutrients better and lowers the possibility of a possible nutrient runoff.

6.5.5 Soil health improvement

By boosting organic matter and encouraging good microbes, liquid fertilizers made from agricultural waste help to create healthier soil (Singh *et al.*, 2022). As a result, the soil's long-term fertility, structure, and water retention all increases. In Vietnam, rice is a staple food. Farmers there take rice straws and mulch them in the soil, which improves its structure and nutrient content (Dinh *et al.*, 2024).

6.6 Drawbacks

6.6.1 Variability in nutrient composition

Depending on the kind of waste utilized, liquid fertilizers made from agricultural waste frequently have different nutritional contents. Farmers may find it difficult to apply the proper amounts of fertilizer to satisfy the demands of a given crop due to inconsistent nutrient content, which can result in either over- or underfertilizing (Case *et al.*, 2017).

For example, in Brazil, the amounts of potassium, phosphorus, and nitrogen in liquid fertilizers derived from sugarcane bagasse and molasses have reportedly been shown to vary (Ajala *et al.*, 2021), and uneven growth has resulted from farmers' inability to determine the proper application rates for sugarcane crops.

6.6.2 Shorter shelf life

Liquid organic fertilizers are known to have a shorter shelf life due to active ingredients. Therefore, they break down more quickly and need to be used shortly after production (Shaji *et al.*, 2021).

6.6.3 Storage challenges

Compared to solid fertilizers, liquid fertilizers have a thicker and heavier consistency, which makes delivery more costly and difficult (Meaclem, 2019) In rural regions of India, inadequate roads make it complicated to move fertilizers produced from cow dung from one place to another (Shaibbur *et al.*, 2021).

6.6.4 Limited availability and production scale

Infrastructure and the right waste substances must be available for agricultural waste to be converted into liquid fertilizer (Koul *et al.*, 2022). There will be places that do not have the appropriate facilities to turn agricultural waste into liquid fertilizer or where insufficient agricultural waste is available. When vineyards in Lutheran France tried to turn grape pomace—a by-product of winemaking—into liquid fertilizers, they discovered that the equipment needed for fermentation and processing was way too costly for smaller vineyards owners (Kosseva, 2013).

6.7 Industrial Perspective on Liquid Fertilizers

Liquid fertilizers have become a cornerstone of modern agricultural input industries due to their efficiency, adaptability, and compatibility with precision farming systems. From an industrial standpoint, their appeal lies in their rapid nutrient availability, suitability for fertigation and foliar application, and ability to be custom-blended with macro- and micronutrients. Their market demand is rising steadily, driven by global shifts toward sustainable and high-efficiency farming practices. According to Grand View Research (2022), the global liquid fertilizer market was valued at over US$13 billion in 2021 and is projected to grow significantly by 2030, propelled by increasing demand for high-yield crops and expanding adoption of precision agriculture technologies.

Unlike granular fertilizers that must dissolve before becoming bioavailable, liquid fertilizers deliver nutrients in readily soluble forms, allowing for faster absorption by plants. This feature is particularly beneficial during critical growth stages, especially in horticulture and cash crops where timing is crucial (Havlin *et al.*, 2014). Their compatibility with irrigation systems such as drip and sprinkler setups enhances uniform distribution and supports site-specific nutrient management, which is fundamental in reducing nutrient runoff and improving fertilizer use efficiency (Fernández and Ebert, 2005). For this reason, fertigation-based delivery of liquid fertilizers has become widespread in commercial farming operations globally.

From a manufacturing perspective, producing liquid fertilizers involves advanced formulation techniques including nutrient solubilization, chelation, pH stabilization, and blending under sterile conditions. Chelation, particularly with EDTA, DTPA (diethylenetriaminepentaacetic acid), and EDDHA (ethylenediamine-N,N′-bis(2-hydroxyphenylacetic acid)) agents, is used to stabilize micronutrients like iron, zinc, and manganese, improving shelf life and uptake efficiency (Mikkelsen, 2018). These processes ensure that nutrient elements remain available even under variable soil pH conditions. Moreover, the production is increasingly automated with inline monitoring systems to control parameters such as pH, viscosity, and nutrient concentration, helping standardize quality across batches (Rodríguez and López, 2020). The ability to produce highly specific formulations tailored to crop and soil conditions is a major industrial advantage that solidifies liquid fertilizers' relevance in modern agronomic strategies.

Industrially, liquid fertilizers are also being formulated with biostimulants such as humic acids, seaweed extracts, and microbial consortia. These enhance not only nutrient uptake but also plant stress tolerance and soil health, adding multifunctionality to the product (Calvo *et al.*, 2014). The integration of biostimulants aligns with sustainable agriculture trends and has opened new market segments focused on eco-friendly and organic farming inputs. This diversification has allowed fertilizer companies to broaden their product portfolios and appeal to environmentally conscious farmers and regulators.

Despite these advantages, liquid fertilizers also present specific challenges for industries, particularly related to logistics and shelf stability. Since they contain water as the primary carrier, their nutrient density is lower compared to solid fertilizers, leading to higher transportation and storage costs per unit of nutrient (Rietra *et al.*, 2017). In regions with inadequate storage infrastructure, this can be a significant limitation. Additionally, organic liquid fertilizers made from compost, manure, or crop residues are prone to microbial contamination, which necessitates pasteurization or the use of preservatives during production (Pant *et al.*, 2012). Manufacturers must invest in filtration, sterilization, and stabilization systems to maintain product integrity over extended shelf life, especially in warm climates.

Environmental concerns also drive the industrial adoption of liquid fertilizers. They are considered more eco-efficient due to reduced nutrient losses, especially when applied through fertigation systems that deliver nutrients directly to the root zone. This targeted approach helps minimize leaching, volatilization, and runoff—common issues associated with overuse of granular fertilizers (Zhang *et al.*, 2015). Furthermore, liquid formulations can be tailored to deliver micronutrients that are typically required in trace quantities but are critical for crop health, such as iron, zinc, and manganese (Hettiarachchi *et al.*, 2010). By providing these nutrients in chelated, plant-available forms, liquid fertilizers help address both visible and latent deficiencies, contributing to improved crop yield and quality.

One of the more significant developments in the liquid fertilizer industry is the role of organic liquid fertilizers, particularly those made from agricultural waste and biodegradable inputs. These formulations not only supply nutrients but also contribute to soil organic matter, enhancing soil texture, water retention, and microbial diversity. Industries have started commercializing vermiwash, compost tea, and fermented plant extracts as liquid biofertilizers, recognizing their role in circular economy models (Bhattacharyya *et al.*, 2016). These products support sustainable waste management while reducing reliance on synthetic inputs.

Application versatility further supports industrial interest in liquid fertilizers. They can be applied through various methods, including foliar sprays, soil drenching, and fertigation, depending on the crop and developmental stage. Foliar application, in particular, is widely used for micronutrient delivery, providing fast correction of deficiencies and allowing nutrient uptake even when root function is impaired (Niu *et al.*, 2021). Industries are increasingly designing crop- and stage-specific foliar blends, enhancing product specialization and market segmentation. Foliar liquid fertilizers also serve as carriers for pesticides, fungicides, or plant growth regulators, enabling tank-mix compatibility and reducing labor and application costs (Fernández and Ebert, 2005).

Digital farming innovations are also influencing the liquid fertilizer industry. With the rise of smart farming tools, liquid fertilizers are being integrated into variable-rate application (VRA) systems, drones, and precision sprayers. These technologies enable real-time nutrient application based on plant health indices captured through sensors and satellite imagery, improving overall input efficiency (Bhatt and Mehta, 2021). Consequently, industries are investing in the development of sensor-compatible and drone-friendly liquid formulations to meet the evolving needs of precision agriculture.

Looking toward the future, nanotechnology and controlled-release mechanisms are being researched to improve the performance of liquid fertilizers. Nanofertilizers can enhance nutrient penetration, reduce leaching losses, and increase uptake efficiency (Dimkpa and Bindraban, 2016). These innovations could redefine the liquid fertilizer sector by improving cost-efficiency and environmental safety. Regulatory acceptance of these advanced technologies will be pivotal in determining the scale and speed of adoption.

Industries are also working toward standardizing and certifying their liquid fertilizer products under eco-labels and organic certification systems. Increasing consumer awareness about food safety and environmentally friendly farming is pushing industries to comply with stricter regulatory norms and to provide transparency regarding product composition, origin, and environmental impact (Savci, 2012). As such, life cycle assessments and environmental impact analyses are becoming routine parts of the product development pipeline for many fertilizer manufacturers.

6.8 Applications of Liquid Fertilizers

Versatility and efficiency are provided by liquid fertilizers in a range of horticultural and agricultural applications. The following are some important applications (see also Fig. 6.2):

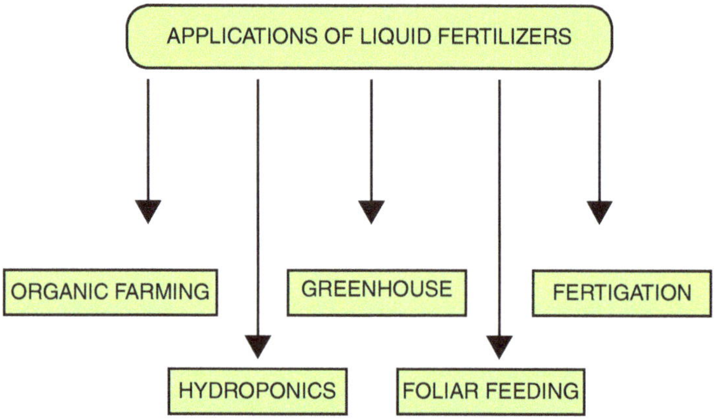

Fig. 6.2. Applications of liquid biofertilizers. Figure author's own.

1. **Fertigation:** in fertigation systems, where nutrients are delivered directly through irrigation water, liquid fertilizers are frequently utilized (Ashrafi *et al.*, 2020). By distributing water and nutrients evenly, this technique boosts productivity and lowers nutrient runoff.
2. **Hydroponics:** Liquid fertilizers give vital nutrients straight to plant roots in a controlled environment in soilless growing methods like hydroponics (El-Kazzaz and El-Kazzaz, 2017). This is very helpful for producing crops such as herbs, tomatoes, and lettuces.
3. **Foliar feeding:** Plants may quickly absorb nutrients via their foliage by simply spraying liquid fertilizer onto their leaves. This is particularly helpful for treating nutritional deficits and encouraging the development of plants (Niu *et al.*, 2021).
4. **Greenhouse and controlled environment:** Liquid fertilizers are used in greenhouses to precisely manage the supply of nutrients, thereby maximizing growing conditions for a variety of crops in controlled surroundings (Savvas *et al.*, 2024).
5. **Organic farming:** In organic farming, liquid organic fertilizers made from natural extracts or composted waste are frequently utilized as a greener substitute for chemical fertilizer (Durán-Lara *et al.*, 2020). Organic farming has been growing in popularity due to its contamination-free and minimized biomagnification effects.

6.9 Efficacy of Liquid Biofertilizer

Biofertilizer efficacy is the ability of microbial inoculants to improve plant nutrition, development, and yield by increasing nutrient availability, absorption, and accessibility. In terms of plant nutrition and yield, liquid biofertilizers have demonstrated the ability to reduce chemical fertilizer usage by 20–30% while maintaining comparable yields, through mechanisms such as nitrogen fixation, phosphate and potassium solubilization, and production of plant-growth substances (Verma and Kumar, 2024). Recent studies continue to highlight the positive impact of liquid biofertilizers (LBFs) on plant productivity, nutrient uptake, and soil health. In a field experiment on fodder sorghum during Rabi 2022–2023, the application of *Azospirillum* + phosphate-solubilizing bacteria (PSB) + potassium-solubilizing bacteria (KSB) in liquid form—combined with 75% of the recommended fertilizer dose—achieved significantly higher green and dry fodder yields and improved NPK uptake compared to control treatments (Vamsi *et al.*, 2023). In finger millet, combining 100% RDF (recommended dose of fertilizer) with liquid biofertilizers via seed inoculation and soil application notably enhanced soil nutrient levels (N: ~185 kg/ha; P: ~19.2 kg/ha; K: ~132.8 kg/ha) while maintaining comparable yields to full RDF alone (Yadav *et al.*, 2025a).

In field pea cultivation on laterite soils, integrating a liquid biofertilizer consortium with 75% RDF significantly improved growth parameters—including plant height, leaf area index, and dry matter accumulation—and led to seed yields (~1624 kg/ha) statistically equivalent to 100% RDF treatments (Kumari *et al.*, 2024b). In tomatoes, liquid inoculants containing

Azotobacter, PSB, and KSB combined with full RDF resulted in superior residual soil fertility (e.g. N: ~386 kg/ha; P: ~71 kg/ha; K: ~208 kg/ha) and higher plant uptake (N: ~164 kg/ha; P: ~26 kg/ha; K: ~200 kg/ha) compared to organic fertilizer treatments or controls (Lakshmikala *et al.*, 2022). Beyond yield and nutrient enhancements, LBF application has been shown to boost secondary plant metabolites and soil microbial activity. For instance, trials with Chinese cabbage revealed LBF treatment improved leaf size, biomass, and antioxidant activity (total phenolics and flavonoids), and enriched soil organic matter, total N, available P, pH, and cation-exchange capacity over control treatments (Allouzi *et al.*, 2022).

A comprehensive review notes that liquid biofertilizers offer eco-friendly, low-input options that can reduce chemical fertilizer dependency; however, challenges like shorter shelf life, susceptibility to climate and soil variability, and the need for optimized additives and application methods remain (Allouzi *et al.*, 2022). Thus, while liquid formulations demonstrate strong efficacy in enhancing crop performance and soil fertility across various crops (sorghum, millet, pea, tomato, cabbage), the consistency of results depends on formulation quality, additives, storage conditions, and integration with nutrient management practices.

6.10 Future Perspectives and Conclusion

The sustainability, efficiency, and compatibility of liquid fertilizers with contemporary farming practices bode well for the future of these products, especially when it comes to agricultural waste. They lessen their influence on the environment and dependency on synthetic fertilizer by turning waste into useful nutrients and promoting circular economies. The focused delivery of liquid fertilizer using techniques like hydroponics and fertigation helps precision agriculture, which maximizes input uses. Liquid fertilizers are being included in sustainable farming methods in areas such as the US and Europe, where they are increasing crop yields and soil health. Developing nations such as Ghana and others are investigating its application to convert local agricultural waste into reasonably priced fertilizers, boosting local economies and lowering reliance on imports. The scalability of liquid fertilizers will be improved by technological advancements in nutrients extraction and microbial treatments, which will be crucial in global agricultural systems going forward, particularly in addressing water constraints and advancing regenerative agriculture.

In conclusion, the production of liquid fertilizer from agricultural waste offers a practical and environmentally responsible solution to two pressing challenges: the excessive use of chemical fertilizers and the disposal of organic farm waste. This study demonstrated that through simple fermentation processes, common agricultural by-products can be converted into nutrient-rich liquid fertilizers that enhance soil health and crop productivity. Moreover, the use of such organic inputs contributes to sustainable farming practices by reducing greenhouse gas emissions, lowering production costs, and minimizing soil degradation. Encouraging wider adoption of this method can lead to significant environmental and economic benefits, especially in rural and resource-limited settings. Further research on scaling up the process and optimizing nutrient profiles can improve its effectiveness and adaptability across various agroecological zones.

References

Ademe, A.A.A., Hirpa, G.H., and Biratu, W.B. (2019). Waste management and utilization in coffee industry: a review. *International Journal of Agriculture and Biosciences* 8(5), 276–281.

Ajala, E.O., Ighalo, J.O., Ajala, M.A., Adeniyi, A.G., and Ayanshola, A.M. (2021) Sugarcane bagasse: a biomass sufficiently applied for improving global energy, environment and economic sustainability. *Bioresources and Bioprocessing* 8(1), 87. DOI: 10.1186/s40643-021-00440-z

Allouzi, M.M.A., Allouzi, S.M.A., Keng, Z.X., Supramaniam, C.V., Singh, A. *et al.* (2022) Liquid biofertilizers as a sustainable solution for agriculture. *Heliyon* 8(12), e12609. DOI: 10.1016/j.heliyon.2022.e12609

Alloway, B.J. (2008) *Zinc in Soils and Crop Nutrition*, 2nd ed. International Zinc Association, Durham, North Carolina.
Anjali, Sharma, P., and Nagpal, S. (2019). Effect of liquid and charcoal based consortium biofertilizers amended with additives on growth and yield in chickpea (*Cicer arietinum* L.). *Legume Research* 44*(5)*, 527–538. DOI: 10.18805/LR-4131
Ashrafi, M.R., Raj, M., Shamim, S., Lal, K., and Kumar, G. (2020) Effect of fertigation on crop productivity and nutrient use efficiency. *Journal of Pharmacognosy and Phytochemistry* 9(5), 2937–2942.
Bednarek, W., Dresler, S., Tkaczyk, P., and Hanaka, A. (2012) Available forms of nutrients in soil fertilized with liquid manure and NPK. *Journal of Elementology* 17(2). DOI: 10.5601/jelem.2012.17.2.01
Bernal, M.P., Alburquerque, J.A., and Moral, R. (2009) Composting of animal manures and chemical criteria for compost maturity assessment. A review. *Bioresource Technology* 100(22), 5444–5453. DOI: 10.1016/j.biortech.2008.11.027
Bhatt, P. and Mehta, T. (2021) Artificial intelligence and machine learning in biosimilar development: a frontier in precision biologics 39(9), 891–893.
Bhattacharyya, P., Chakrabarti, K., Chakraborty, A., and Tripathy, S. (2016) Recycling of organic wastes for sustainable soil health and crop productivity. *Compost Science & Utilization* 24(4), 241–249. DOI: 10.1080/1065657X.2016.1195740
Calvo, P., Nelson, L., and Kloepper, J.W. (2014) Agricultural uses of plant biostimulants. *Plant and Soil* 383(1–2), 3–41. DOI: 10.1007/s11104-014-2131-8
Case, S.D.C., Oelofse, M., Hou, Y., Oenema, O., and Jensen, L.S. (2017) Farmer perceptions and use of organic waste products as fertilisers – a survey study of potential benefits and barriers. *Agricultural Systems* 151, 84–95. DOI: 10.1016/j.agsy.2016.11.012
Chew, K.W., Chia, S.R., Yen, H.W., Nomanbhay, S., Ho, Y.C. *et al.* (2019) Transformation of biomass waste into sustainable organic fertilizers. *Sustainability* 11(8), 2266. DOI: 10.3390/su11082266
Chojnacka, K., Moustakas, K., and Mikulewicz, M. (2022) Valorisation of agri-food waste to fertilisers is a challenge in implementing the circular economy concept in practice. *Environmental Pollution* 312, 119906. DOI: 10.1016/j.envpol.2022.119906
Dimkpa, C.O. and Bindraban, P.S. (2016) Fortification of micronutrients for efficient agronomic production: a review. *Agronomy for Sustainable Development* 36(1), 7. DOI: 10.1007/s13593-015-0346-6
Dinh, V.-P., Tran-Vu, H.-A., Tran, T., Duong, B.-N., Dang-Thi, N.-M. *et al.* (2024) Improving soil quality and crop yields using enhancing sustainable rice straw management through microbial enzyme treatments. *Environmental Health Insights* 18, 11786302241283001. DOI: 10.1177/11786302241283001
Dotaniya, M.L., Datta, S.C., Biswas, D.R., Dotaniya, C.K., Meena, B.L. *et al.* (2016) Use of sugarcane industrial by-products for improving sugarcane productivity and soil health. *International Journal of Recycling of Organic Waste in Agriculture* 5(3), 185–194. DOI: 10.1007/s40093-016-0132-8
Durán-Lara, E.F., Valderrama, A., and Marican, A. (2020) Natural organic compounds for application in organic farming. *Agriculture* 10(2), 41. DOI: 10.3390/agriculture10020041
El-Kazzaz, K.A. and El-Kazzaz, A.A. (2017) Soilless agriculture a new and advanced method for agriculture development: an introduction. *Agricultural Research and Technology Open Access Journal* 3, 63–72. DOI: 10.19080/ARTOAJ.2017.03.555610
El-Ramady, H.R., Alshaal, T.A., Shehata, S.A., Domokos-Szabolcsy, É., Elhawat, N. *et al.* (2014) Plant nutrition: from liquid medium to micro-farm. *Sustainable Agriculture Reviews* 14, 449–508. DOI: 10.1007/978-3-319-06016-3_12
European Commission (2018) *A Sustainable Bioeconomy for Europe: Strengthening the Connection between Economy, Society and the Environment*. European Commission, Brussels.
Fageria, N.K. (2009) *The Use of Nutrients in Crop Plants*. CRC Press, Boca Raton, Florida.
Fernández, V. and Ebert, G. (2005) Foliar iron fertilization: a critical review. *Journal of Plant Nutrition* 28(12), 2113–2124. DOI: 10.1080/01904160500320954
Fernández-Delgado, M., Del Amo-Mateos, E., Lucas, S., García-Cubero, M.T., and Coca, M. (2022) Liquid fertilizer production from organic waste by conventional and microwave-assisted extraction technologies: techno-economic and environmental assessment. *The Science of the Total Environment* 806(Pt 4), 150904. DOI: 10.1016/j.scitotenv.2021.150904
Gatabazi, A., Botha, M., and Mvondo-She, M.A. (2023) Assessing liquid inoculant formulation of biofertilizer (*Sinorhizobium meliloti*) on growth, yield, and nitrogen uptake of lucerne (*Medicago sativa*). *Nitrogen* 4(1), 125–134. DOI: 10.3390/nitrogen4010009
Gopinath, K.A., Supradip, S., and Mina, B.L. (2008) Influence of organic amendments on growth, yield and soil health in organic farming. *Agricultural Research* 42(1), 25–34.

Grand View Research (2022) Liquid fertilizers market size. Available at: https://www.grandviewresearch.com/industry-analysis/liquid-fertilizers-market-report (accessed September 6, 2025).

Havlin, J.L., Tisdale, S.L., Nelson, W.L., and Beaton, J.D. (2014) *Soil Fertility and Fertilizers*, 8th ed. Pearson, London.

Hazra, K.K. and Santosa, D.A. (2024) Combined use of chemical fertilizers and biofertilizers in rice (*Oryza sativa* L.): yield and economic evaluation. *Agricultural Research Journal* 61(2), 105–112.

Hettiarachchi, G.M., Pierzynski, G.M., and Ransom, M.D. (2010) Nutrient management in soils of the tropics. In: Aulakh, M.S. and Grant, C.A. (eds) *Nutrient Management for Sustainable Crop Production*. CRC Press, Boca Raton, Florida, pp. 85–98.

Horta, C., Riaño, B., Anjos, O., and García-González, M.C. (2022) Fertiliser effect of ammonia recovered from anaerobically digested orange peel using gas-permeable membranes. *Sustainability* 14(13), 7832. DOI: 10.3390/su14137832

Indrawati, S., Anggoro, D., Sukamto, H., Puspitasari, N., Indarto, B. *et al.* (2020) The effectiveness of the addition of EM4 and molasses in increasing levels of N, P and K in environmentally friendly liquid fertilizers made from banana pseudostem. In: De Oliveira, A.M., Susanto, H., Madlazim, M., Yundra, E., Muslim, S. *et al.* (eds), *International Joint Conference on Science and Engineering (IJCSE 2020)*, Atlantis Press, Surabaya, East Java, Indonesia, pp. 365–369. DOI: 10.2991/aer.k.201124.066

Kosseva, M.R. (2013). Sources, characterization, and composition of food industry wastes. In: Kosseva, M.R. and Webb, C. (eds) *Food Industry Wastes*. Amsterdam, Elsevier, 37–88.

Koul, B., Yakoob, M., and Shah, M.P. (2022) Agricultural waste management strategies for environmental sustainability. *Environmental Research* 206, 112285. DOI: 10.1016/j.envres.2021.112285

Kovačić, Đ., Lončarić, Z., Jović, J., Samac, D., Popović, B, *et al.* (2022) Digestate management and processing practices: a review. *Applied Sciences* 12(18), 9216. DOI: 10.3390/app12189216

Kumari, A., Singh, B., and Prasad, R. (2024a) Effect of liquid biofertilizer with or without vermicompost on growth and yield of field pea (*Pisum sativum* L.). *Journal of Scientific Research and Reports* 30(9), 62–68. DOI: 10.9734/jsrr/2024/v30i92330

Kumari, S., Banerjee, M., Malik, G.C., and Duvvada, S.K. (2024b) Effect of liquid biofertilizer with or without vermicompost on growth and yield of field pea (*Pisum sativum*) grown under laterite soil of West Bengal. *Journal of Scientific Research and Reports* 30(9), 62–68. DOI: 10.9734/jsrr/2024/v30i92330

Lakshmikala, K., Babu, B.R., Babu, M.R., and Devi, P.R. (2022) Studies on efficacy of liquid and carrier based biofertilizers on residual soil and plant NPK uptake and microbial count in tomato (*Solanum lycopersicum* L.). *Journal of Eco-Friendly Agriculture* 17(2), 211–218. DOI: 10.5958/2582-2683.2022.00043.0

Majee, S., Halder, G., Mandal, D.D., Tiwari, O.N., and Mandal, T. (2021) Transforming wet blue leather and potato peel into an eco-friendly bio-organic NPK fertilizer for intensifying crop productivity and retrieving value-added recyclable chromium salts. *Journal of Hazardous Materials* 411, 125046. DOI: 10.1016/j.jhazmat.2021.125046

Malhotra, S.K. (2016) Water soluble fertilizers in horticultural crops—an appraisal. *The Indian Journal of Agricultural Sciences* 86(10), 1245–1256. DOI: 10.56093/ijas.v86i10.62095

Meaclem, T.F. (2019) Investigation into the production and release behaviour of controlled release fertilisers. Doctoral thesis, University of Canterbury, UK.

Mekki, A., Dhouib, A., Aloui, F., and Sayadi, S. (2006) Olive wastewater as an ecological fertiliser. *Agronomy for Sustainable Development* 26(1), 61–67. DOI: 10.1051/agro:2005061

Mikkelsen, R. (2018) Micronutrient Fertilizers: Types, Sources, and Application Methods. In: International Plant Nutrition Institute (ed.) *IPNI Plant Nutrition Manual*. International Plant Nutrition Institute, Peachtree Corners, Georgia, pp. 20–30.

Molla, A.A., Singh, V., Singh, A.C., and George, S.G. (2024) Effect of biofertilizer and liquid organic nutrient sources on growth and yield of black gram (*Vigna mungo* L.). *International Journal of Agriculture Sciences* 7(5), 129–132. DOI 10.33545/2618060X.2024.v7.i5b.660

Niu, J., Liu, C., Huang, M., Liu, K., and Yan, D. (2021a) Effects of foliar fertilization: a review of current status and future perspectives. *Journal of Soil Science and Plant Nutrition* 21(1), 104–118. DOI: 10.1007/s42729-020-00346-3

Niu, X., Song, L., Xiao, Y., and Ge, D. (2021b) Review on foliar fertilization: mechanism, technology, and application. *Agronomy* 11(8), 1604. DOI: 10.3390/agronomy11081604

Osman, K.T. (2013) Plant nutrients and soil fertility management. In: Osman, K.T. (ed.) *Soils: Principles, Properties and Management*. Springer, Dordrecht, Netherlands, pp. 129–159.

Pant, A.P., Radovich, T.J., Hue, N.V., Talcott, S.T., and Krenek, K.A. (2012) Vermicompost extracts influence growth, mineral nutrients, phytonutrients, and antioxidant activity in pak choi. *Journal of the Science of Food and Agriculture* 92(13), 2557–2564. DOI: 10.1002/jsfa.5655

Prado, J., Ribeiro, H., Alvarenga, P., and Fangueiro, D. (2022) A step towards the production of manure-based fertilizers: disclosing the effects of animal species and slurry treatment on their nutrients content and availability. *Journal of Cleaner Production* 337, 130369. DOI: 10.1016/j.jclepro.2022.130369

Reig, M., Vecino, X., Gibert, O., Valderrama, C., and Cortina, J.L. (2021) Study of the operational parameters in the hollow fibre liquid-liquid membrane contactors process for ammonia valorisation as liquid fertiliser. *Separation and Purification Technology* 255, 117768. DOI: 10.1016/j.seppur.2020.117768

Richa (2023) Liquid bio-fertilizers: prospects and challenges. In: Kaur, S., Dwibedi, V., Sahu, P.K., and Singh Kocher, G. (eds) *Metabolomics, Proteomes and Gene Editing Approaches in Biofertilizer Industry*. Springer, Singapore, pp. 77–99.

Rietra, R.P.J.J., Heinen, M., Dimkpa, C.O., and Bindraban, P.S. (2017) Effects of nutrient antagonism and synergism on fertilizer use efficiency. *Frontiers in Plant Science* 8, 444. DOI: 10.1080/00103624.2017.1407429

Rodríguez, H. and López, M.A. (2020) Formulation strategies of fertilizers: efficiency and safety concerns. In: *Fertilizer Technology and Use*. Elsevier, Amsterdam, pp. 31–54. DOI: 10.1016/B978-0-12-819648-5.00002-7

Sanadi, N.F.A., Lee, C.T., Sarmidi, M.R., Klemeš, J.J., and Zhang, Z. (2019) Characterisation of liquid fertiliser from different types of bio-waste compost and its correlation with the compost nutrients. *CET Journal-Chemical Engineering Transactions* 72, 253–258. DOI: 10.3303/CET1972043

Savci, S. (2012) An agricultural pollutant: chemical fertilizer. *International Journal of Environmental Science and Development* 3(1), 73–80. DOI: 10.7763/IJESD.2012.V3.191

Savvas, D., Giannothanasis, E., and Ntatsi, G. (2024) Precise fertilization in greenhouse crops using modern decision support systems to save fertilisers and mitigate pollution of water resources. *Acta Horticulturae* 1391, 723–734. DOI: 10.17660/ActaHortic.2024.1391.97

SCRiP Research Team (2022) Organic liquid biofertilizer boosts corn (*Zea mays*) growth under reduced chemical input. *Sustainable Crop Research and Innovation Proceedings* 2(1), 44–52.

Shaibur, M.R., Husain, H., and Arpon, S.H. (2021) Utilization of cow dung residues of biogas plant for sustainable development of a rural community. *Current Research in Environmental Sustainability* 3, 100026. DOI: 10.1016/j.crsust.2021.100026

Shaji, H., Chandran, V., and Mathew, L. (2021) Organic fertilizers as a route to controlled release of nutrients. In: Lewu, F.B., Volova, T., Thomas, S., and Rakhimol, K.R.: *Controlled Release Fertilizers for Sustainable Agriculture*. Academic Press, London, pp. 231–245.

Shao, Z., Bu, X., Gou, Q., Guo, X., Zhou, L., Wang, J., and Li, Y. (2023) Effects of organic carbon fertilizer from waste alcohol mash on root growth, yield and quality of cherry tomatoes. *Journal of Henan Agricultural Sciences,* 52(3), 109–117. DOI: 10.15933/j.cnki.1004-3268.2023.03.012

Sharma, B., Vaish, B., Monika, Singh, U.K., Singh, P. *et al.* (2019) Recycling of organic wastes in agriculture: an environmental perspective. *International Journal of Environmental Research* 13(2), 409–429. DOI: 10.1007/s41742-019-00175-y

Singh, V.K. (2021) Environmental and health consequences of distillery wastewater and ways to tackle: a review. *International Journal for Research in Applied Sciences and Biotechnology* 8(6), 127–135.

Singh, V.K., Malhi, G.S., Kaur, M., Singh, G., and Jatav, H.S. (2022) Use of organic soil amendments for improving soil ecosystem health and crop productivity. In: Singh Jatav, H. and Rajput, V.D. (eds) *Ecosystem Services*, Nova Science Publishers, New York, Vol. 12. p. 45.

Timilsena, Y.P., Adhikari, R., Casey, P., Muster, T., Gill, H. *et al.* (2015) Enhanced efficiency fertilisers: a review of formulation and nutrient release patterns. *Journal of the Science of Food and Agriculture* 95(6), 1131–1142. DOI: 10.1002/jsfa.6812

Unnisa, S.A. (2015) Liquid fertilizer from food waste: a sustainable approach. *International Research Journal of Environment Sciences* 4(8), 22–25.

Vamsi, Y., Tirumala Reddy, S., Raveendra Reddy, M., and Prathima, T. (2023) Performance of liquid biofertilizers in enhancing productivity, nutrient availability and uptake by fodder sorghum. *Journal of Research ANGRAU* 51(4), 81–87. DOI: 10.58537/jorangrau.2023.51.4.10

Verma, L. and Kumar, V. (2024) Use of biofertilizers in modern agriculture. *Agriculture and Biological Research* 40(4), 1191–1193. DOI: 10.35248/0970-1907.24.40.1191-1193

White, P.J. and Broadley, M.R. (2009) Biofortification of crops with seven mineral elements often lacking in human diets—Iron, zinc, copper, calcium, magnesium, selenium and iodine. *The New Phytologist* 182(1), 49–84. DOI: 10.1111/j.1469-8137.2008.02738.x

Yadav, B., Bharati, V., Yadav, T.K., Kumar, A., Kumar, A. *et al.* (2025a) Impact of liquid biofertilizers on nutrition of finger millet and residual soil fertility status. *Journal of Experimental Agriculture International* 47(4), 314–322. DOI: 10.9734/jeai/2025/v47i43380

Yadav, B., Bharati, V., Yadav, T.K., Kumar, A., Kumar, A. *et al.* (2025b) Effect of liquid biofertilizer practices on the growth and yield of finger millet (*Eleusine coracana* L.) *International Journal of Agriculture Sciences* 8(3B), 132–136. DOI: 10.33545/2618060X.2025.v8.i3b.2605

Yadav, S.K. and Meena, R.H. (2024) Response of black gram to liquid biofertilizers under rainfed conditions. *International Journal of Current Microbiology and Applied Sciences* 13(2), 119–126.

Zhang, X., Davidson, E.A., Mauzerall, D.L., Searchinger, T.D., Dumas, P. *et al.* (2015) Managing nitrogen for sustainable development. *Nature* 528(7580), 51–59. DOI: 10.1038/nature15743

7 Applications and Limitations of Biofertilizers

Anirudh Sharma*
Department of Biotechnology, Jaypee Institute of Information Technology, Noida, Uttar Pradesh, India

Abstract
This chapter explores the critical role of microbial biofertilizers as sustainable alternatives to chemical fertilizers in modern agriculture. With the growing concerns over soil degradation, nutrient runoff, water contamination, and ecological imbalance caused by excessive synthetic fertilizer use, biofertilizers offer eco-friendly solutions to enhance crop productivity while preserving soil health. The chapter details the diverse categories of biofertilizers, including nitrogen-fixing bacteria (e.g. *Rhizobium*, *Azospirillum*), phosphate- and potassium-solubilizing microorganisms, arbuscular mycorrhizal fungi, and siderophore producers. These microbes improve nutrient availability through biological nitrogen fixation, solubilization of phosphorus and potassium, mobilization of iron, phytohormone production, and enhanced rhizospheric interactions. Their applications extend beyond nutrient efficiency to fostering soil biodiversity, structure, and long-term fertility. Despite these advantages, widespread adoption of biofertilizers faces biological, technical, and socio-economic limitations. Field performance often varies depending on soil type, climatic conditions, crop species, and farming practices. Other challenges include limited shelf life, inadequate storage infrastructure, inconsistent quality standards, and low farmer awareness. The absence of robust regulatory frameworks further hampers their commercialization and scalability. The chapter concludes by highlighting recent innovations aimed at overcoming these barriers, such as improved microbial formulations, consortia-based inoculants, and integrated nutrient management strategies. It emphasizes that policy support, farmer training, and continued research into microbial mechanisms are essential for maximizing their potential. Ultimately, biofertilizers represent a vital tool in transitioning toward sustainable, resilient, and environmentally conscious agricultural systems capable of meeting global food security challenges.

7.1 Introduction

For decades, modern agriculture has heavily depended on chemical fertilizers to enhance crop productivity and meet the nutritional demands of a rapidly expanding global population. Although these synthetic inputs have played a crucial role in boosting food production, their overuse has led to significant ecological and environmental issues. Concerns such as soil quality deterioration, nutrient runoff, contamination of water bodies, and disruption of indigenous microbial communities have highlighted the pressing need for more sustainable nutrient management approaches (Mahanty *et al.*, 2017). In response to these challenges, biofertilizers—microbial

*Corresponding author: anirudhsharma172@gmail.com

DOI: 10.1079/9781836991021.0007

formulations that enhance plant growth by increasing nutrient availability—are emerging as eco-friendly alternatives for crop improvement (Vessey 2003; Afzal and Bano, 2008). Biofertilizers operate through multiple biological mechanisms, including atmospheric nitrogen fixation, solubilization of phosphorus and potassium, synthesis of phytohormones, and facilitation of nutrient uptake by plants. Commonly utilized microorganisms include nitrogen-fixing bacteria such as *Rhizobium*, *Azospirillum*, and *Azotobacter*; phosphate-solubilizing bacteria; potassium-mobilizing strains; and beneficial fungi like mycorrhizae (Mazid and Khan, 2014). Beyond improving nutrient efficiency, these microbes contribute to better soil texture, increased microbial biodiversity, and enhanced long-term soil fertility, positioning biofertilizers as a key component of sustainable and organic farming practices (Pandey and Singh, 2012; Mazid and Khan, 2014).

However, despite their ecological and agronomic potential, the widespread adoption of biofertilizers faces several hurdles. Their effectiveness under field conditions often varies, depending on factors such as soil properties, climate, crop type, and agricultural practices. Moreover, limitations such as the short shelf life of microbial products, inadequate cold-chain or storage facilities, and limited farmer awareness further restrict their usage. In addition, the lack of uniform regulatory guidelines in many regions complicates quality control and hinders access to reliable biofertilizer products.

This chapter aims to offer a detailed exploration of the role and limitations of biofertilizers in current agricultural systems. It begins by outlining the different categories of biofertilizers and their functional attributes across diverse agroecological settings. The discussion then shifts to examining major constraints—biological, technical, and socio-economic—that affect their field performance and scalability. The chapter concludes with insights into recent innovations, possible solutions to overcome existing challenges, and the future prospects of biofertilizers in the context of sustainable agricultural transformation.

7.2 Biofertilizers

The continuous application of conventional chemical fertilizers has been shown to degrade soil health and fertility over time. As a result, there is an increasing need to explore sustainable alternatives that can fulfill the nutrient demands of crops while reducing environmental harm.

Among these alternatives, microorganisms and their associative or symbiotic interactions with plants are gaining attention for their role in developing effective biofertilizers. These microbial bio-inputs are generally classified based on their physiological properties and their specific contributions to plant growth promotion.

1. **Nitrogen-fixing microorganisms**: Nitrogen is a vital nutrient for all life forms, as it constitutes a major part of nucleic acids, proteins, and other biologically important molecules. Although nitrogen is abundant in the atmosphere, its concentration in soil is relatively low and its biologically available form often limits plant productivity in both natural ecosystems and agricultural settings. This limitation is primarily due to the complex process required to convert atmospheric nitrogen into a usable form. Certain microorganisms possess the ability to fix atmospheric nitrogen either freely in the soil or through symbiotic and associative relationships with plants. These nitrogen-fixing microbes enhance nitrogen availability in the rhizosphere, making them promising agents for use as biofertilizers due to their potential to contribute significantly to soil nitrogen content.
2. **Phosphate-solubilizing microorganisms**: While soils generally contain large reserves of phosphorus, most of it exists in insoluble forms that plants cannot directly utilize. This phosphorus may occur as inorganic minerals such as apatite or in various organic forms, including inositol phosphate (phytate), phosphomonoesters, and phosphotriesters (Hamid and Ahmad, 2012). Phosphate-solubilizing bacteria (PSB), such as *Bacillus* and *Pseudomonas*, play a key role in converting these unavailable forms into plant-accessible phosphorus. They achieve this by secreting organic acids that acidify their surroundings, thereby releasing bound phosphorus through chelation and solubilization processes.

7.3 Nitrogen-fixing Biofertilizers

Nitrogen is a fundamental element required for the survival of all living organisms, serving as a key component in the synthesis of proteins, nucleic acids, and various other nitrogen-containing organic compounds (Barker and Bryson, 2016). However, plants lack the enzymatic machinery necessary to convert atmospheric dinitrogen (N_2) into ammonia, which is the bioavailable form required for their growth. This vital transformation is carried out exclusively by certain prokaryotic microorganisms, including specific bacteria and cyanobacteria, that possess the natural ability to fix atmospheric nitrogen.

Organisms capable of nitrogen fixation, often referred to as "nitrogen fixers" or "N-fixing microorganisms," are commonly used in the development of biofertilizers. These biofertilizers consist of living microbial inoculants that can convert atmospheric nitrogen into forms usable by plants. Nitrogen-fixing microbes are generally classified into free-living forms—such as *Azotobacter* and *Azospirillum*—blue-green algae (cyanobacteria), and symbiotic types like *Rhizobium*, *Frankia*, and *Azolla*.

The diversity of nitrogen-fixing bacteria associated with non-leguminous plants is quite broad and includes genera such as *Acetobacter*, *Achromobacter*, *Alcaligenes*, *Arthrobacter*, *Azomonas*, *Bacillus*, *Beijerinckia*, *Campylobacter*, *Clostridium*, *Corynebacterium*, *Derxia*, *Desulfovibrio*, *Enterobacter*, *Erwinia*, *Herbaspirillum*, *Klebsiella*, *Lignobacter*, *Methylosinus*, *Mycobacterium*, *Rhodopseudomonas*, *Rhodospirillum*, and *Xanthobacter*.

Nitrogen-fixing prokaryotes are capable of forming a wide range of beneficial associations with plants, from loose interactions in the rhizosphere to more intimate intercellular symbioses. Although both the plant host and the microbial partner can exist independently, their interaction tends to be mutually advantageous. Studies have shown that up to 25% of a plant's total nitrogen content can be derived through biological nitrogen fixation. The efficiency of this process is strongly influenced by the availability of carbon sources and a low concentration of combined nitrogen in the surrounding environment (Fig. 7.1).

Nitrogen exists in various chemical forms throughout nature. To harness nitrogen from the atmosphere, the highly stable triple covalent bond between two nitrogen atoms must be broken to yield ammonia (NH_3) or nitrate (NO_3^-). This transformation, known as nitrogen fixation, can occur through both industrial methods and naturally

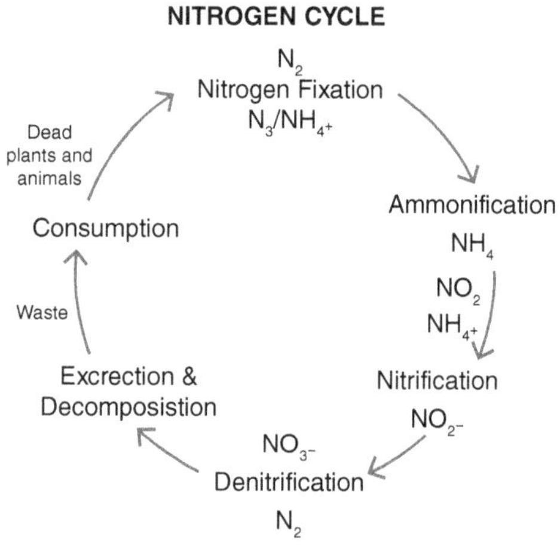

Fig. 7.1. Nitrogen fixation mechanism. Figure author's own.

occurring processes. Among the natural pathways, biological nitrogen fixation is particularly significant—it relies on specific microorganisms capable of converting atmospheric nitrogen into usable forms, and these microbes are collectively referred to as nitrogen-fixing biofertilizers.

This process is reductive and biosynthetic in nature. Only certain prokaryotic organisms possess the enzymatic machinery necessary for biological nitrogen fixation, whereas eukaryotes lack this capability. During the biochemical reaction, at least one mole of hydrogen gas (H_2) and two moles of ammonia (NH_3) are produced per mole of nitrogen (N_2) fixed. As a result, the complete reduction of a single nitrogen molecule requires the input of eight electrons:

$$N_2 + 8H^+ + 8e^- \rightarrow 2NH_3 + H_2$$

However, the actual reduction of a nitrogen molecule (N_2) into ammonia, without the concurrent formation of molecular hydrogen, necessitates only six electrons. As nitrogen fixation is fundamentally a reductive biochemical process, it requires a suitable electron donor. In most nitrogen-fixing microorganisms, reduced ferredoxin typically serves as the primary electron donor. This reaction is energetically demanding due to the presence of a strong triple covalent bond between the two nitrogen atoms in N_2, rendering the molecule extremely stable and unreactive. Consequently, a considerable amount of energy is needed to cleave this bond and to incorporate three hydrogen atoms into each nitrogen atom. The overall process requires a minimum of 16 ATP (adenosine triphosphate) molecules to reduce a single N_2 molecule. The enzymatic transformation of atmospheric nitrogen into ammonia is catalyzed by nitrogenase, a highly conserved, oxygen-sensitive enzyme complex found in both free-living and symbiotic diazotrophic organisms. The most widely studied and common variant is Mo-nitrogenase (also known as conventional nitrogenase), which incorporates a molybdenum-containing prosthetic group known as FeMoCo (iron-molybdenum cofactor). Certain microorganisms, including *Azotobacter* and various photosynthetic nitrogen fixers (such as cyanobacteria), possess alternative forms of nitrogenase. These include V-nitrogenase, which utilizes vanadium, and Fe-nitrogenase, which contains only iron (Fig 7.2).

Purified nitrogenase consists of two metalloprotein components. The first, referred to as the MoFe protein (Component I), is a tetrameric complex with a molecular mass of approximately 220 kDa, composed of two distinct α and β subunits. The second component, known as the Fe protein (Component II), is a homodimer of 68 kDa. Each MoFe protein harbors two FeMoCo cofactors associated with the α subunits. Additionally, it contains two types of prosthetic groups: P-clusters, which are 4Fe–4S clusters covalently bound to cysteine residues that bridge the α and β subunits, and a third type of iron-sulfur (Fe–S) cluster found within the Fe protein itself.

Nitrogen fixation involves two key metalloproteins: a larger molybdenum-iron (Mo-Fe) protein and a smaller iron (Fe) protein. The Fe protein binds ATP and Mg^{2+} and accepts electrons from oxidized ferredoxin or flavodoxin. These electrons are then transferred to the Mo-Fe protein, which interacts with the reducible substrate, nitrogen (N_2), ultimately producing two molecules of ammonia (NH_3) (Fig. 7.3).

The nitrogenase enzyme complex is highly sensitive to oxygen and becomes irreversibly inactivated upon exposure. As a result, nitrogen fixation must occur under anaerobic conditions. This poses no issue for obligate anaerobic prokaryotes, while facultative anaerobes carry out nitrogen fixation only when oxygen is absent. In aerobic nitrogen-fixing organisms, such as cyanobacteria, specialized cells known as heterocysts provide a micro-anaerobic environment necessary for nitrogen fixation. These cells lack photosystem II—the component responsible for oxygen evolution during photosynthesis—thereby preventing internal oxygen generation. The development of heterocysts is considered an

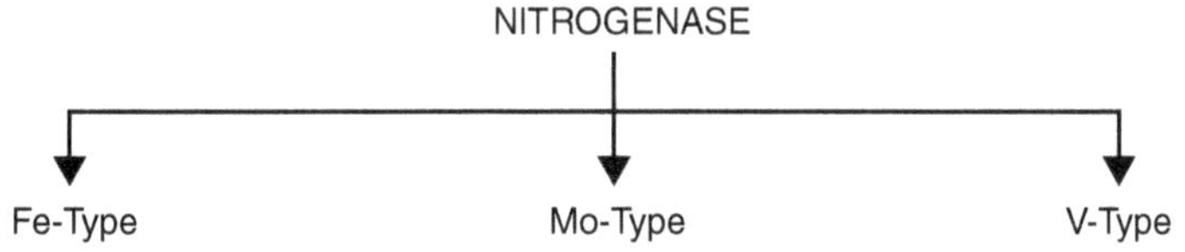

Fig 7.2. Alternative forms of nitrogenase. Figure author's own.

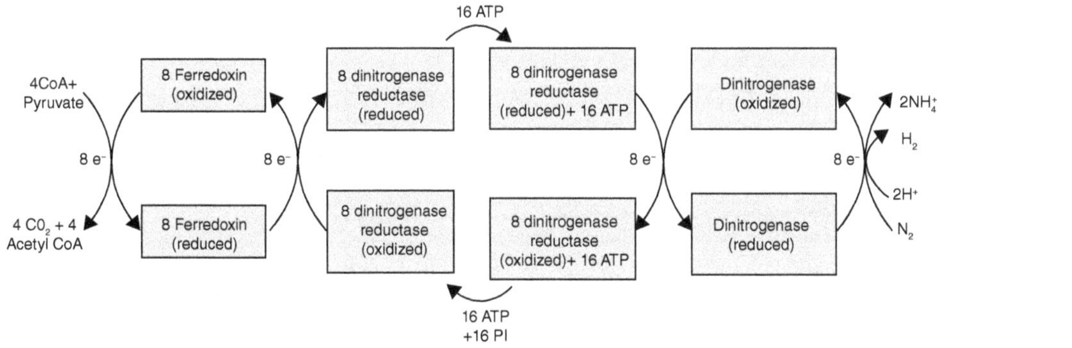

Fig 7.3. Cycle of nitrogenase enzyme. Figure author's own.

evolutionary adaptation that enables nitrogen fixation in oxygen-rich environments.

7.4 Phosphate-solubilizing Biofertilizers

Phosphorus is a key macronutrient that often limits plant growth and productivity. It constitutes approximately 0.2–0.8% of the plant's dry weight and is an essential component of nucleic acids, enzymes, coenzymes, nucleotides, and phospholipids. In plant systems, phosphorus is irreplaceable, and unlike nitrogen, it cannot be synthesized artificially. Plants absorb it primarily as orthophosphate ions. Phosphorus plays a crucial role at various biological scales—right from molecular functions to broader physiological and biochemical processes. These include, but are not limited to, photosynthesis, root system development, stem and stalk strengthening, flowering and seed formation, crop maturity and overall quality, intracellular energy transfer, cell division and elongation, nitrogen fixation in leguminous crops, conversion of sugars to starch, transmission of genetic information, and resistance against plant pathogens. Additionally, during the early stages of plant development, sufficient phosphorus levels are vital for initiating the primordia of reproductive structures.

Following nitrogen, phosphorus ranks as the second most critical nutrient for plant growth. Despite its importance, phosphorus availability in soil is typically low due to its strong tendency to become fixed in forms that plants cannot absorb. This occurs through the formation of insoluble complexes with metals such as iron, aluminum, and calcium. Although soils generally contain significant quantities of phosphorus, most of it exists in forms bound to soil minerals and particles, limiting its bioavailability. Even when phosphorus fertilizers are applied to soils with low native phosphorus levels, their retention is poor. A considerable portion—approximately 75–90%—of the added inorganic phosphorus fertilizers becomes rapidly immobilized by metal cations and thus unavailable to plants. This not only results in poor phosphorus use efficiency but also contributes to long-term environmental issues, including reduced soil fertility, eutrophication of water bodies, and increased carbon emissions.

A diverse range of soil microorganisms, particularly phosphate-solubilizing microorganisms (PSM), are known to participate actively in phosphorus transformation. Some of the most studied examples include species from bacterial genera such as *Pseudomonas*, *Bacillus*, *Micrococcus*, *Flavobacterium*, and fungal genera like *Aspergillus*, *Penicillium*, *Fusarium*, and *Sclerotium*. Among these, *Pseudomonas*, *Bacillus*, *Rhizobium*, and *Enterobacter*, along with *Penicillium* and *Aspergillus*, have shown high efficiency in solubilizing phosphorus. Notable strains include *Bacillus megaterium*, *B. circulans*, *B. subtilis*, *B. polymyxa*, *B. sircalmous*, *Pseudomonas striata*, and *Enterobacter* spp. Bacilli are frequently present in soils, whereas spirilla are relatively rare in natural settings. PSB are widespread and exhibit diversity in population and form across different soil types. Their abundance and activity are influenced by multiple factors, including soil physicochemical properties, organic matter content, phosphorus levels, and agronomic practices. Higher populations of PSB are commonly reported in cultivated and pasture lands.

Microorganisms are central to the natural phosphorus cycling process. The potential of PSM to serve as biofertilizers has been widely investigated over the past several decades. These microbes can mobilize phosphate from insoluble sources, thereby improving phosphorus availability to crops. Similarly, microbes capable of degrading complex carbohydrates such as starch and cellulose also contribute to enhanced soil fertility. The application of PSB as biofertilizers dates back to the 1950s, with records of naturally occurring rhizospheric phosphate solubilizers traced to as early as 1903. Bacteria generally outperform fungi in terms of phosphate solubilization capacity. Within the total microbial community in soil, PSB account for about 1–50%, while phosphate-solubilizing fungi (PSF) contribute only around 0.1–0.5%. Some bacterial strains isolated from alkaline soils have shown the ability to solubilize phosphate under extreme conditions such as high salinity, elevated pH, and increased temperature. Furthermore, several PSB have been recognized as plant growth-promoting rhizobacteria (PGPR) in crops like rice and tomato. These microorganisms are distributed widely across various soil types, although the composition of PSM communities differs significantly based on environmental and edaphic conditions. Most PSM are commonly isolated from plant rhizospheres, where they exhibit active metabolic behavior.

Microorganisms also contribute to phosphorus immobilization processes. After uptake, phosphorus is assimilated into microbial cellular structures, including nucleic acids, coenzymes, free phosphate ions, and phosphorus esters. Surplus phosphorus within microbial cells is frequently stored in the form of polyphosphates. These microbial biomass pools act as temporary phosphorus reservoirs and play a significant role in long-term nutrient cycling. Upon microbial turnover, stored phosphorus can be mineralized and gradually released into the soil, making it accessible to plants. In most microorganisms, inorganic polyphosphates serve as key phosphorus reserves. These are linear chains composed of multiple phosphate units and can range from a few to several hundred phosphate residues. Many bacterial species are known to accumulate polyphosphates under stress conditions. The biosynthesis of these compounds is catalyzed by specific enzymes known as polyphosphate kinases (polyP kinases). The functional significance of polyphosphates as energy-rich phosphate reservoirs has been demonstrated across various microbial taxa, from archaea to fungi. Due to their high-energy anhydride bonds, polyphosphates can serve as both phosphate donors and alternative energy sources through the release of inorganic phosphate (Pi).

Microbial biomass serves as a significant phosphorus sink within the soil matrix, limiting immediate plant uptake but ensuring long-term nutrient availability. As microbes undergo turnover during biogeochemical cycles, phosphorus locked within their cells is gradually released and made available to plants. This has been substantiated by studies showing a direct correlation between microbial phosphorus biomass and phosphorus uptake by crops.

7.4.1 Inorganic phosphate solubilization mechanism

Phosphate solubilizing-biofertilizers are believed to enhance the availability of phosphorus by releasing low molecular weight organic acids. These acids play a crucial role in dissolving otherwise insoluble phosphate compounds. In alkaline soils, phosphorus often becomes immobilized through the formation of calcium phosphate complexes, which are less soluble at higher pH levels. By secreting organic acids, these biofertilizer help lower the pH of the surrounding soil environment, thereby improving the solubility of these phosphate forms. As soil pH decreases, the balance shifts toward more soluble phosphorus species, while at higher pH levels, less bioavailable forms such as HPO_4^{2-} and PO_4^{3-} tend to dominate (Timofeeva *et al.*, 2022).

7.4.2 Organic acid production

Organic acids are generated within the periplasmic space of microbial cells through direct oxidative pathways. Their subsequent release into the external environment is often associated with a drop in pH levels (De Freitas *et al.*, 1997; Rashid *et al.*, 2004; Gulati *et al.*, 2010). Interestingly, studies have shown that this acidification does not necessarily correlate with the extent of phosphorus solubilization. One proposed explanation for this phenomenon involves the release of protons (H^+), which is linked to the uptake of cations by microbial cells. For instance, during the assimilation of ammonium (NH_4^+), the simultaneous excretion of protons may facilitate the solubilization of phosphate (Illmer and Schinner, 1995; Collavino *et al.*, 2010).

In addition to this, another potential mechanism for mineral phosphate solubilization involves the extrusion of protons across the cell membrane—either through cation exchange or via ATPase-mediated H^+ transport (Rodríguez and Fraga, 1999). Notably, ammonium assimilation alone has also been reported to trigger proton release, leading to phosphate solubilization even in the absence of organic acid production.

The effectiveness of phosphate solubilization largely depends on both the type and strength of the organic acids involved. Tri- and dicarboxylic acids tend to be more efficient than monobasic or aromatic acids. Moreover, aliphatic acids generally exhibit a higher solubilization capacity compared to phenolic compounds, as well as citric and fumaric acids. A variety of organic acids have been implicated in phosphate solubilization, including but not limited to citric, lactic, glycolic, oxalic, gluconic, 2-ketogluconic, acetic, malic, succinic, tartaric, malonic, and propionic acids (Kalayu, 2019). Among these, gluconic acid and 2-ketogluconic acid are particularly prominent for their consistent and potent ability to solubilize mineral-bound phosphates.

7.4.3 Inorganic acid production

Phosphate solubilization has also been attributed to the action of inorganic acids such as sulfuric, nitric, and carbonic acids. Nonetheless, these acids generally appear to be less efficient in releasing orthophosphate when compared to the solubilization potential of organic acids. Inorganic acids are typically generated by sulfur-oxidizing and nitrifying microorganisms during the biochemical oxidation of inorganic sulfur or nitrogen-containing compounds (Vazquez *et al.*, 2000). Once produced, these acids can react with insoluble phosphate minerals, facilitating their conversion into more soluble forms.

7.4.4 Chelation

Chelating agents also contribute significantly to the solubilization of phosphate. Organic acids such as 2-keto-gluconic acid, humic acid, and fulvic acid are recognized for their strong ability to bind with metal cations like calcium, iron, and aluminum (Sperber, 1958; Katznelson and Bose, 1959). This chelating action enhances the availability of phosphate that is otherwise locked in insoluble complexes with these metals. Humic and fulvic acids, in particular, are naturally released by microorganisms as they break down plant-derived organic matter (Stevenson and Cole, 1999).

7.4.5 Exopolysaccharide production

Exopolysaccharides (EPS) produced by microorganisms are believed to contribute indirectly to phosphate solubilization. These polysaccharides are typically synthesized as a microbial response to environmental stress. Several studies have highlighted their ability to interact with metal ions present in the soil, which can, in turn, affect the solubility of metal-associated phosphates (Ochoa-Loza *et al.*, 2001). In particular, microbial EPS have been observed to enhance the solubilization of tricalcium phosphate, often in combination with organic acid anions. Moreover, a positive relationship has been identified between the quantity of EPS secreted and the degree of phosphate solubilization (Yi *et al.*, 2008). Despite these findings, the precise role of high-molecular weight EPS in mobilizing phosphate from soil minerals remains an area requiring further research.

7.4.6 Phosphate-solubilizing biofertilizer

Phosphate-solubilizing biofertilizers consist of microorganisms that can transform insoluble phosphate compounds, such as aluminum phosphate ($AlPO_4$), iron phosphate ($FePO_4$), and tricalcium phosphate ($Ca_3(PO_4)_2$), into plant-accessible forms like HPO_4^{2-} and $H_2PO_4^-$ [21]. The involvement of rhizospheric microbes in mineral phosphate solubilization has been recognized for over a century, with initial observations reported as early as 1903. Since then, significant research has focused on the natural phosphate-solubilizing potential of these soil-associated microorganisms. Among the key bacterial genera with this capability are *Bacillus*, *Pseudomonas*, and *Micrococcus*, while *Aspergillus* and *Penicillium* are well-known fungal counterparts (Sperber, 1958; Illmer and Schinner, 1992). Additionally, the nematode-trapping fungus *Arthrobotrys oligospora* has demonstrated phosphate-solubilizing activity in both laboratory and field conditions, particularly with various forms of rock phosphate.

These microorganisms can be identified and isolated by incubating them on solid media containing insoluble phosphate, where clear zones are produced around their colonies. Variability is commonly observed among different strains in terms of their phosphate-solubilizing efficiency. To ensure stability of this trait, strains are often re-cultured multiple times and re-evaluated for consistent solubilizing activity. The most effective strains are then assessed in liquid culture systems to confirm their ability to solubilize insoluble phosphate under controlled conditions. Following this, the promising candidates are formulated into microbial inoculants, which are subsequently evaluated under greenhouse or field trials for their effectiveness. Prior to commercial application, these inoculants undergo biosafety evaluations to ensure they do not pose risks to human, animal, or environmental health.

7.5 Phosphate Mobilization by AM Fungi

Phosphorus (P) plays a crucial role in sustaining life, particularly in supporting plant growth and development. It forms an integral part of key macromolecules involved in various biological functions such as genetic regulation, cellular signaling, structural integrity, and metabolic activity. Beyond its presence as orthophosphate, phosphorus is also incorporated into molecules like nucleic acids and ATP/ADP (adenosine diphosphate), which are essential for vital energy-related processes, including photosynthesis and respiration. Additionally, phosphorus is found in storage forms such as phytate and pyrophosphate, and in structural components like membrane phospholipids and phosphoproteins.

Soil microorganisms play a pivotal role in nutrient cycling and contribute significantly to plant health and productivity (de Bruijn, 2013). Among them, PGPR enhance the plant's ability to access phosphorus from the soil through several mechanisms. These include: (i) modifying the soil's sorption dynamics to improve phosphorus mobility, (ii) releasing phosphorus from less-available sources, (iii) promoting root surface expansion, (iv) stimulating root branching and the development of root hairs, and (v) modifying root surface characteristics to optimize phosphorus absorption. Through these processes, microorganisms facilitate the conversion of phosphorus from insoluble inorganic forms (via solubilization) and complex organic forms (via mineralization) into bioavailable forms. This microbial mediation is especially important from a sustainability standpoint, as it contributes to maintaining soil fertility and enhancing the long-term productivity of both managed and natural ecosystems (Velázquez and Rodriguez-Barrueco, 2007).

Arbuscular mycorrhizal fungi (AMF) are known to colonize over 80% of terrestrial plant species, including a wide range of agricultural crops, and significantly influence plant growth and productivity (Smith and Read, 2010). These fungi extend their extraradical hyphae into the surrounding soil, and these serve dual functions: they expand the rhizosphere to enhance nutrient and water uptake (Li *et al.*, 1991), and simultaneously provide a niche for colonization by other soil microorganisms (Artursson, 2006). A key role of AMF is to assist in the supply of phosphorus (P) to plants (Smith and Read, 2010). However, AMF alone are generally inefficient in utilizing organic forms of phosphorus, likely due to their limited ability to secrete phosphatase enzymes into the soil (Tisserant *et al.*, 2013). As a result, they depend on cooperative interactions with specific soil microbes to fulfill this function.

Recent research has explored the synergistic interaction between AMF and hyphae-associated bacteria in enhancing organic phosphorus utilization and improving plant phosphorus acquisition. In particular, PSB that associate with AMF hyphae are recognized for their crucial role in phytate phosphorus mineralization. Under phosphorus-deficient conditions, AMF can also exert a priming effect, accelerating the mineralization and cycling of organic phosphorus within the hyphosphere, ultimately increasing the phosphorus content in host plants (Zhang, 2014).

AMF may also influence the process of organic phosphorus mineralization by modulating the composition of the microbial community. Since different bacterial genera exhibit varying capacities to mineralize organic phosphorus, shifts in microbial populations can substantially affect the mineralization process (Rodríguez and Fraga, 1999; Stone *et al.*, 2014).

Given their limited ability to excrete phosphatases externally, AMF are reliant on the recruitment of other soil microbes for efficient organic phosphorus mobilization. To facilitate this, AMF release hyphal exudates that attract PSB—organisms typically constrained by carbon scarcity in bulk soil environments. These bacteria colonize the surface of AMF hyphae (Wang *et al.*, 2016), enhancing phosphatase activity and stimulating the breakdown of organic phosphorus. The resulting inorganic phosphorus is then accessible to AMF through their hyphal networks (Tarafdar and Marschner, 1994). Although such interactions have been well-documented under controlled conditions like microcosms and in vitro systems (Rodríguez and Fraga, 1999; Stone *et al.*, 2014;

Wang *et al.*, 2016), their occurrence and functional relevance under natural field conditions remain to be fully validated.

7.6 Siderophore Production

Iron is a vital micronutrient required by almost all living organisms. As a transition metal, it can readily alternate between oxidation states, which is central to many biological processes. Although iron is relatively plentiful in the environment, its bioavailable form is extremely scarce—typically less than 10^{-18} M—due to its rapid oxidation in the presence of atmospheric oxygen, leading to the formation of poorly soluble ferric oxyhydroxides (Haas, 2003). In response to iron deficiency, many microorganisms release specific molecules known as siderophores to enhance iron solubilization. These compounds, produced by bacteria, fungi, and plants, are low-molecular-weight chelators (ranging from 200 to 2000 Da) that facilitate iron uptake (Schalk *et al.*, 2011).

In oxygen-rich soil environments, iron is predominantly found in the form of insoluble oxide-hydrate complexes. Similarly, during host-pathogen interactions, pathogenic microbes face restricted access to iron, as it is tightly bound to host proteins and heme groups (Miethke and Marahiel, 2007). Siderophores exhibit exceptionally high binding affinity for Fe^{3+}, with stability constants often approaching 10^{30} M^{-1}, thereby playing a critical role in mobilizing extracellular iron from mineral sources. These compounds display a wide array of chemical structures, with more than 500 variants identified to date (Schalk *et al.*, 2011). Functional groups commonly involved in Fe^{3+} coordination include catechol and hydroxamate moieties. Typically, siderophore production is triggered under iron-limiting conditions, and the molecules are secreted into the surrounding environment to scavenge and solubilize iron for microbial use.

7.6.1 Synthesis of siderophores

Siderophores are primarily synthesized within the cytoplasm and peroxisomes of microbial cells (Miethke and Marahiel, 2007). Once produced, these molecules are secreted into the surrounding environment. Upon binding with ferric ions (Fe^{3+}), the resulting siderophore–iron complexes are transported into the periplasmic space via specific receptor-mediated mechanisms. This transport is energy-dependent rather than driven by a simple concentration gradient, suggesting a functional link between ATP generation, nitrogen fixation, and phosphate solubilization within the bacterial system.

Iron becomes available for intracellular metabolic activities through two main pathways: either by the reduction of Fe^{3+} to its more soluble Fe^{2+} form or through the dissociation of the siderophore–iron complex, often facilitated by the presence of a more competitive iron-binding molecule (Hider and Kong, 2010). Several proteins and enzymes involved in this process have been identified, including siderophore-binding proteins, siderophore reductases, and hydrolases capable of cleaving iron–siderophore complexes (Miethke and Marahiel, 2007). Of these, reductive release of iron is generally more favourable for the cell, as it allows the siderophore to be recycled—an advantage not possible when the molecule is degraded enzymatically.

In the rhizosphere, siderophores produced by beneficial bacteria enhance plant iron acquisition and support growth, especially under iron-limiting conditions. Two primary mechanisms have been described by which plants can access iron from microbial siderophores. The first involves the transport of siderophore-bound iron into the apoplastic space, followed by reduction of Fe^{3+} and uptake of Fe^{2+} by the root system. The second mechanism entails direct exchange of Fe^{3+} between microbial siderophores and those produced by the plant itself (Ahmed and Holmström, 2014).

7.7 Conclusions

The integration of biofertilizers into modern agriculture represents a meaningful step toward creating more sustainable and environmentally conscious food production systems. Drawing on the natural capabilities of microorganisms to fix atmospheric nitrogen, solubilize essential minerals like phosphorus, and improve soil biological activity, biofertilizers offer a promising alternative to chemical-based inputs. Their ability to promote plant growth while preserving soil health makes them an essential component in the transition toward low-input, ecologically balanced farming systems.

Despite their benefits, the application of biofertilizers is not without challenges. Field performance is often influenced by a variety of factors, including climatic conditions, soil characteristics, and compatibility with specific crops. Inconsistent results, limited shelf life, and infrastructural constraints related to storage and distribution continue to hinder their widespread adoption. Moreover, gaps in farmer awareness, extension services, and regulatory frameworks further complicate efforts to scale their use across diverse agricultural settings.

Nonetheless, ongoing scientific advancements—ranging from improved microbial strain selection and formulation technologies to synergistic consortia and smart delivery systems—are steadily addressing many of these obstacles. Emerging research also highlights the potential of combining biofertilizers with other sustainable practices such as integrated nutrient management and organic amendments to enhance their efficacy.

As agriculture grapples with the twin pressures of feeding a growing global population and mitigating environmental degradation, biofertilizers stand out as a biologically inspired solution with long-term potential. Continued investment in research, farmer education, and policy support will be crucial in translating this potential into practice. With appropriate strategies in place, biofertilizers could become a cornerstone of resilient and regenerative agriculture in the years to come.

References

Afzal, A. and Bano, A. (2008) Rhizobium and phosphate solubilizing bacteria improve the yield and phosphorus uptake in wheat (*Triticum aestivum*). *International Journal of Agriculture and Biology* 10(1), 85–88.

Ahmed, E. and Holmström, S.J.M. (2014) Siderophores in environmental research: roles and applications. *Microbial Biotechnology* 7(3), 196–208. DOI: 10.1111/1751-7915.12117

Artursson, V., Finlay, R.D., and Jansson, J.K. (2006) Interactions between arbuscular mycorrhizal fungi and bacteria and their potential for stimulating plant growth. *Environmental Microbiology* 8(1), 1–10. DOI: 10.1111/j.1462-2920.2005.00942.x

Barker, A.V. and Bryson, G.M. (2016) Nitrogen. In: Barker, A.V. and Pilbeam, D.J. (eds) *Handbook of Plant Nutrition*. CRC Press, Boca Raton, Florida, pp. 37–66.

Collavino, M.M., Sansberro, P.A., Mroginski, L.A., and Aguilar, O.M. (2010) Comparison of *in vitro* solubilization activity of diverse phosphate-solubilizing bacteria native to acid soil and their ability to promote *Phaseolus vulgaris* growth. *Biology and Fertility of Soils* 46(7), 727–738. DOI: 10.1007/s00374-010-0480-x

de Bruijn, F.J. (2013) *Molecular Microbial Ecology of the Rhizosphere*. Wiley, Oxford, UK.

de Freitas, J.R., Banerjee, M.R., and Germida, J.J. (1997) Phosphate-solubilizing rhizobacteria enhance the growth and yield but not phosphorus uptake of canola (*Brassica napus* L.). *Biology and Fertility of Soils* 24(4), 358–364. DOI: 10.1007/s003740050258

Gulati, A., Sharma, N., Vyas, P., Sood, S., Rahi, P. *et al.* (2010) Organic acid production and plant growth promotion as a function of phosphate solubilization by *Acinetobacter rhizosphaerae* strain BIHB 723 isolated from the cold deserts of the trans-Himalayas. *Archives of Microbiology* 192(11), 975–983. DOI: 10.1007/s00203-010-0615-3

Haas, H. (2003) Molecular genetics of fungal siderophore biosynthesis and uptake: the role of siderophores in iron uptake and storage. *Applied Microbiology and Biotechnology* 62(4), 316–330. DOI: 10.1007/s00253-003-1335-2

Hamid, A. and Ahmad, L. (2012) Soil phosphorus fixation chemistry and role of phosphate solubilizing bacteria in enhancing its efficiency for sustainable cropping—a review. *Journal of Pure and Applied Microbiology* 66(4), 1905–1911.

Hider, R.C. and Kong, X. (2010) Chemistry and biology of siderophores. *Natural Product Reports* 27(5), 637–657. DOI: 10.1039/b906679a

Illmer, P. and Schinner, F. (1992) Solubilization of inorganic phosphates by microorganisms isolated from forest soils. *Soil Biology and Biochemistry* 24(4), 389–395. DOI: 10.1016/0038-0717(92)90199-8

Illmer, P. and Schinner, F. (1995) Solubilization of inorganic calcium phosphates—solubilization mechanisms. *Soil Biology and Biochemistry* 27(3), 257–263. DOI: 10.1016/0038-0717(94)00190-C

Kalayu, G. (2019) Phosphate solubilizing microorganisms: promising approach as biofertilizers. *International Journal of Agronomy* 2019(1), 1–7. DOI: 10.1155/2019/4917256

Katznelson, H. and Bose, B. (1959) Metabolic activity and phosphate-dissolving capability of bacterial isolates from wheat roots, rhizosphere, and non-rhizosphere soil. *Canadian Journal of Microbiology* 5(1), 79–85. DOI: 10.1139/m59-010

Li, X., George, E., and Marschner, H. (1991) Phosphorus depletion and pH decrease at the root–soil and hyphae–soil interfaces of VA mycorrhizal white clover fertilized with ammonium. *New Phytologist* 119(3), 397–404. DOI: 10.1111/j.1469-8137.1991.tb00039.x

Mahanty, T., Bhattacharjee, S., Goswami, M., Bhattacharyya, P., Das, B. *et al.* (2017) Biofertilizers: a potential approach for sustainable agriculture development. *Environmental Science and Pollution Research International* 24(4), 3315–3335. DOI: 10.1007/s11356-016-8104-0

Mazid, M. and Khan, T.A. (2014) Future of bio-fertilizers in Indian agriculture: an overview. *International Journal of Agricultural and Food Research* 3(3), 10–23. DOI: 10.24102/ijafr.v3i3.132

Miethke, M. and Marahiel, M.A. (2007) Siderophore-based iron acquisition and pathogen control. *Microbiology and Molecular Biology Reviews* 71(3), 413–451. DOI: 10.1128/MMBR.00012-07

Ochoa-Loza, F.J., Artiola, J.F., and Maier, R.M. (2001) Stability constants for the complexation of various metals with a rhamnolipid biosurfactant. *Journal of Environmental Quality* 30(2), 479–485. DOI: 10.2134/jeq2001.302479x

Pandey, J. and Singh, A. (2012) Opportunities and constraints in organic farming: an Indian perspective. *Journal of Scientific Research* 56(1), 47–72.

Pradhan, N. and Sukla, L. (2006) Solubilization of inorganic phosphates by fungi isolated from agriculture soil. *African Journal of Biotechnology* 5(10), 850–854.

Rashid, M., Khalil, S., Ayub, N., Alam, S., and Latif, F. (2004) Organic acids production and phosphate solubilization by phosphate solubilizing microorganisms (PSM) under *in vitro* conditions. *Pakistan Journal of Biological Sciences* 7(2), 187–196. DOI: 10.3923/pjbs.2004.187.196

Rodríguez, H. and Fraga, R. (1999) Phosphate solubilizing bacteria and their role in plant growth promotion. *Biotechnology Advances* 17(4–5), 319–339. DOI: 10.1016/s0734-9750(99)00014-2

Schalk, I.J., Hannauer, M., and Braud, A. (2011) New roles for bacterial siderophores in metal transport and tolerance. *Environmental Microbiology* 13(11), 2844–2854. DOI: 10.1111/j.1462-2920.2011.02556.x

Smith, S.E. and Read, D.J. (2010) *Mycorrhizal Symbiosis*. Academic Press, London.

Sperber, J.I. (1958) Solution of apatite by soil microorganisms producing organic acids. *Australian Journal of Agricultural Research* 9(6), 782–787. DOI: 10.1071/AR9580782

Stevenson, F.J. and Cole, M.A. (1999) *Cycles of Soils: Carbon, Nitrogen, Phosphorus, Sulfur, Micronutrients*. Wiley, Oxford, UK.

Stone, M.M., DeForest, J.L., and Plante, A.F. (2014) Changes in extracellular enzyme activity and microbial community structure with soil depth at the Luquillo Critical Zone Observatory. *Soil Biology and Biochemistry* 75, 237–247. DOI: 10.1016/j.soilbio.2014.04.017

Tarafdar, J.C. and Marschner, H. (1994) Phosphatase activity in the rhizosphere and hyphosphere of VA mycorrhizal wheat supplied with inorganic and organic phosphorus. *Soil Biology and Biochemistry* 26(3), 387–395. DOI: 10.1016/0038-0717(94)90288-7

Timofeeva, A., Galyamova, M., and Sedykh, S. (2022) Prospects for using phosphate-solubilizing microorganisms as natural fertilizers in agriculture. *Plants* 11(16), 2119. DOI: 10.3390/plants11162119

Tisserant, E., Malbreil, M., Kuo, A., Kohler, A., Symeonidi, A. *et al.* (2013) Genome of an arbuscular mycorrhizal fungus provides insight into the oldest plant symbiosis. *Proceedings of the National Academy of Sciences* 110(50), 20117–20122. DOI: 10.1073/pnas.1313452110

Vazquez, P., Holguin, G., Puente, M.E., Lopez-Cortes, A., and Bashan, Y. (2000) Phosphate-solubilizing microorganisms associated with the rhizosphere of mangroves in a semiarid coastal lagoon. *Biology and Fertility of Soils* 30(5–6), 460–468. DOI: 10.1007/s003740050024

Velázquez, E. and Rodriguez-Barrueco, C. (eds) (2007) *First International Meeting on Microbial Phosphate Solubilization*, Vol. 102. Springer Science & Business Media, Dordrecht, Netherlands.

Vessey, J.K. (2003) Plant growth promoting rhizobacteria as biofertilizers. *Plant and Soil* 255(2), 571–586. DOI: 10.1023/A:1026037216893

Wang, F., Shi, N., Jiang, R., Zhang, F., and Feng, G. (2016) *In situ* stable isotope probing of phosphate-solubilizing bacteria in the hyphosphere. *Journal of Experimental Botany* 67(6), 1689–1701. DOI: 10.1093/jxb/erv561

Yi, Y., Huang, W., and Ge, Y. (2008) Exopolysaccharide: a novel important factor in the microbial dissolution of tricalcium phosphate. *World Journal of Microbiology and Biotechnology* 24(7), 1059–1065. DOI: 10.1007/s11274-007-9575-4

Zhang, L., Fan, J., Ding, X., He, X., Zhang, F. *et al.* (2014) Hyphosphere interactions between an arbuscular mycorrhizal fungus and a phosphate solubilizing bacterium promote phytate mineralization in soil. *Soil Biology and Biochemistry* 74, 177–183. DOI: 10.1016/j.soilbio.2014.03.004

8 Organic Farming and Green Manuring

Anirudh Sharma*
Department of Biotechnology, Jaypee Institute of Information Technology, Noida, Uttar Pradesh, India

Abstract

This chapter explores the principles, practices, and benefits of organic farming and green manuring as sustainable alternatives to conventional agriculture. Organic farming emphasizes ecological balance and the exclusion of synthetic fertilizers, pesticides, and genetically modified inputs, instead relying on compost, crop rotation, biological pest control, and green manures. Rooted in traditional practices and reinforced by modern sustainability goals, organic farming enhances soil fertility, improves biodiversity, reduces environmental pollution, and provides healthier food. The chapter highlights the four principles of organic farming-health, ecology, fairness, and care—as outlined by IFOAM, and discusses their role in shaping a holistic farming system that supports environmental and human wellbeing. Green manuring, an integral part of organic farming, involves incorporating leguminous and non-leguminous crops into the soil to improve organic matter, enhance nutrient cycling, fix atmospheric nitrogen, and reduce soil erosion. Historical and contemporary perspectives are presented, tracing its practice from ancient civilizations to its revival in modern sustainable agriculture. The chapter also examines global and Indian scenarios of organic farming, institutional support, certification systems, and government schemes promoting adoption. Techniques such as composting, vermicomposting, and crop rotation are described alongside the agronomic, ecological, and socio-economic benefits. Despite challenges like yield gaps, certification hurdles, and policy limitations, organic farming and green manuring are gaining momentum globally due to increasing awareness of food safety, climate change, and soil degradation. By integrating traditional wisdom with scientific advances, these practices promise to strengthen soil health, ensure food security, and foster resilient agricultural systems for future generations.

8.1 Introduction

Organic farming has grown outstandingly in recent years and is considered a sustainable alternative to agriculture that relies on chemical methods. Organic farming is not a new notion in India; it has been practiced since ancient times. The main aim of the United Nations' Sustainable Development Goals (SDGs) is to eradicate hunger and achieve food security (Lu & Wu, 2022). In modern agriculture, large amounts of nutrients are removed from the soil during crop production (Mamiev et al., 2019). The world's population has been projected to reach approximately 11 billion by 2100 (Ekins *et al.*, 2019). As a result, to keep pace with the escalating population, the production of the food must increase. Apart from meeting the basic daily need for food, the

*Corresponding author: anirudhsharma172@gmail.com

DOI: 10.1079/9781836991021.0008

agriculture sector also fulfills the demands of a country's economic development. Organic farming depends on the usage of biological fertilizers as well as some ancient practices such as the use of crop rotation, green manure, and animal manure, and a number of approaches to biological pest control, to improve soil quality and control insects. Most nations rely on the idea of the green revolution, which led to food production in surplus amounts, higher income levels, and self-sufficiency (Kansanga et al., 2019). The adoption of the green revolution has led to agricultural development and farming practices involving chemical fertilizers, high-yielding crops, irrigation water, and herbicides (Rubenstein *et al.*, 2021). The green revolution has resulted in some negative impacts, such as reduction of soil fertility, deterioration of the soil structure, eradication of beneficial insects and microbes, air pollution, and depletion of ground water resources and natural minerals (Rubenstein *et al.*, 2021).

Long before the contemporary period, agricultural technology was developed in China, India, and Japan. These nations recognized the need for leguminous green manure crops as a source of nitrogen and other nutrients (Yang *et al.*, 2014). Green manuring is a traditional and sustainable agricultural practice, where green plants, typically leguminous crops, are incorporated into the soil to improve the organic matter and nutrient quality present in the soil. Green manure can be beneficial for promoting soil health, reducing soil erosion, improving soil fertility, and enhancing the nutrient content in soil (Akbarian *et al.*, 2021). Furthermore, green manure serves as a good and effective source of root microbes as well as leachable nutrients and root bacteria. However, with the passing of time, chemical fertilizer became the main source of nutrients for the growth of crops (Meng *et al.*, 2021).

8.2 Organic Farming

8.2.1 Definitions of organic farming

Many administrations and scientists have elaborated on the perception of organic farming, providing different descriptions, as follows:

- According to the United States Department of Agriculture (USDA), organic farming is a system that to the greatest extent possible avoids the use of synthetic inputs like fertilizers, hormones, pesticides, feed additives, etc., and relies on crop rotation, crop residues, animal manures, organic waste from farms, mineral grade rock additives, etc.
- According to Organics International (IFOAM, formerly International Federation of Organic Agriculture Movements), organic farming is rearing animals and cultivating plants by natural methods. This reduces the pollutants and waste from the environment by using organic components, and avoiding the use of synthetic ones, to maintain soil fertility and ecological balance.
- According to Food and Agricultural organization (FAO), organic farming is a unique production management system that supports and improves the condition of the ecosystem. This excludes all artificial off-farm inputs and encompasses biodiversity, soil biological activity, and biological cycles, all of which are accomplished using biological, mechanical, and agronomic practices.

8.2.2 Principles of organic farming

There are four major principles of organic farming, as introduced by IFOAM. These principles are taken from the IFOAM Basic Standards and IFOAM Accreditation Criteria, Version 2005 (Fig. 8.1; IFOAM, n.d.).

1. Principle of health: Organic farming must sustain and increase the health of soil, plants, animals, humans, and the planet as one. This idea emphasizes the close connection between individual health and communal health. Healthy soil produces healthy crops, which support the health of people and animals. Organic farming promotes the use of natural methods and ecological diversity to maintain health and wellbeing at all levels.

2. Principle of ecology: This principle aims at using organic to for enhance the ecological balance of the environment. Organic farming should maintain and sustain the ecological balance and cycle. It is engaged in ecological processes like nutrient cycling, soil preservation, and biodiversity. It also supports sustainable agricultural practices like crop rotation, composting, and green manuring. The main goal is to respect the natural balance of the ecosystem.

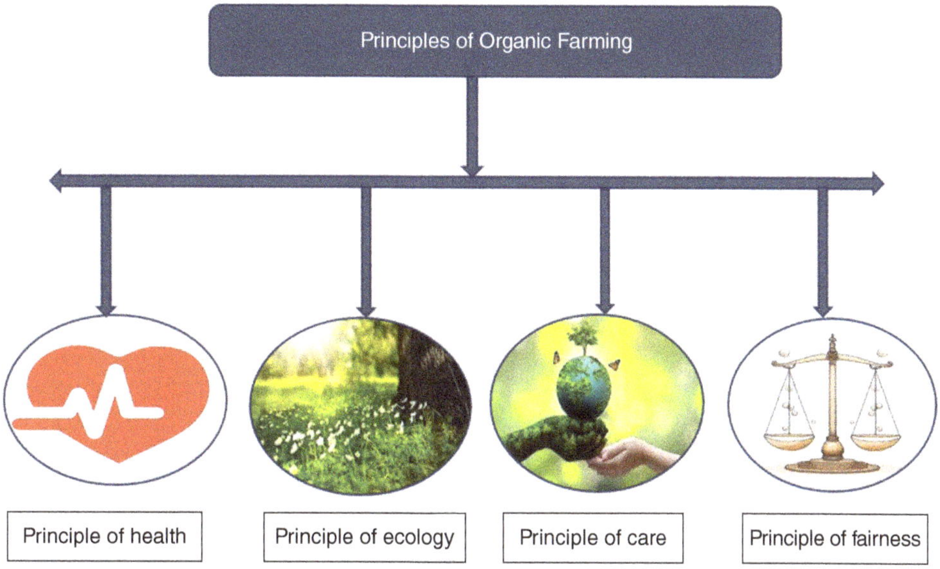

Fig. 8.1. Major principles of organic farming. Figure author's own.

3. Principle of care: Organic farming must protect the health of present and future generations and also protect the environment and preserve biodiversity. According to this principle, preventive measures and accountability are the primary concerns in organic agriculture's management and development. The principle emphasizes employing caution in the adoption of new organic inputs or technologies. It focuses on the long-term impacts of farming decisions and sustainability and supports innovation and research that align with organic principles.

4. Principle of fairness: To remain sustainable, the environment and organic farming should have a fairly equitable relationship. According to this principle, equality is upheld at every stage of production and the delivery system. It promotes ethical treatment of animals and ensures fair wages and fair working conditions for every individual. It encourages a transparent food system for every individual and supports rural livelihoods.

8.2.3 Benefits of organic farming

1. Healthier food: Due to the absence of toxic chemicals and pesticides, organic crops are believed to provide heathier and nutritionally superior food (Kumar *et al.*, 2023).

2. Improved soil condition: Organic farming enhances soil fertility and structure through compost, green manure, and crop rotation (Reganold and Wachter, 2016). Organic manures provide all the essential nutrients required by the plant. It also strengthens the physical properties of soil. Soil under organic management shows 15–30% higher microbial biomass and enzyme activity. Compost, green manure, and crop rotation also help in the regeneration of humus.

3. Nutritional and health safety benefits: Organically grown food, particularly green vegetables, contain higher levels of dry mass, according to research. Organic plants exhibit minerals such as magnesium, iron, calcium, and phosphorous (Rembiałkowska, 2007).

4. Socio-economic impact: Organic products are often accessible at low prices, increasing demand and thus farmers' revenue. Also, the growing demand supports economic growth in a sustainable way. Organic farming can be adopted by small farms, reducing the dependency on costly machines and technologies. This not only benefits the farmers, but also contributes to employment generation at community level.

5. Ecological benefits: The energy used by the organic farming is way less than in conventional farming methods, making it more eco-friendly.

It also maintains diversity and does not disturb the habitats of any species (Asopa, n.d.). It releases much lower amounts of CO_2 as compared to other farming methods. It also helps to prevent environmental degradation.

6. No synthetic chemicals: Organic farming avoids use of synthetic chemical fertilizers, pesticides, herbicides, etc. (Pimentel & Burgess, 2014).

7. Biodiversity: As organic farming is a natural method without the use of any synthetic chemicals, it provides a pesticide-free environment that supports the process of pollination, resulting in the presence of more diverse flora and fauna above and below ground, as there are no harmful chemicals involved. Meta-analysis shows 30% higher species richness in organic fields.

8.2.4 Evolution of the organic agriculture movement

According to the report of the Research Institute of Organic Culture (FiBL), there were 4.3 million organic producers in the world in 2023, increasing from 2.3 million in 2022. How did we get here?

At the beginning of the 20th century, there were some issues that needed to be addressed—soil erosion and depletion, lack of crop varieties, and inadequate food quality—so the organic movement began to address these issues. But as time progressed, the mechanization of agriculture evolved very quickly, making farming easier and more affordable and increasing yields. This had negative effects on the environment and incited the birth of the organic farming movement (Fig. 8.2).

1. Traditional agricultural practices (pre-1900s): At the beginning of the 19th century, farming practices were integrally organic; they were done using compost, manure, crop rotation, and natural pest control. There was no practice of using artificial fertilizers, and soil fertility was preserved using indigenous knowledge systems.

2. Industrial agriculture (1900–1940s): In the middle of the 19th century, the use of synthetic pesticides and fertilizers started to become more prevalent. Nonetheless, a number of trailblazers, such as Sir Albert Howard, started experimenting with composting and other natural techniques after realizing the worth of organic farming. He developed a method of composting, which he called

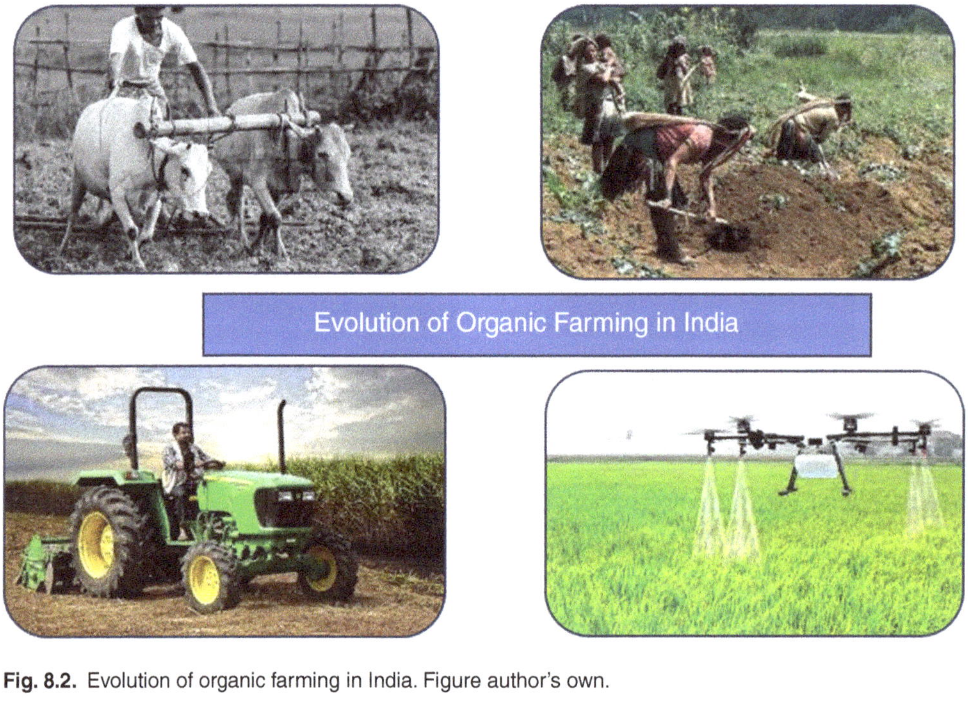

Fig. 8.2. Evolution of organic farming in India. Figure author's own.

Indore Farming, and while residing in India he emphasized the role of organic farming.

3. Birth of the organic movement (1940s–1970s): Walter James was the first person to use the phrase "organic farming," in his book *Look to the Land*, in which he discussed natural and ecological way of farming (Northbourne, 2005). His theories were essential in the development of the global organic farming movement. Additionally, in 1940s, J.I. Rodale, the founder of the Rodale Institute, offered his own insights on chemical-free farming practices. Rodale gained inspiration from Sir Albert Howard, a British scientist who for years observed the agricultural system of India that used green manure and waste as fertilizers. Howard wrote a book, *An Agricultural Testament*, published in 1943, in which he emphasized the importance of using animal waste to preserve soil fertility (Howard, 1943).

4. Institutionalization of organic farming (1970s–1990): The 1970s witnessed growing awareness of the environmental impact of industrial agricultural practices. In 1972, the most important body, IFOAM, was founded in Versailles, France, to raise awareness about sustainable agriculture. In addition, the certification system began in Europe and the USA. In 1990, the US Congress passed the Organic Foods Production Act (OFPA) to create standards for organic food production. The OFPA ultimately resulted in the creation of a board, called the National Organic Standards Board, which would make recommendations for which ingredients could be used in organic farming practices.

5. Global expansion and regulation of organic farming (beyond the 2000s): The development of regulations under the OFPA over the course of a decade in the USA culminated, in 2002, in the implementation of final organic standards under the USDA National Organic Program (NOP). This aligned with a global surge in the organic food market during the 2000s. From 1999 to 2014, the area of certified organic farming expanded significantly, from 11 million hectares to approximately 43.7 million hectares.

6. Current scenario in organic farming (India, 2025): Organic food is produced in India in large amounts, and within a decade India is set to become the largest producer of organic food in world. Currently, India has an area of 4.9 million hectares under organic farming. Major organic producer states in India are Rajasthan, Madhya Pradesh, Maharashtra, Uttarakhand, and Sikkim. Sikkim became India's first organic state in 2016. Popular organic products in India are spices, wheat, tea, fruit and vegetables, rice, and pulses (Fig. 8.3).

In Madhya Pradesh, 9.6 million hectares of land come under organic farming, which is the largest area of any state in India. The major crops grown in are soybean, wheat, and pulses. Rajasthan is the second largest organic producer state, with 5.5 million hectares of land (Fig. 8.4).

8.2.5 Institutionalization of organic farming in India

The following are some of the programs launched by the Government of India to provide support for organic farming in north-eastern India.

1. Rashtriya Krishi Vikas Yojna (RKVY): This scheme was launched in 2007 by the Ministry of Agriculture and Farmers' Welfare; it is a central government scheme. It emphasizes achieving development in agriculture and allied sectors by encouraging states to increase public investment, agricultural productivity, and farmers' incomes. It promotes the usage of biofertilizers and biopesticides and aims to provide training and organic certification. It provides enhanced support to north-eastern India, where farming is more traditional and feasible (Darjee, 2023).

2. National Project on Organic Farming (NPOF): NPOF is a central government scheme that comes under the Ministry of Agriculture and Farmers' Welfare. It was introduced during the tenth five-year plan. It aims to support and promote organic farming in India through technical support, capacity building, and training. It also aims to provide the distribution of biofertilizers, biopesticides, compost, etc. (Ministry of Agriculture and Farmers' Welfare 2023).

3. Mission for Integrated Development of Horticulture: This scheme, launched in 2014, is sponsored by central government and comes under the Ministry of Agriculture and Farmers' Welfare. It aims to promote growth and development in the horticultural sector, including vegetables, flowers, spices, coconut, and medicinal plants. It provides support to organic farming practices in horticulture and enhances farm income and employment generation (Darjee, 2023).

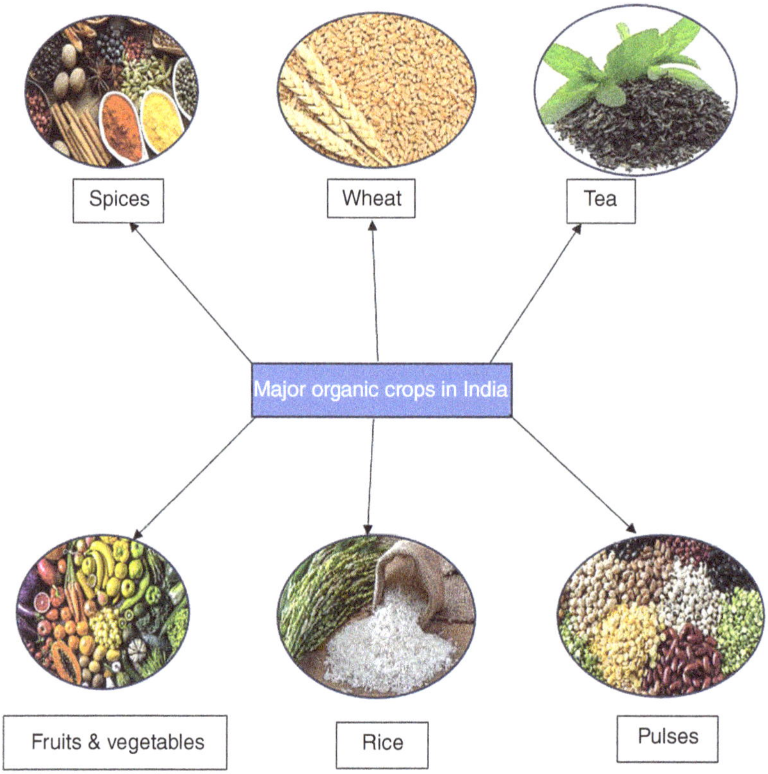

Fig. 8.3. Major organic crops in India. Figure author's own.

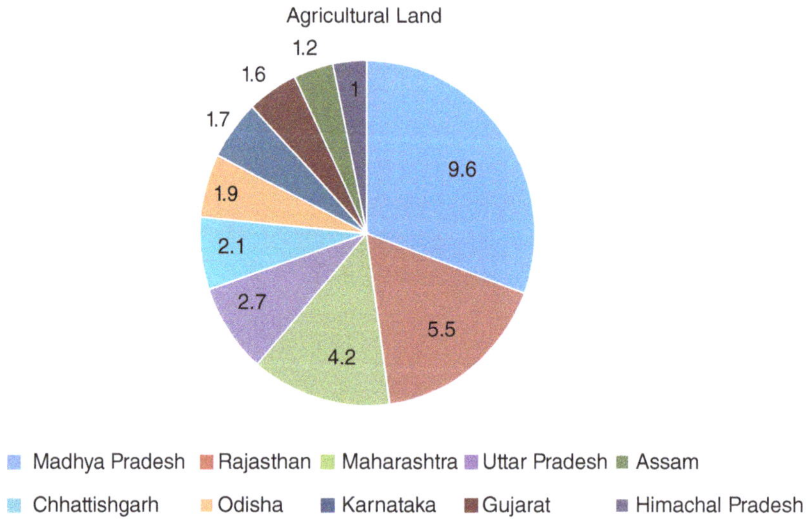

Fig. 8.4. Top 10 Indian states by agricultural area under organic farming. Figure author's own.

4. Bhartiya Prakritik Krishi Paddhati (BPKP): This scheme was launched in 2020 under the National Mission for Sustainable Agriculture (NMSA). It aims to promote traditional chemical-free methods of farming that are economically viable. It reduces dependency on chemical fertilizers and pesticides and improves soil fertility and crop productivity as well as the farmer income. It encourages the use of cow dung/urine, green manuring, and other bio-inputs. As of 2025, it had been implemented in 13 Indian states, including Himachal Pradesh, Andhra Pradesh, Tamil Nadu, and Maharashtra. It covers nearly 1 million hectares of land. It also aims to supports crop diversity.

8.2.6 Organic certification system in India

The demand for the organic products has been increasing in recent years, and people's knowledge about sustainability has also grown. In India, organic certification ensures that agricultural products are grown in a natural way without the use of synthetic fertilizers or any genetically modified organisms (GMOs). Certification is the process of verifying producers of organic food, seeds, and other organic products. The main certification body in India is the National Programme for Organic Production (NPOP), which is managed by the Agricultural and Processed Food Products Export Development Authority (APEDA), which comes under Ministry of Commerce. NPOP provides a third-party assurance that products comply with given organic standards.

8.2.7 World scenario of organic farming

To satisfy the need for both food security and food safety, organic agriculture has become more and more acceptable throughout the world following the COVID-19 pandemic. The European market has been particularly affected. In 2021, nearly 4.7 million tonnes of organic products were imported into the European Union and the United States. Ecuador was the largest exporter, but there was a significant decline in imports from China, Ukraine, and Russia. The top three most imported organic products were bananas, sugar, and soybeans. At least 3.4 million farmers practiced organic farming in 191 nations, covering 74.9 million hectares of land (Rani, 2023) (Fig. 8.5). Leading organic farming countries by area are:

- **Australia:** 53 million hectares
- **India:** 4.5 million hectares
- **Argentina:** 4 million hectares

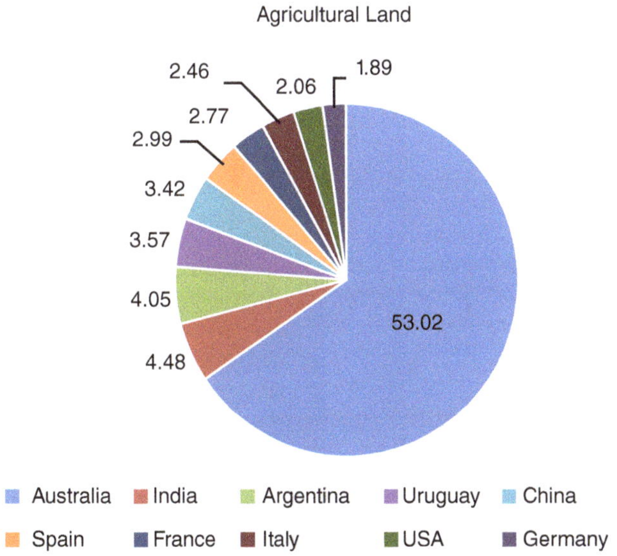

Fig. 8.5. World's top 10 countries by agricultural area under organic farming. Figure author's own.

The universal organic farming market will be worth US$175,215.6 million in 2024, and is projected to expand between 2025 and 2031, according to the *World of Organic Agriculture: Statistics and Emerging Trends 2025* (Willer *et al.*, 2025). According to IFOAM and FiBL, there were 4.3 million organic producers in the world in 2023, up from 2.3 million in 2022, and organic farming accounted for 2.1% of global agricultural land. These organic producers manage farming on 98.9 million hectares of land. Africa saw the biggest relative percentage growth (24%), reaching 3.4 million hectares, while Latin America saw the largest rise, adding 1 million hectares, a growth rate of 10.8%. With 53.2 million hectares, or more than half of the world's organic land, Oceania continues to be the largest organic farming region. Europe comes in second place, with 19.5 million hectares, followed by Latin America with 10.3 million hectares.

8.3 Types of Organic Farming

Two types of organic farming are practiced in India: pure organic farming and integrated organic farming.

8.3.1 Pure organic farming

Pure organic farming methods are 100% natural and involve no chemical fertilizers or pesticides. They use only natural constituents like manure, compost, natural seeds, etc. Soil fertility is maintained through crop rotation, mulching, and composting. Pure organic farming practices include vermicomposting, mulching, crop rotation, and mixed cropping. For biological pest control, natural methods are used, such as pheromone traps, neem and garlic extracts, etc. Pure organic farming emphasizes the use of compost, cow dung, vermicompost, and manure and supports biodiversity and water conservation.

8.3.2 Integrated organic farming

Integrated organic farming is a sustainable method that combines crop production, poultry, aquaculture, beekeeping, animal husbandry, and other activities. The main aim is to maximize resource use efficiency by minimizing waste, recycling nutrients, reducing external inputs, etc. This method maintains soil fertility, improves biodiversity, reduces water pollution, and converts dung and waste into energy. The approach enables the conversion of organic waste into rich compost. It also helps in the process of beekeeping, aiding in pollination and the production of honey. Soil fertility is maintained by the crop rotation and organic input. Integrated organic farming engages rural people in various farm-related activities.

8.4 Techniques in Organic Farming

8.4.1 Composting

Composting involves a biochemical reaction in which aerobic and anaerobic bacteria are employed for the decomposition of organic matter under a controlled environment to produce compost. It employs the breakdown of organic materials such as plant residues, food scraps, and animal manure by microbes. It also enhances soil structure and fertility.

There are two methods for the preparation of compost: the Indore method and the Bangalore method. The Indore method was developed by A. Howard and Y.D Wad at the Institute of Plant Industry in Indore between 1924 and1931. Compost is ready within 4 months in aerobic conditions using this method. A pit approximately 3 m long and 2 m deep is prepared. Plants, weeds, sugarcane residue, and wood ash are used as raw materials. The pit is filled with different layers: for example, in the first layer, waste from cattle spread, in the second, a thick layer of dung is sprinkled (Gamage *et al.*, 2023).

The second method is the Bangalore method, which was developed by C.N. Acharya at the Indian Institute of Science, Bangalore. It is an anaerobic method for the preparation of manure in urban areas. In this method trenches are prepared—the size of trench depends upon the population size of the city. At the bottom of the trench, waste products are spread, and above these, human feces is spread evenly. After that, a thick layer of soil is added. The quantity of manure by this method is better compared to the Indore method.

8.4.2 Vermicomposting

The most common method of composting is vermicomposting. In this method, worms are used to decompose organic waste, creating vermicast (or worm castings), a rich soil amendment containing nutrients and beneficial microorganisms. There are approximately 700 species of earthworm that can be used for this process, which requires regular moisture maintenance at 60–70%. The main benefits of vermicomposting are that it enhances soil fertility and adds essential nutrients like NPK (nitrogen, phosphorous, and potassium) and other micronutrients. It also increases aeration of the soil and its water holding capacity. It is an eco-friendly practice as it reduces farm, municipal, and kitchen waste. Vermicomposting promotes plant growth and leads to better root development and seed germination. It also minimizes the need for synthetic fertilizers as it acts as a natural fertilizer. It is suitable for both large-scale and small-scale operations (Fig. 8.6).

8.4.3 Green manuring

Green manuring is the process of growing crops and deliberately inserting them into the soil at the flowering stage (Bista and Dahal, 2018; see also Section 8.5). It helps preserve soil fertility and improves the physical and chemical properties of the soil. It also enhances water retention capability. Some green manures help in the eradication of soil-borne pathogens. Green manuring crops also incorporate essential nutrients into the soil.

Fig. 8.6. Picture showing vermicomposting. Figure author's own.

8.4.4 Crop rotation

This is a method in which different crops, like maize, barley, wheat, and vegetables (Fig. 8.7), are grown in sequence on the same land to prevent pest buildup and improve soil health. It aims to enhance crop yields. It involves rotating crops with different nutrients needs and pest and disease profiles. It maintains the natural cycles of plant growth and nutrient uptake to retain soil fertility and eradicate pests and disease (Weil and Brady, 2017) Different crops have different nutrient requirements. Farmers can optimize the soil nutrients to reduce the risk of nutrient depletion. For example, deep-rooted crops can access nutrients from deeper layers of soil. Crop rotation helps improve soil structure by enhancing aeration and increasing organic matter (Magdoff & Van Es, 2021). This method promotes biodiversity both above and below, supporting a plethora of beneficial organisms and improving ecosystem and nutrient cycling.

8.5 Green Manure

In recent years, sustainable agricultural methods that support ecosystems and promote cover crops and green manure cultivation have become increasingly important. Among these methods, green manure has become the most popular of all (Haring and Hanson, 2022). In recent years, the use of organic crops has increased considerably due to environmental awareness and efforts to reduce input costs (Iderawumi and Kamal, 2022). Green manuring is an economically effective technique that aims to reduce use of chemical fertilizers and preserve soil (Sharma *et al.*, n.d.). The crops are designed to be integrated into the soil, adding nutrients and organic matter to improve soil fertility. The

Fig. 8.7. Different crops, like wheat, barley, and maize, are grown in rotation on a single piece of land. Figure author's own.

Table 8.1. Different nitrogen, phosphorous pentoxide, and potassium oxide compositions in green manure. Table author's own.

Crop name	Scientific name	N content (kg/ha)	P_2O_5 content (kg/ha)	K_2O content (kg/ha)
Dhaincha	*Sesbania aculeata*	2.51	0.92	0.92
Mung bean	*Vigna radiata*	0.80	0.46	1.15
Cowpea	*Vigna unguiculata*	0.70	0.34	1.15
Maize stover	*Zea mays*	15.5	0.71	3.00
Black Gram	*Vigna mungo*	0.80	0.46	1.15
Sun Hemp	*Crotalaria juncea*	0.70	0.27	1.15
Acacia	*Acacia arabica*	2.61	0.39	2.75

green manure method has been practiced since ancient time in parts of India and Japan, where the leguminous green manure crop is valued as an important source of nutrients for rice and other agriculture (Table 8.1). In recent decades, the increasing use of synthetic fertilizers has had a detrimental effect on crops, so farmers have been moving toward using green crops, which benefit from high nutrient content and rapid decomposition (Xie *et al.*, 2016). In India, leguminous species of crops like *Crotalaria juncea* and *Sesbania aculeate* are most frequently used in green manuring, and *Trifolium* spp. is frequently used in milder regions. There are two types of green crops, leguminous and non-leguminous, and both are frequently found in India.

8.5.1 Historical background of green manuring

8.5.1.1 Ancient period

Green manuring is one of the oldest agricultural practices used to enhance soil fertility and nutrient composition in soil. The practice of green manuring was first begun in China in 1134 BC. In particular, Chinese farmers used a water fern called *Azolla* in rice fields. *Azolla* is able to fix nitrogen in association with *Anabaena*. There is a mention of organic inputs and green manuring in the ancient Indian text the *Rigveda*. The Greeks and Romans also integrated leguminous cover crops into the soil to enhance its fertility.

8.5.1.2 Pre-modern era (1600s–1900s)

European colonies introduced systematic farming practices. In India, green manuring was acknowledged by British scientists in the early 20th century. Arab agriculturists also promoted practices like crop rotation and green manuring.

8.5.1.3 Modern era (20th century onwards)

Green manuring declined in 1950s during the period of green revolution and use of synthetic fertilizers. However, post-2000, green manuring started regaining importance in organic farming and sustainable agriculture.

8.5.1.4 Current status (2020–2025)

There was significant use of green manuring in the practice of organic farming after the COVID-19 pandemic. The government of India also launched schemes like Pradhan Mantri Krishi Vikas Yojna (PMKVY) and Bhartiya Prakritik Krishi Paddhati (BPKP). These schemes aim to enhance the use of green manuring crops.

8.5.2 Benefits of green manuring

The main benefits of green manuring are that it:

- adds organic matter to soil and increases the nutrient content;
- preserves and enhances the fertility of soil;
- lessens the chances of soil erosion;
- improves soil texture and aeration;
- increases nitrogen content in soil;
- reduces leaching;
- suppresses weeds;
- increases microbial population in soil;
- is an eco-friendly practice, as it has no adverse effect on soil or the environment;
- helps in the reduction of greenhouse gas emissions (particularly, carbon dioxide, methane, and nitrous oxide);
- helps in the preservation of soil fertility, which leads to enhanced crop diversity (Bista and Dahal, 2018); and
- helps to enhance carbon sequestration by increasing the nutrient content of soil.

8.5.3 Types of green manure crops in India

There are two types of green manure, leguminous and non-leguminous.

8.5.3.1. Leguminous green manure

Leguminous green manure is considered to be most effective in fixing nitrogen and improving soil fertility by forming a symbiotic association with *Rhizobium* bacteria. These are found naturally in the soil, but inoculating the seed prior to sowing may benefit some species. It must be kept in mind that some of these plants can result in bloat or reproductive issues in livestock if they are grazed without being diluted with feed (Fig. 8.8).

8.5.3.1.1 CHARACTERISTICS OF LEGUMINOUS GREEN MANURE CROPS. The key characteristics of leguminous green manure crops are that they:

- belong to the family *Fabaceae*;
- improve soil structure and preserve the fertility of soil;
- have nitrogen-fixing bacteria;
- improve microbial activity in soil;
- have high moisture content, which helps in rapid breakdown of biomass; and
- reduce dependency on synthetic fertilizers and pesticides.

8.5.3.2 Non-leguminous green manure

Non-leguminous green manure includes non-leguminous plants that are very helpful at preventing nitrate leaching and adding organic matter to the soil. Unlike legumes, these manures will not fix atmospheric nitrogen. They are incorporated into the soil to add organic matter and improve soil structure (Fig. 8.9).

8.5.3.2.1 CHARACTERISTICS OF NON-LEGUMINOUS GREEN MANURE CROPS. The key characteristics of non-leguminous green manure crops are that they:

- suppress weeds;
- have high biomass production;
- do not fix nitrogen but tend to fix nutrient cycling;
- improve water holding capacity in soil; and
- improve soil structure and humus content.

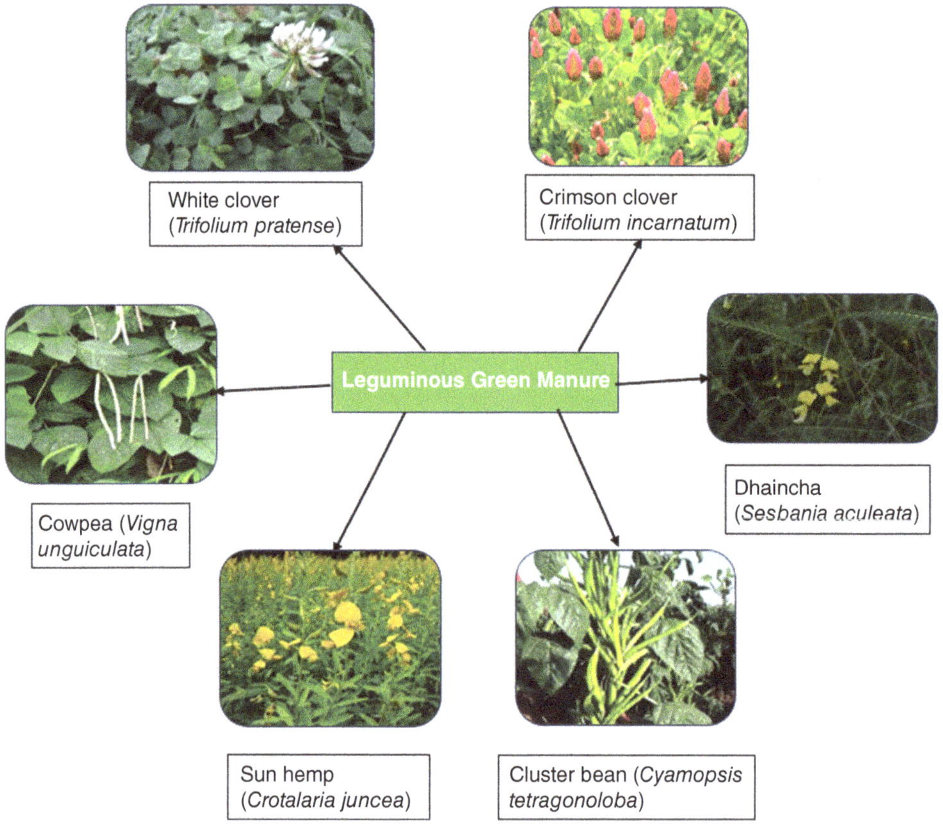

Fig. 8.8. Examples of leguminous green manure crops. Figure author's own.

8.5.4 Steps for the growing and incorporation of green manure crops

8.5.4.1 Well-minced soil

Before we incorporate green manure into the field, the field must be finely ploughed and the soil pulverized. This is done to ensure a good seed–soil ratio, better aeration and moisture content, and proper germination of green manure seeds. One of the main aims of green manuring is to ensure 100% ground cover so maximum germination can be achieved.

8.5.4.2 Green manure seed

It is important to select the appropriate green manure seed. The seed can be selected on the basis of soil type, climate type, and cropping pattern. Some common seeds include Sun hemp, dhaincha, and cowpea. These seeds are usually fast growing and able to fix nitrogen quickly (Ramanjaneyulu *et al.*, 2021).

8.5.4.3 Distribution of green manure seed

Seeds are distributed either manually or using seed drills. Uniform spreading is vital to ensure the even density of plant growth, effective and rapid biomass production, and eradication of weeds.

8.5.4.4 Green manure at seeding stage

Some 7–10 days after the sowing of the seed, they start to germinate and reach the seedling stage. At this point, the plants are able to establish a well-developed root system. The early suppression of weeds begins and nitrogen fixation starts.

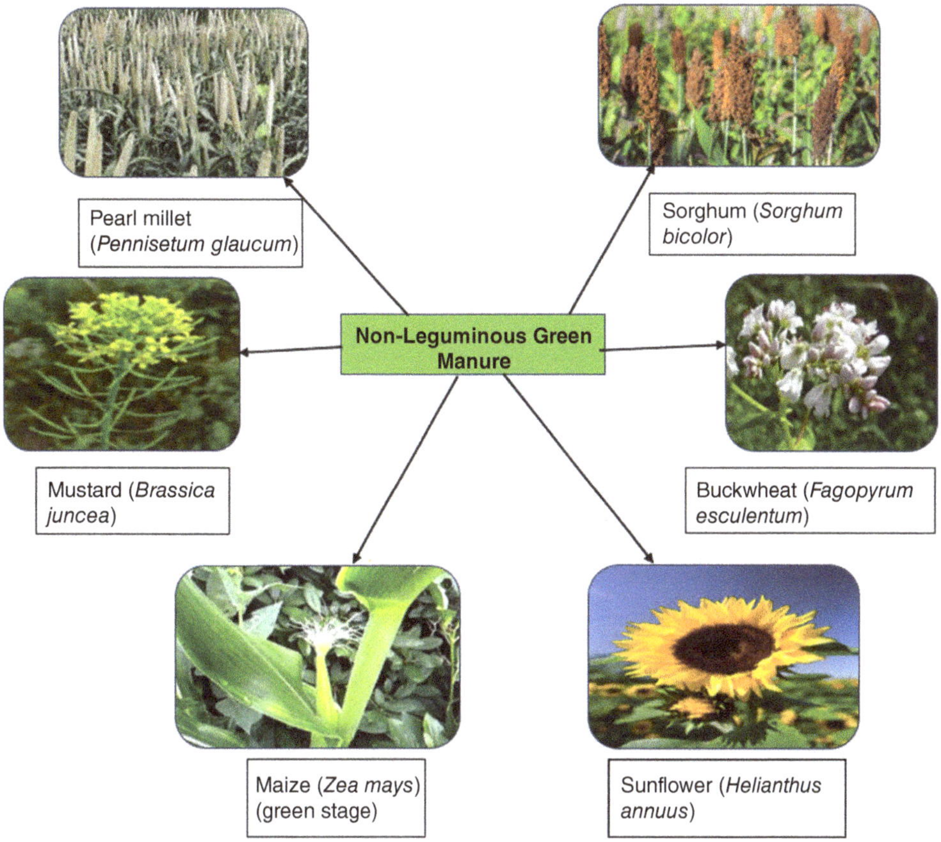

Fig. 8.9. Examples of non-leguminous green manure crops. Figure author's own.

8.5.4.5 Vegetative growth state of green manure crops

Plants are growing rapidly and able to produce leafy biomass and developed a strong root system. This is essential for biomass production, which contributes to soil organic matter and access to maximum nitrogen fixation benefits.

8.5.4.6 Flowering stage of green manure plants

When the flowering stage is 50% complete, the amount of biomass and nitrogen are at their peak. Plants are tender and decompose easily and timely incorporation ensures nutrient release.

8.5.4.7 Incorporation of green manure plants into rice fields

Green manure crops are mainly incorporated into the soil using tractors. Green manure is ploughed into the soil and mixed evenly into the upper layer. After ploughing, it is left to decompose for 2–3 weeks before transplanting.

8.5.4.8 Rice transplanting

Once the green manure has decomposed, rice seed are transplanted into the soil. Benefits of sowing a rice crop into the field include increased nitrogen availability, improved soil structure, and enhanced microbial activity (Fig. 8.10).

8.5.5 Biological nitrogen fixation by green manure

One of the most beneficial effects of green manure crops is the addition of nutrients to the soil. This is often the main source of nitrogen. The nitrogen-fixing bacteria *Rhizobium*, which

Fig. 8.10. Growing and incorporation of green manure crops. Figure author's own.

is present into the root nodule, is often used by legumes to fix atmospheric nitrogen.

8.5.5.1 Why nitrogen fixation?

Although nitrogen is the most abundant element present on Earth, due to its complex physical and biological transformations in the soil, its vulnerability to gaseous emission and leaching makes it is the most difficult nutrient to manage (Wysokinski and Lozak, 2021). Ample amounts of nitrogen are added to the soil after green manuring. These nutrients are then used by the next crop sown into that soil. A process called mineralization is carried out by bacteria and fungi, where complex molecules are broken down and converted into ammonium and nitrate ions. Mineralization proceeds faster in moist soil. Sometimes nitrogen does not fix even after the symbiosis process. So, green manure has different nitrogen fixation abilities. Some of the methods for nitrogen fixation are N-accumulation, N-balance, N-difference, and acetylene reduction. Some of the benefits of nitrogen fixation include eliminating the need for synthetic fertilizers and supporting the healthy life of microbes in the soil.

8.5.6 General requirement for green manure crops

It is important to know the different qualities of green manure crops in order to get the benefits from them. These include which family they belong to, their growth habits, their life cycle, and their ability to control weeds, fix nitrogen, and recycle nutrients. Their selection for green manure depends upon environmental conditions, the cropping pattern, and specific soil needs. The following are some of the criteria for selection of green manure.

8.5.6.1 Rapid growth and biomass production

Crops should produce ample green biomass rapidly for effective soil coverage. Ample amounts of biomass mean ample amounts of organic matter to be incorporated into the soil. For example, the Sun hemp and *Sesbania aculeata* can produce 12–15 tonnes of biomass per hectare in every 45–60-day cycle (Palaniappan and Annadurai, 2018).

8.5.6.2 Nitrogen fixing ability (especially for legumes)

Leguminous green manure has the ability to fix atmospheric nitrogen through the process of symbiosis and the nitrogen-fixing bacteria *Rhizobium* found in the root nodules. For example, sesbania, commonly called dhaincha, can fix up to 120 kg N/ha in a single season.

8.5.6.3 Deep rooting system

One of the pivotal roles played by green manure or cover crops involves the deep and extensive root system. In many agricultural areas, important nutrients like nitrogen (N), Potassium (K), Magnesium (Mg), Calcium (Ca), and Sulphur (S) can be leached, especially in areas with heavy rainfall. Deep-rooted green manures such as Sun hemp, sesbania, and mustard can penetrate deep into the soil and absorb these nutrients. This process is called pumping, because these plants act like a natural nutrient pump.

8.5.6.4 Fast initial growth

Soil is extremely vulnerable to natural phenomena like erosion, nutrient loss, weed invasion, etc. After the harvest period, green manure crops are immediately sown into the soil, providing rapid vegetative cover on the surface of the soil. This acts as a natural barrier that protects the soil from the direct impact of raindrops, sunlight, and wind. As a result, both the layer of top soil and moisture are preserved. Sun hemp, cowpea, mustard, and buckwheat are fast-growing green manure crops that play an initial role in the suppression of weeds.

8.5.6.5 Easy decomposition

At a very specific growth stage, preferably when they are young and not woody, green manure crops are ploughed into the soil. At this stage, the plant tissues are soft, tender, and juicy, and easy for the microbes to decompose. The microbes, like bacteria and fungi, are present in the soil and they find it easy to digest these soft tissues, which have a low amount of lignin present. This means they are easy to break down compared to mature and fibrous woody plants. As the process of decomposition proceeds, the nutrients, such as nitrogen, phosphorous, and potassium, are released into the soil and are available to be taken up by the next crop (Palaniappan and Annadurai, 2018).

8.5.7 Global market report 2025 for green manure

The green manure market has grown strongly in the past decade. In 2024, it was US$2.32 billion, projected to increase to $2.49 billion in 2025. The size of the green manure market is forecast to increase further in the coming years, to $3.32 billion in 2029, with a compound annual growth rate (CAGR) of 7.5%. However, this growth forecast depends upon climate change, regenerative agricultural practices, water quality management, etc. Major companies that operate green markets include BioGreens, BioStar Organics, and BioAtlantis. These companies are mainly focusing on adapting collaborations with one another to promote sustainability in green manuring. These collaborations are intended to promote the use of green manure in agriculture to enhance soil health and improve crop yield. This initiative emphasizes the use of green manure. It is also planned to distribute 10 lakh plant samples of neem, which is known for the pest control properties (Koul, 2004).

8.5.8 Status of green manure in India

Green manuring continues to gain importance as a key and sustainable practice in India. The government also provides consolidated support for these practices. In the states of Sikkim, Uttarakhand, and Himachal Pradesh, adoption is moderate but growing progressively. Paramparagat Krishi Vikas Yojana (PKVY) and natural farming initiatives promote green manuring in situ. In 2024, the Tamil Nadu government launched the Chief Minister's Mannuyir Kaathu Mannuyir Kaappom scheme. This scheme was launched with a budget of 206 crore (US$2.6 billion), with the initial allocation of 20 crore specifically for distributing green manure seeds, which will cover more than 80,000 ha (200,000 acres) of land in the year 2024–2025. The market value for green manure was defined as the revenue gained from sales and goods or services within the specified market and through sales, grants, and donations in terms of currency. The specified currency is US dollars unless otherwise specified. The area under green manuring is estimated to be 2.0–2.5 million hectares for the year 2025. In India, the practice of green manuring is most common in rice growing states like Andhra Pradesh, Uttar Pradesh, Karnataka, Punjab, and Odisha. The common growing crops include dhaincha (*Sesbania aculeata*), Sun hemp (*Crotalaria juncea*), cowpea, and berseem (*Trifolium alexandrinum*).

8.5.9 Green manuring in organic farming

Green manuring is a fundamental practice in organic farming, which is used to improve soil fertility and enhance nutrient content in soil. Various leguminous green manure crops, such as *Sesbania aculeata*, *Crotalaria juncea*, and *Vigna* spp., fix atmospheric nitrogen by symbiotic association with *Rhizobium*, which is present in the root nodule. These green manure crops are able to produce 60–150 kg N/ha of nitrogen. Green manuring avoids the chances of soil erosion during the rainy season and can maintain the population of soil mycorrhiza. Surface roughness is increased by green manure, which lessens the wind speed and helps minimize wind erosion. Green manure improves soil texture, aeration, and moisture retention. A major advantage of green manure is the suppression of weeds. If organic matter is not used to replenish the negative balance of humus, a top layer of soil is created that negatively affects the physicochemical properties and biological activity of the soil and deteriorates its fertility (Mamiev *et al.*, 2019). Some green manure has the ability to secrete chemicals into the soil that inhibit the growth of weeds. Special care must be taken to make sure that manures do not turn into weeds themselves and worsen the weed infestation in a rotation. Green manure also minimizes the dependency on chemical fertilizers.

8.6 Conclusions

Organic farming is sustainable and eco-friendly. Unlike conventional farming methods, which are heavily dependent upon synthetic fertilizers and pesticides, organic farming emphasizes the use of natural resources and methods such as green manure, compost, crop rotation, etc. These help to improve and preserve the fertility of the soil. Organic farming also promotes ecological balance and reduces environmental pollution. As global awareness about food safety, climate change, and soil degradation increases, so does awareness about organic farming. Organic farming and green manuring emerge as an alternative for conventional farming methods, ensuring long-term agricultural sustainability and food security and empowering farmers to produce healthy and chemical-free food. In 2016, Sikkim became the first organic state of India. Various schemes, such as RKVY and NPOF, have been launched by the Indian government to increase awareness about organic farming and green manuring. To maximize its benefits, strengthened policy support, farmer education, and greater awareness about the benefits of organic farming are needed.

This chapter focuses on the benefits of organic farming and green manuring for maintaining an ecological balance. The practice not only reduces the dependency on synthetic fertilizers but also enhances soil fertility, nutrient composition, and microbial flora in soil. Embracing organic farming and green manuring is a crucial step toward ecological sustainability so that future generations can inherit a more balanced life. It is our collective responsibility to sustain organic approaches so that what we leave behind is care, health, and harmony with nature.

References

Akbarian, M.M., Mojaradi, T., and Shirzadi, F. (2021) Effects of *Hedysarum coronarium* L. (sulla) as a green manure along with nitrogen fertilizer on maize production. *agriTECH* 41(1), 95. DOI: 10.22146/agritech.58944

Asopa, R. (n.d.) Environmental benefits of organic or natural farming. *International Journal of Education and Science Research Review* 11(2), 44–54. Available at: https://ijesrr.org/publication/99/575.%20april%202024%20ijesrr.pdf (accessed August 25, 2025).

Bista, B. and Dahal, S. (2018) Cementing the organic farming by green manures. *International Journal of Applied Sciences and Biotechnology* 6(2), 87–96. DOI: 10.3126/ijasbt.v6i2.20427

Darjee, D.K. (2023) A review on policy initiatives, institutional mechanism and support for organic farming in India with special reference to north-east states: an exploratory study. *International Journal of Research and Analytical Reviews* 10(4), 193–211.

Ekins, P., Gupta, J., and Boileau P. (eds) (2019) *Global Environment Outlook–GEO-6: Healthy Planet, Healthy People*. Cambridge University Press, Cambridge, UK.

Gamage, A., Gangahagedara, R., Gamage, J., Jayasinghe, N., Kodikara, N., Suraweera, P., and Merah, O. (2023) Role of organic farming for achieving sustainability in agriculture. *Farming System* 1(1), 100005. DOI: 10.1016/j.farsys.2023.100005

Haring, S.C. and Hanson, B.D. (2022) Agronomic cover crop management supports weed suppression and competition in California orchards. *Weed Science* 70(5), 595–602. DOI: 10.1017/wsc.2022.48

Howard, A. (1943) *An Agricultural Testament*. Oxford University Press, New York.

Iderawumi, A.M. and Kamal, T.O. (2022) Green manure for agricultural sustainability and improvement of soil fertility. *Farming & Management* 7(1), 1–8. DOI: 10.31830/2456-8724.2022.FM-101

IFOAM (n.d.) Standards and Certification. Available at: https://ifoam.bio/our-work/how/standards-certification (accessed September 9, 2025).

Kansanga, M., Andersen, P., Kpienbaareh, D., Mason-Renton, S., Atuoye, K. *et al.* (2019) Traditional agriculture in transition: examining the impacts of agricultural modernization on smallholder farming in Ghana under the new green revolution. *International Journal of Sustainable Development & World Ecology* 26(1), 11–24. DOI: 10.1080/13504509.2018.1491429

Koul, O. (2004) Neem: a global perspective. In: Koul, O. and Wahab, S. (eds) *Neem: Today and in the New Millennium*. Springer, Dordrecht, Netherlands, pp. 1–19.

Kumar, D., Ravisankar, N., and Panghal, A. (eds) (2023) *Transforming Organic Agri-Produce into Processed Food Products: Post-COVID-19 Challenges and Opportunities*. Apple Academic Press, Palm Bay, Florida.

Lu, C. and Wu, A. (2022) The impact of migration characteristics on rural migrant households' farmland use arrangements in China. *PLOS One* 17(8), e0273624. DOI: 10.1371/journal.pone.0273624

Magdoff, F. and Van Es, H. (2021) *Building Soils for Better Crops: Ecological Management for Healthy Soils*. Sustainable Agriculture Research & Education, College Park, Maryland.

Mamiev, D., Abaev, A., Tedeeva, A., Khokhoeva, N., and Tedeeva, V. (2019) Use of green manure in organic farming. *IOP Conference Series* 403(1), 012137. DOI: 10.1088/1755-1315/403/1/012137

Meng, T., Chen, X., Ge, J., Zhang, X., Zhou, G. *et al.* (2021) Reduced nitrogen rate with increased planting density facilitated grain yield and nitrogen use efficiency in modern conventional japonica rice. *Agriculture* 11(12), 1188. DOI: 10.3390/agriculture11121188

Northbourne, Lord (2005) *Look to the Land*. Sophia Perennis, Brooklyn, New York.

Palaniappan, S.P. and Annadurai, K. (2018) *Organic Farming: Theory & Practice*. Scientific Publishers, Jodhpur, India.

Pimentel, D. and Burgess, M. (2014) An environmental, energetic and economic comparison of organic and conventional farming systems. In: Pimentel, D. and Peshin, R. (eds) *Integrated Pest Management: Pesticide Problems*, Vol. 3. Springer, Dordrecht, Netherlands, pp. 141–166.

Ramanjaneyulu, A.V., Sainath, N., Swetha, D., Reddy, R.U., and Jagadeeshwar, R. (2021) Green manure crops: a review. *Biological Forum – An International Journal* 13(2), 445. Available at: www.researchtrend.net (accessed September 9, 2025).

Rani, A. (2023) Green farming in India: issues and policy perspective. Available at: https://www.researchsquare.com/article/rs-2962228/v1 (accessed September 9, 2025).

Reganold, J.P. and Wachter, J.M. (2016) Organic agriculture in the twenty-first century. *Nature Plants* 2(2), 15221. DOI: 10.1038/nplants.2015.221

Rembiałkowska, E. (2007) Quality of plant products from organic agriculture. *Journal of the Science of Food and Agriculture* 87(15), 2757–2762. DOI: 10.1002/jsfa.3000

Rubenstein, J.M., Hulme, P.E., Buddenhagen, C.E., Rolston, M.P., and Hampton, J.G. (2021) Weed seed contamination in imported seed lots entering New Zealand. *PLOS One* 16(8), e0256623. DOI: 10.1371/journal.pone.0256623

Sharma, S., Singh, Y., Sekhon, B., Sharma, S., Sran, R.S. *et al.* (n.d.) Underutilized crops and organic farming: to combat climate change and hidden hunger. *Bulletin of Environment, Pharmacology and Life Sciences* 7(6), 92–97. Available at: https://www.researchgate.net/publication/327831169_Underutilized_Crops_and_Organic_Farming_To_Combat_Climate_Change_and_Hidden_Hunger (accessed August 25, 2025).

Weil, R.R. and Brady, N.C. (2017) *The Nature and Properties of Soils*, 15th edn. Pearson Education, London. Available at: https://www.Researchgate.Net/Publication/301200878_The_Nature_and_Properties_of_Soils_15th_edition (accessed August 25, 2025).

Willer, H., Trávníček, J., and Schlatter, B. (eds) (2025) *The World of Organic Agriculture: Statistics and Emerging Trends 2025*. Research Institute of Organic Agriculture FiBL and IFOAM–Organics International. Available at: https://www.fibl.org/fileadmin/documents/shop/1797-organic-world-2025.pdf (accessed September 9, 2025).

Wysokinski, A. and Lozak, I. (2021) The dynamic of nitrogen uptake from different sources by pea (*Pisum sativum* L.). *Agriculture* 11(1), 81. DOI: 10.3390/agriculture11010081

Xie, Z., Tu, S., Shah, F., Xu, C., Chen, J. *et al.* (2016) Substitution of fertilizer-N by green manure improves the sustainability of yield in double-rice cropping system in South China. *Field Crops Research* 188, 142–149. DOI: 10.1016/j.fcr.2016.01.006

Yang, Z., Zheng, S., Nie, J., Liao, Y., and Xie, J. (2014) Effects of long-term winter planted green manure on distribution and storage of organic carbon and nitrogen in water-stable aggregates of reddish paddy soil under a double-rice cropping system. *Journal of Integrative Agriculture* 13(8), 1772–1781. DOI: 10.1016/S2095-3119(13)60565-1

9 Application Methods of Agrifood Waste-based Biofertilizers and their Role in the Promotion of Sustainable Agriculture and a Clean Environment

Anirudh Sharma*
Department of Biotechnology, Jaypee Institute of Information Technology, Noida, Uttar Pradesh, India

Abstract
This chapter explores the application methods of agrifood waste-based biofertilizers and their critical role in advancing sustainable agriculture and environmental protection. It begins by tracing the historical development of biofertilizers, from early organic amendments to modern biotechnological innovations, emphasizing their growing importance as eco-friendly alternatives to chemical fertilizers. The chapter highlights the valorization of agrifood waste as a substrate for biofertilizer production, offering dual benefits of waste management and soil fertility enhancement. Various preparation techniques, including solid-state fermentation, liquid fermentation, and immobilization, are detailed alongside factors influencing microbial efficacy, such as substrate choice, strain selection, and formulation strategies. Application methods—seed treatment, soil incorporation, foliar sprays, and fertigation—are comprehensively examined, with discussion on their respective advantages, limitations, and suitability under different agroecological contexts. Emerging approaches such as nanoencapsulation and controlled-release formulations are identified as promising avenues to improve nutrient delivery, crop productivity, and environmental compatibility. The ecological benefits of agrifood waste-based biofertilizers are underscored, including improved soil structure, enhanced nutrient cycling, reduced greenhouse gas emissions, and contributions to climate change mitigation through carbon sequestration. Their role in phytoremediation and reduction of agricultural pollution further strengthens their environmental value. The chapter concludes by outlining future prospects, stressing the need for research on plant–microbe interactions, consortia optimization, and site-specific protocols, while also calling for robust legal and policy frameworks to ensure quality assurance and scalability. Ultimately, the adoption of these biofertilizers represents a transformative step toward resilient farming systems and clean, sustainable food production.

9.1 Introduction

The concept of biofertilizers, though not termed as such, has roots stretching back to the dawn of agriculture. Ancient farmers observed the benefits of adding organic matter to soil, unknowingly fostering microbial growth. The scientific foundation for biofertilizers was laid in the late 1800s with the discovery of *Rhizobium* bacteria's role in nitrogen fixation. This breakthrough

*Corresponding author: anirudhsharma172@gmail.com

DOI: 10.1079/9781836991021.0009

opened doors to a new realm of agricultural microbiology, setting the stage for future developments. The turn of the millennium brought a paradigm shift, with biotechnology enabling the creation of more robust and effective microbial formulations. Recent years have seen an explosion in biofertilizer research and adoption, driven by the need for sustainable agricultural practices and the growing organic food market. As we face global challenges like climate change and food security, biofertilizers are increasingly recognized as a crucial component of future farming systems (Zhao *et al.*, 2024a).

The burgeoning interest in agrifood waste-based biofertilizers necessitates a comprehensive elucidation of their application methodologies and their pivotal role in fostering sustainable agricultural practices and environmental stewardship (Kavitha Shree *et al.*, 2024). This chapter meticulously delineates the multifarious techniques employed in the application of these biofertilizers, elucidating their efficacy and environmental implications. The discourse commences with a thorough explication of biofertilizers, encompassing their composition, microbial constituents, and biochemical mechanisms. Subsequently, it delves into the myriad approaches utilized in their preparation, encompassing solid-state fermentation, liquid fermentation, and immobilization techniques (Mishra *et al.*, 2023). The chapter expounds on the critical parameters influencing biofertilizer efficacy, including substrate selection, microbial strain optimization, and formulation strategies (Arora *et al.*, 2020).

An exhaustive analysis of application methods follows, encompassing seed inoculation, soil application, foliar sprays, and fertigation techniques. The discourse elucidates the merits and limitations of each approach, providing a nuanced understanding of their suitability for diverse agronomic contexts. Furthermore, the chapter explores novel application technologies, such as nanoencapsulation and controlled-release formulations, which promise enhanced efficacy and environmental compatibility (Karuvelan *et al.*, 2025). The environmental ramifications of agrifood waste-based biofertilizers are scrutinized, with particular emphasis on their role in soil health amelioration, nutrient cycling enhancement, and mitigation of agricultural pollution. The chapter examines the potential of these biofertilizers in phytoremediation and their capacity to augment soil carbon sequestration, thereby contributing to climate change mitigation efforts. The discourse culminates in a prognostication of future trajectories in biofertilizer development and application (Sarker *et al.*, 2024). It delineates research imperatives, including the elucidation of plant–microbe interactions, optimization of consortia formulations, and development of site-specific application protocols.

Recommendations for policy frameworks and industrial scaling are proffered, aimed at facilitating the widespread adoption of these sustainable agricultural inputs (Malusá and Vassilev, 2014). The promulgation of a comprehensive legal framework for biofertilizers necessitates a multifaceted approach, encompassing myriad aspects of production, quality assurance, and market deployment. Such a framework must deftly navigate the intricate balance between fostering innovation and safeguarding environmental and public health interests. At its core, the legislative edifice must be predicated upon a precise, scientifically grounded definition of biofertilizers, delineating them from conventional agrochemicals and other biological products. This definitional clarity serves as the cornerstone for subsequent regulatory provisions, ensuring consistent application across diverse jurisdictions and mitigating potential legal ambiguities (Khan *et al.*, 2024).

This chapter provides a comprehensive treatise on the application methods of agrifood waste-based biofertilizers, their environmental implications, and their potential to revolutionize agricultural practices. It serves as a cornerstone for researchers, agronomists, and policy makers striving to actualize the vision of sustainable, environmentally benign food production systems.

9.2 Seed Treatment

Seed treatment via biofertilizers and botanical extracts is effectuated to engender multifarious agronomic benefits. Microbial inoculant ameliorations and phyto-derived substances are adhibited to seeds, whereupon growth-promoting and protective moieties are imparted. The process entails the infusion of bioactive compounds into the seed integument, thereby modifying its physicochemical properties and microenvironment. Such treatments aim to augment seed viability, enhance seedling vigor, and confer resistance against biotic and abiotic stressors (Bakonyi *et al.*, 2013). The intricate interplay between seed physiology and biofertilizer-mediated microbial interactions engenders a complex network of biochemical cascades and symbiotic associations within the pedosphere. Upon imbibition and subsequent metabolic reactivation, quiescent embryonic structures exude a diverse array of rhizodeposits, including low-molecular-weight organic acids, amino acids, and flavonoids, which function as chemoattractants for soil-dwelling microorganisms. These rhizosphere-competent microbes, often formulated as biofertilizers, exhibit remarkable colonization efficacy, establishing intricate biofilms on the emergent radicle and root hairs (Khalequzzaman and Hossain, 2007) (Fig. 9.1).

Rhizobial and mycorrhizal symbionts, ensconced within the developing rhizosphere, engage in reciprocal nutrient exchange paradigms with their plant hosts. Through sophisticated molecular dialogues, these microorganisms modulate plant gene expression, instigating morphological and physiological alterations that augment nutrient acquisition efficiency (Chiquito-Contreras *et al.*, 2024). Concomitantly, the microbial partners receive photosynthetically derived carbon compounds, facilitating their proliferation and metabolic activities. This mutualistic association engenders enhanced seedling vigor, manifested through augmented biomass accumulation, elevated stress tolerance, and amplified resistance to phytopathogens. The multifaceted interactions between seed physiology and biofertilizer-associated microbes precipitate a panoply of soil health amelioration effects. The burgeoning microbial populations contribute to soil structure enhancement through the production of extracellular polymeric substances, fostering aggregate stability and improving water retention capacity (Dal Cortivo *et al.*, 2020). Moreover, the diverse metabolic repertoire of these microorganisms facilitates biogeochemical cycling of essential nutrients, enhancing their bioavailability to subsequent plant generations. This complex, self-reinforcing ecosystem dynamic underpins the conceptual framework of regenerative agriculture, wherein the synergistic interactions between seeds, microbes, and soil engender sustainable agroecosystems characterized by enhanced productivity and resilience (Bayu, 2024).

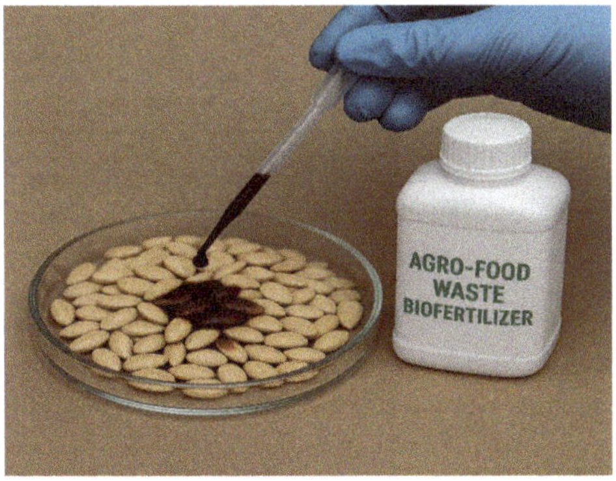

Fig. 9.1. Seed treatment using agrifood waste biofertilizers. Figure author's own.

Diverse plant materials are exploited for seed treatment purposes (Ahmed and Agnihotri, 1977). Leaf powders, obtained through desiccation and pulverization of foliar tissues, serve as repositories of phytochemical constituents. Rhizome preparations, derived from subterranean stem structures, are utilized for their potent bioactive compounds (Punj and Gupta, 1988). Seed and plant oils, extracted through various methodologies, are employed for their antifungal and insecticidal properties (Kazmi *et al.*, 1993). These botanical derivatives are applied through techniques such as seed coating, priming, and pelleting. The efficacy of such treatments is manifested through enhanced seedling establishment and field performance. Treated seeds exhibit superior emergence rates and heightened resilience against adverse edaphic and climatic conditions (Khan *et al.*, 2003). The biofertilizer components facilitate nutrient mobilization and rhizosphere colonization, while phytochemical constituents confer antimicrobial and insecticidal properties. Systemic resistance is induced, bolstering the plant's defensive capabilities against pathogens and pests. This synergistic approach engenders a holistic enhancement of plant health and productivity. The integration of nanotechnology into seed treatment represents a significant advancement in agricultural biotechnology, offering novel solutions to longstanding challenges in crop production and stress management. This approach leverages the unique properties of nanoparticles (NPs) to enhance seed performance and plant resilience (Khilare and Gangawane, 1997).

Nanoparticles, characterized by their diminutive size and distinctive physicochemical attributes, possess the capacity to traverse biological barriers with unprecedented efficiency. This ability facilitates rapid penetration into seed structures, enabling direct modulation of cellular metabolic processes (Zhao *et al.*, 2024b). The catalytic properties inherent to many NPs further augment their potential to influence seed physiology at the molecular level. Recent empirical evidence demonstrates that nano-enabled seed treatments yield multifaceted benefits. These include enhanced germination rates, accelerated emergence, and improved resilience to both abiotic and biotic stressors. The mechanistic underpinnings of these improvements are attributed to the priming effect, wherein pre-exposure to specific stimuli—in this case, nanoparticles—engenders epigenetic, metabolic, and transcriptomic modifications. These alterations effectively "prime" the seed for rapid and robust activation of defense responses upon encountering subsequent stressors (Tiwari and Park, 2024).

The development of seed-specific nanoformulations represents a frontier in this field. Various nanostructures, including nanoencapsulations, nanoemulsions, nanocomposites, and nanogels, are being engineered to optimize delivery and efficacy. These formulations incorporate a diverse array of materials, ranging from polymeric and non-metallic substances to metallic and mesoporous structures. The tailored design of these nanoformulations addresses critical parameters such as stability, solubility, and controlled release of active compounds (Eevera *et al.*, 2023). This nanoscale approach to seed treatment offers several advantages over conventional chemical priming methods. Notably, it circumvents issues of instability and poor penetration associated with low molecular weight compounds traditionally used in seed priming. Moreover, the scalability and cost-effectiveness of nano-based treatments render them particularly promising for adoption by smallholder producers, potentially democratizing access to advanced agricultural technologies in the face of climate change (Shelar *et al.*, 2023).

Storage deterioration of seeds is markedly attenuated through the application of these bioactive formulations. The antifungal and insecticidal properties imbued to the seed coat serve to preclude pest infestation and pathogen proliferation during storage (Shivpuri *et al.*, 1997). Moisture uptake is regulated, thereby mitigating the deleterious effects of hydration-dehydration cycles. The overall seed viability is prolonged, ensuring maintenance of germination potential over extended periods. Thus, the integrative approach of melding biological agents with phytochemical constituents optimizes seed performance from storage through field establishment, culminating in enhanced agronomic outcomes and sustainable crop production.

9.3 Soil Treatment

Soil application via biofertilizers and botanical extracts, particularly those synthesized from agrifood waste, is executed to induce a series of agronomic enhancements that contribute to sustainable agriculture and environmental conservation (Chojnacka *et al.*, 2022). This method involves the strategic incorporation of organic biofertilizer materials into the edaphic substrate, where they undergo decomposition, thereby liberating essential nutrients that augment soil physicochemical properties, including structure, porosity, and moisture retention (Vurukonda *et al.*, 2024). Composting, inclusive of vermicomposting, accelerates the biodegradation of agrifood residues, yielding a humus-rich amendment replete with beneficial microbial consortia, such as nitrogen-fixing bacteria, phosphate-solubilizing organisms, and mycorrhizal fungi. These microbial agents colonize the rhizosphere, facilitating nutrient cycling, enhancing bioavailability, and fostering a dynamic and resilient soil ecosystem (Vyas *et al.*, 2022) (Fig. 9.2).

Liquid biofertilizers, when applied as a soil drench or integrated within drip irrigation systems, deliver a concentrated supply of nutrients and microbial inoculants directly to the root zone, thereby promoting root morphogenesis and optimizing nutrient uptake (Bradáčová *et al.*, 2019). Furthermore, botanical extracts, when infused into organic mulches, confer additional protective and nutritive benefits, including the suppression of soil-borne pathogens and pests, while enhancing the overall fertility of the soil matrix (Tóthné Bogdányi *et al.*, 2021).

This method substantially enhances soil health by promoting microbial diversity, improving

Fig. 9.2. Soil treatment using agrifood waste biofertilizers. Figure author's own.

soil aggregation, and mitigating compaction. The recycling of nutrients through the application of agrifood waste-based biofertilizers diminishes reliance on synthetic fertilizers, thereby curtailing nutrient leaching and reducing environmental pollution (Álvarez Salas *et al.*, 2024). Consequently, this approach alleviates the ecological burden imposed by conventional agricultural practices, fostering biodiversity and bolstering the sustainability of agricultural ecosystems (Prabhakar and Brar, 2024).

Moreover, biofertilizer applications contribute to the mitigation of greenhouse gas emissions by reducing the dependency on energy-intensive synthetic fertilizers and enhancing carbon sequestration within the soil matrix. This method concurrently mitigates the risk of eutrophication in adjacent aquatic ecosystems by minimizing nutrient runoff, thereby preserving water quality and protecting biodiversity (Sharma *et al.*, 2023).

Soils treated with these bioactive formulations exhibit enhanced resilience to abiotic stressors, including drought, salinity, and temperature fluctuations, owing to improved soil water retention and nutrient mobilization (Rahim *et al.*, 2024). The synergistic interaction between biofertilizers and botanical extracts not only bolsters crop yields and enhances produce quality but also aligns with the overarching principles of sustainable agriculture, ensuring long-term food security while conserving natural resources and safeguarding environmental integrity. The soil application of agrifood waste-based biofertilizers and botanical extracts engenders a comprehensive enhancement of soil health, promoting a robust, sustainable agricultural paradigm and contributing to a cleaner, more resilient environment (Amanullah, 2024).

9.4 Foliar Treatment

Foliar treatment is a method of applying nutrients directly to plant leaves, offering an effective alternative to traditional soil-based fertilization techniques. This approach is particularly beneficial for addressing nutrient deficiencies quickly and efficiently, as nutrients are absorbed directly through the leaf surface. By bypassing the soil, this method mitigates issues such as nutrient fixation, leaching, or poor soil fertility (Kaya and Ashraf, 2024). When coupled with biofertilizers derived from agrifood waste, foliar treatment not only enhances plant growth but also aligns with sustainable agricultural practices, reducing the environmental footprint of farming (Voss *et al.*, 2024).

The primary advantage of foliar treatment lies in its ability to deliver nutrients rapidly, ensuring a swift response to deficiencies. Within hours of application, plants begin absorbing the nutrients, leading to visible improvements in health and vitality. This immediacy is particularly crucial in cases where soil applications would take longer to show results (Ray *et al.*, 2024). Foliar treatments are also more resource-efficient, as they minimize nutrient wastage that can occur through soil processes like leaching or immobilization. Additionally, because this method involves lower quantities of fertilizers compared to soil amendments, it is more environmentally friendly, reducing the risk of nutrient runoff into water bodies and subsequent eutrophication (Voss *et al.*, 2024) (Fig. 9.3).

In sustainable agriculture, foliar treatment is especially valuable for delivering micronutrients such as zinc (Zn), boron (B), and manganese (Mn). These nutrients are essential for various physiological processes in plants, including photosynthesis, enzyme activation, and reproductive development (Singh *et al.*, 2013). For example, zinc plays a critical role in carbohydrate metabolism, while boron is vital for cell wall structure and pollen tube formation. Manganese, on the other hand, is crucial for photosynthesis and

Fig. 9.3. Foliar treatment using agrifood waste biofertilizers. Figure author's own.

nitrogen metabolism. Foliar application ensures that these micronutrients are directly available to plants, even under adverse soil conditions such as high pH or low organic matter content, where nutrient availability is naturally limited (Karim *et al.*, 2012).

The use of foliar-applied biofertilizers derived from agrifood waste is a sustainable innovation in modern agriculture. These biofertilizers not only provide essential nutrients but also improve soil health over time by reducing dependency on synthetic chemicals (Stewart *et al.*, 2021). For instance, research has demonstrated that foliar applications of zinc and manganese significantly enhance grain yield and quality in crops like rice, wheat, and barley. Similarly, boron foliar sprays have been shown to improve flowering and fruit setting, particularly under conditions of water stress. These benefits make foliar treatment a critical tool in promoting crop productivity and resilience in the face of climate challenges (Dass *et al.*, 2022).

Despite its advantages, foliar treatment requires careful consideration of application timing, concentration, and environmental conditions to maximize its efficacy. For optimal results, foliar sprays should be applied during critical growth stages, such as flowering or grain filling, when plants have the highest nutrient demand (Fageria *et al.*, 2009). Over-application or improper formulation can lead to leaf damage, emphasizing the need for precise management. Environmental factors such as temperature, humidity, and wind speed also play a significant role in the absorption and effectiveness of foliar-applied nutrients (Rehman *et al.*, 2019).

Foliar treatment represents an innovative and sustainable approach to fertilization, particularly when used with biofertilizers made from agrifood waste. By delivering nutrients directly to plants, this method addresses deficiencies efficiently, enhances crop productivity, and minimizes environmental harm. As agriculture continues to evolve toward sustainability, integrating foliar treatment into farming practices offers a promising pathway to achieving both high yields and ecological balance (Vurukonda *et al.*, 2024).

Foliar application of biofertilizers and botanical extracts, particularly those synthesized from agrifood waste, is a method employed to induce a spectrum of agronomic enhancements that align with sustainable agricultural practices and environmental stewardship. This approach entails the direct application of liquid biofertilizer formulations and botanical extracts onto the aerial parts of plants, enabling the rapid assimilation of essential nutrients and bioactive compounds through the plant's cuticle and stomatal apertures. The foliar route is particularly effective for addressing acute nutrient deficiencies, amplifying plant vigor, and imparting protective properties against biotic and abiotic stressors (Zanzotti *et al.*, 2023).

Biofertilizers derived from agrifood waste incorporate a diverse array of beneficial microorganisms and phytonutrients that are swiftly absorbed by the foliage, leading to increased photosynthetic capacity, elevated chlorophyll synthesis, and enhanced overall metabolic processes in the plant. These formulations frequently include vital micronutrients—such as iron, zinc, and manganese—essential for enzymatic activities and more effectively utilized when administered via the foliar route. Botanical extracts, obtained from various plant parts including leaves, stems, and roots, are integrated into these foliar applications due to their pesticidal, antifungal, and growth-regulatory properties. The phytochemical constituents in these extracts synergize with biofertilizer components to strengthen the plant's innate defense systems and enhance its resilience to pathogenic attacks and environmental stress (Laishram *et al.*, 2024).

The foliar application technique is particularly beneficial in scenarios where soil conditions hinder nutrient uptake, such as in saline or waterlogged environments, or during drought conditions when root absorption is compromised. By bypassing the soil medium, foliar application ensures the direct delivery of nutrients to the photosynthetically active tissues, thereby sustaining plant growth and development even under adverse conditions. Additionally, this method allows for the prompt rectification of nutrient imbalances, which is critical during crucial phenological stages such as flowering and fruiting (Hu *et al.*, 2023).

Employing agrifood waste-based biofertilizers in foliar applications significantly advances sustainable agriculture by diminishing the dependence on synthetic foliar fertilizers, which are often energy-intensive to produce and may contribute to environmental degradation through runoff and volatilization. This approach not only conserves natural resources by recycling organic waste into valuable agricultural inputs but also reduces the ecological footprint of farming practices (Chew *et al.*, 2019). The inclusion of botanical extracts further augments sustainability by providing a natural alternative to chemical pesticides, thereby decreasing the chemical burden on the environment and fostering biodiversity within agroecosystems.

The application of these bioactive foliar treatments also plays a pivotal role in enhancing crop yield and quality. Plants receiving foliar treatments with biofertilizers and botanical extracts exhibit superior tolerance to abiotic stresses, including extreme temperatures, ultraviolet radiation, and oxidative stress, due to the presence of antioxidants and stress-alleviating compounds in the formulations. Furthermore, the rapid uptake and systemic distribution of these bioactive substances within the plant contribute to uniform crop development, resulting in increased yields and improved quality of produce (Ngegba *et al.*, 2022).

Thus, the foliar application of agrifood waste-based biofertilizers and botanical extracts emerges as a crucial component in the paradigm of sustainable agriculture. This method offers an efficient means of nutrient delivery, stress mitigation, and environmental conservation, while simultaneously promoting enhanced plant health, improved agricultural productivity, and a reduced ecological impact (Calabi-Floody *et al.*, 2018). Such a comprehensive approach is indispensable for advancing sustainable agricultural practices and ensuring food security amidst the global challenges posed by environmental changes (Selvaraju *et al.*, 2011).

9.5 Integration with Other Farming Practices

The synergistic integration of foliar, soil, and seed treatment methods utilizing agrifood waste-based biofertilizers and botanical extracts into broader agricultural practices is essential for enhancing sustainable farming and environmental protection (Zhao *et al.*, 2024a). When these techniques are applied in tandem, they significantly improve nutrient management, pest resistance, and stress tolerance, thereby advancing agricultural productivity and ecological balance (Zhou *et al.*, 2024).

Seed treatment with biofertilizers and botanical extracts lays a strong foundation for crop growth by enhancing seed germination and seedling vigor, and by providing early protection against various stress factors. This initial treatment is effectively complemented by soil application, where biofertilizers and botanical extracts are incorporated into the soil to enrich soil fertility, enhance soil structure, and promote beneficial microbial populations (Husen, 2024). These microbes colonize the root zone, facilitating nutrient cycling and establishing optimal conditions for root development. The combination of seed treatment and soil application ensures that crops are provided with an ideal growth environment from the outset, leading to improved establishment and early-stage growth (Oyedoh *et al.*, 2024).

As crops develop, foliar application of biofertilizers and botanical extracts serves as a precise method for addressing nutrient deficiencies and mitigating stress during critical growth phases. By applying nutrients and bioactive compounds directly to the plant foliage, foliar application acts as a timely intervention that sustains plant health during key stages such as vegetative growth, flowering, and fruiting. The integration of foliar application with prior seed and soil treatments creates a multi-layered approach to crop management, where each method reinforces the others, resulting in a comprehensive strategy for improving plant resilience and productivity (Shahrajabian *et al.*, 2022).

Incorporating these biofertilizer applications into other farming practices such as crop rotation, intercropping, and conservation tillage, further enhances their effectiveness. In crop rotation, the residual benefits of soil-applied biofertilizers from one crop cycle can improve soil health and nutrient availability for the following crops, thus reducing the need for additional inputs. In intercropping systems, foliar applications can be customized to meet the diverse nutritional needs of multiple crops grown together, optimizing nutrient use and minimizing competition (Wakindiki *et al.*, 2018). Conservation tillage, which reduces soil disturbance, works in harmony with soil biofertilizer applications by maintaining soil structure and organic matter, thereby supporting the long-term health of the soil ecosystem (Bhattacharya *et al.*, 2023).

Moreover, combining these biofertilizer applications with precision agriculture technologies, such as soil sensors and plant health monitoring systems, allows for more accurate management of nutrient levels and plant health. This precision enhances the efficiency of biofertilizer use and reduces environmental impacts, such as nutrient leaching and runoff, which can lead to soil and water contamination (Stafford, 2000).

By integrating seed treatment, soil application, and foliar application with other farming activities, a more resilient agricultural system is established—one that is better equipped to handle the challenges of climate change, resource limitations, and environmental degradation (Lin, 2011). Utilizing agrifood waste-based biofertilizers and botanical extracts in a coordinated manner enables farmers to achieve higher crop yields, improve soil health, and reduce dependency on synthetic agrochemicals. This approach contributes to the advancement of sustainable agriculture and environmental protection, ensuring the long-term sustainability of food production systems and the preservation of natural resources (Zafar *et al.*, 2024).

9.6 Promotion of Sustainable Agriculture

Agro-based biofertilizers are gaining traction in sustainable agriculture as an alternative to chemical fertilizers. These biofertilizers are derived from agricultural and food waste, containing beneficial microorganisms that enhance soil fertility and plant growth (Zhao *et al.*, 2024a). The use of waste materials like fruit peels, vegetable scraps, and food canteen waste in biofertilizer production addresses both waste management issues and soil health concerns in agriculture. The efficacy of agro-based biofertilizers in improving crop yields and soil quality, for instance, solid-state fermentation of fruit waste, has been shown to increase microbial diversity, particularly *Aspergillus* and *Bacillus* species, leading to improved soil macronutrient content and enhanced plant growth (Zaini *et al.*, 2023). Studies have also found that biofertilizers made from fruit peel waste can promote mustard plant growth and increase potassium content in the soil.

The integration of organic fertilization methodologies within vertical farming systems represents a paradigm shift toward circular economic principles, aimed at mitigating resource influx into urban environments. This approach encompasses a diverse array of nutrient sourcing strategies, each with its own merits and limitations. One such avenue involves the utilization of anthropogenic waste streams, including grey water and human urine, as nutrient reservoirs. These materials, rich in essential plant nutrients, offer a promising alternative to conventional fertilizers. Concurrently, the application of biofertilizers, exemplified by *Rhizobium* species in leguminous crop cultivation, presents a symbiotic approach to nitrogen fixation, potentially reducing reliance on exogenous nitrogen inputs. Further exploration has been conducted into the repurposing of municipal waste products, such as sewage sludge and its incinerated ash (Arcas-Pilz *et al.*, 2021).

Biofertilizers, comprising beneficial microorganisms such as *Rhizobium*, *Azospirillum*, and arbuscular mycorrhizal fungi (AMF), serve as pivotal agents in the maintenance of soil biodiversity. These microbial communities act as keystone species, modulating the intricate web of below-ground interactions. For instance,

AMF produce glomalin, a glycoprotein that significantly contributes to soil aggregation and carbon sequestration, thereby fostering a stable soil structure conducive to diverse microbial habitats. The application of biofertilizers has demonstrated remarkable efficacy in augmenting crop yields whilst concomitantly reducing reliance on synthetic fertilizers. This dual benefit not only enhances agronomic productivity but also mitigates the deleterious effects of chemical inputs on non-target organisms (Singh *et al.*, 2016). For example, studies have shown that *Rhizobium* inoculation in pulse crops cultivated under temperate conditions significantly improves yield attributes, while *Azospirillum* application enhances leaf area index and harvest index across various agricultural crops. Moreover, the integration of biofertilizers into agricultural practices has been observed to stimulate soil enzymatic activities. The enhanced alkaline phosphatase activity, particularly notable in *Azotobacter chroococcum* treatments, indicates a more dynamic nutrient cycling process. This biochemical enhancement fosters a more diverse and functionally robust soil ecosystem, supporting a wider array of soil fauna and flora.

The synergistic application of biofertilizers with reduced rates of chemical fertilizers has shown promising results. For instance, the combination of *Glomus fasciculatum* and *Azotobacter chroococcum* with half the recommended phosphorus rate improved bulb size in onion crops (Singh *et al.*, 2016). Such integrated approaches not only optimise resource utilization but also promote a more balanced soil ecosystem, conducive to the proliferation of diverse microbial communities.

The application of these biofertilizers contributes to soil health in multiple ways. They improve soil structure, increase water retention capacity, and enhance microbial activity (Souza *et al.*, 2015). For example, the use of *Paecilomyces lilacinus*, a fungus found in some food waste-based biofertilizers, has been observed to inhibit the growth of harmful nematophagous fungi (Zhao *et al.*, 2024a). Additionally, the organic acids produced during the fermentation of waste materials can help reduce soil pH and increase nitrogen content, creating a more favorable environment for plant growth. Integrating agro-based biofertilizers into farming practices aligns with the principles of sustainable agriculture. This approach reduces reliance on synthetic inputs, promotes nutrient recycling, and supports the development of more resilient agricultural systems. By transforming agricultural by-products and food waste into valuable soil amendments, farmers can reduce external inputs and minimize environmental impacts. This shift toward bio-based fertilizers represents a step toward more sustainable and resource-efficient farming practices, contributing to long-term food security and environmental conservation (Singh *et al.*, 2016).

9.7 Clean Environment

The application of agrifood waste-based biofertilizers induces substantial alterations in soil physicochemical properties and microbial ecology. These amendments facilitate enhanced nutrient cycling, augmenting soil organic matter content and promoting the proliferation of beneficial microorganisms (Ammar *et al.*, 2023). Concomitantly, the reduced application of synthetic fertilizers mitigates the risks of eutrophication and soil acidification, thereby preserving ecosystem integrity. The enhanced soil structure resulting from biofertilizer application improves water retention capacity and reduces erosion susceptibility, contributing to long-term soil health and stability (Sreethu *et al.*, 2024).

Biofertilizers derived from agrifood waste exhibit significant potential in phytoremediation processes. The bioaugmentation of contaminated soils with these products can enhance the degradation of recalcitrant pollutants and facilitate the sequestration of heavy metals (Bhatia and Sindhu, 2024). This bioremediation approach not only addresses legacy contamination issues but also prevents further environmental degradation associated with conventional remediation techniques. The synergistic action of plant–microbe interactions stimulated by biofertilizers accelerates the restoration of polluted ecosystems, promoting the reestablishment of native flora and fauna (da Fonseca and Gaylarde, 2024). The implementation of agrifood waste-based biofertilizers in agricultural systems engenders a reduction in greenhouse gas emissions through multiple pathways. The decreased reliance on energy-intensive synthetic fertilizer production and the enhanced carbon sequestration potential of improved soils contribute to climate change mitigation efforts (Kumar *et al.*, 2022). Furthermore, the efficient recycling of organic waste reduces methane emissions associated with landfill disposal, while the improved nutrient use efficiency minimizes nitrous oxide emissions from agricultural soils. These cumulative effects significantly diminish the carbon footprint of agricultural practices (Yasmin *et al.*, 2022).

The adoption of biofertilizers in agriculture catalyzes a shift toward more sustainable pest management strategies. The enhanced plant vigor and induced systemic resistance resulting from biofertilizer application reduce crop susceptibility to pathogens and pests, thereby decreasing the necessity for chemical pesticides. This reduction in pesticide use mitigates the adverse impacts on non-target organisms, preserves biodiversity, and reduces the bioaccumulation of harmful substances in food chains (Pandiyan *et al.*, 2024). The resultant ecosystem equilibrium promotes natural pest control mechanisms, further reducing the need for external interventions. The integration of agrifood waste-based biofertilizers into agricultural systems facilitates the development of closed-loop nutrient cycles, aligning with circular economy principles (Gamage *et al.*, 2023). This approach not only minimizes waste generation but also optimizes resource utilization, enhancing the overall sustainability of food production systems. The reduced dependence on finite resources for fertilizer production and the valorization of waste streams contribute to resource conservation efforts (Sarangi *et al.*, 2024). Moreover, the improved soil health and reduced environmental externalities associated with biofertilizer use enhance the long-term viability and resilience of agricultural ecosystems, ensuring continued productivity in the face of climate change and other environmental stressors (Ibáñez *et al.*, 2023).

References

Ahmed, S. and Agnihotri, J. (1977) Antifungal activity of some plant extracts. *Indian Journal of Mycology and Plant Pathology* 7(2), 180–181.

Álvarez Salas, M., Sica, P., Rydgård, M., Sitzmann, T.J., Nyang'au, J.O. *et al.* (2024) Current challenges on the widespread adoption of new bio-based fertilizers: insights to move forward toward more circular food systems. *Frontiers in Sustainable Food Systems* 8, 1386680. DOI: 10.3389/fsufs.2024.1386680

Amanullah, K. (2024) *Integrated Agriculture: An Approach for Sustainable Agriculture*, Vol. 3. Walter de Gruyter, Berlin, Germany.

Ammar, E.E., Rady, H.A., Khattab, A.M., Amer, M.H., Mohamed, S.A. *et al.* (2023) A comprehensive overview of eco-friendly bio-fertilizers extracted from living organisms. *Environmental Science and Pollution Research International* 30(53), 113119–113137. DOI: 10.1007/s11356-023-30260-x

Arcas-Pilz, V., Parada, F., Villalba, G., Rufí-Salis, M., Rosell-Melé, A. *et al.* (2021) Improving the fertigation of soilless urban vertical agriculture through the combination of struvite and rhizobia inoculation in *Phaseolus vulgaris*. *Frontiers in Plant Science* 12, 649304. DOI: 10.3389/fpls.2021.649304

Arora, N.K., Fatima, T., Mishra, I., and Verma, S. (2020) Microbe-based inoculants: role in next green revolution. In: Shukla, V. and Kumar, N. (eds) *Environmental Concerns and Sustainable Development: Volume 2: Biodiversity, Soil and Waste Management*. Springer, Singapore, pp. 191–246.

Bakonyi, N., Bott, S., Gajdos, E., Szabó, A., Jakab, A. *et al.* (2013) Using biofertilizer to improve seed germination and early development of maize. *Polish Journal of Environmental Studies* 22(6), 1595–1599.

Bayu, T. (2024) Systematic review on the role of microbial activities on nutrient cycling and transformation implication for soil fertility and crop productivity. *bioRxiv*. DOI: 10.1101/2024.09.02.610905

Bhatia, T. and Sindhu, S.S. (2024) Sustainable management of organic agricultural wastes: contributions in nutrients availability, pollution mitigation and crop production. *Discover Agriculture* 2(1), 1–42. DOI: 10.1007/s44279-024-00147-7

Bhattacharya, U., Naskar, M.K., Venugopalan, V.K., Sarkar, S., Bandopadhyay, P. *et al.* (2023) Implications of minimum tillage and integrated nutrient management on yield and soil health of rice-lentil cropping system – being a resource conservation technology. *Frontiers in Sustainable Food Systems* 7, 1225986. DOI: 10.3389/fsufs.2023.1225986

Bradáčová, K., Sittinger, M., Tietz, K., Neuhäuser, B., Kandeler, E. *et al.* (2019) Maize inoculation with microbial consortia: contrasting effects on rhizosphere activities, nutrient acquisition and early growth in different soils. *Microorganisms* 7(9), 329. DOI: 10.3390/microorganisms7090329

Calabi-Floody, M., Medina, J., Rumpel, C., Condron, L.M., Hernandez, M. *et al.* (2018) Smart fertilizers as a strategy for sustainable agriculture. *Advances in Agronomy* 147, 119–157. DOI: 10.1016/bs.agron.2017.10.003

Chew, K.W., Chia, S.R., Yen, H.-W., Nomanbhay, S., Ho, Y.-C. *et al.* (2019) Transformation of biomass waste into sustainable organic fertilizers. *Sustainability* 11(8), 2266. DOI: 10.3390/su11082266

Chiquito-Contreras, C.J., Meza-Menchaca, T., Guzmán-López, O., Vásquez, E.C., and Ricaño-Rodríguez, J. (2024) Molecular insights into plant–microbe interactions: a comprehensive review of key mechanisms. *Frontiers in Bioscience-Elite* 16(1), 9. DOI: 10.31083/j.fbe1601009

Chojnacka, K., Moustakas, K., and Mikulewicz, M. (2022) Valorisation of agri-food waste to fertilisers is a challenge in implementing the circular economy concept in practice. *Environmental Pollution* 12, 119906. DOI: 10.1016/j.envpol.2022.119906

da Fonseca, E.M. and Gaylarde, C.C.C. (2024) Biofertilization and bioremediation - how can microbiological technology assist the ecological crisis in developing countries? *Environmental and Earth Sciences*. DOI: 10.20944/preprints202411.0003.v1

Dal Cortivo, C., Ferrari, M., Visioli, G., Lauro, M., Fornasier, F. *et al.* (2020) Effects of seed-applied biofertilizers on rhizosphere biodiversity and growth of common wheat (*Triticum aestivum* L.) in the field. *Frontiers in Plant Science* 11, 72. DOI: 10.3389/fpls.2020.00072

Dass, A., Rajanna, G.A., Babu, S., Lal, S.K., Choudhary, A.K. *et al.* (2022) Foliar application of macro- and micronutrients improves the productivity, economic returns, and resource-use efficiency of soybean in a semiarid climate. *Sustainability* 14(10), 5825. DOI: 10.3390/su14105825

Eevera, T., Kumaran, S., Djanaguiraman, M., Thirumaran, T., Le, Q.H. *et al.* (2023) Unleashing the potential of nanoparticles on seed treatment and enhancement for sustainable farming. *Environmental Research* 236(Pt 2), 116849. DOI: 10.1016/j.envres.2023.116849

Fageria, N.K., Filho, M.P.B., Moreira, A., and Guimarães, C.M. (2009) Foliar fertilization of crop plants. *Journal of Plant Nutrition* 32(6), 1044–1064. DOI: 10.1080/01904160902872826

Gamage, A., Gangahagedara, R., Gamage, J., Jayasinghe, N., Kodikara, N. *et al.* (2023) Role of organic farming for achieving sustainability in agriculture. *Farming System* 1(1), 100005. DOI: 10.1016/j.farsys.2023.100005

Hu, Y., Bellaloui, N., and Kuang, Y. (2023) Factors affecting the efficacy of foliar fertilizers and the uptake of atmospheric aerosols, volume II. *Frontiers in Plant Science* 14, 1146853. DOI: 10.3389/fpls.2023.1146853

Husen, A. (2024) *Biostimulants in Plant Protection and Performance*. Elsevier, Amsterdam.

Ibáñez, A., Garrido-Chamorro, S., Vasco-Cárdenas, M.F., and Barreiro, C. (2023) From lab to field: biofertilizers in the 21st century. *Horticulturae* 9(12), 1306. DOI: 10.3390/horticulturae9121306

Karim, M.R., Zhang, Y.Q., Zhao, R.R., Chen, X.P., Zhang, F.S. *et al.* (2012) Alleviation of drought stress in winter wheat by late foliar application of zinc, boron, and manganese. *Journal of Plant Nutrition and Soil Science* 175(1), 142–151. DOI: 10.1002/jpln.201100141

Karuvelan, M., Raj, S., Gururajan, G., Chelliah, R., Sultan, G. *et al.* (2025) Perspectives of nanomaterials for crop plants improvement and practices. In: Husen, A. (ed.) *Agricultural Crop Improvement*. CRC Press, Boca Raton, Florida, pp. 1–22.

Kavitha Shree, G., Arokiamary, S., Kamaraj, M., and Aravind, J. (2024) Biorefinery approaches for converting fruit and vegetable waste into sustainable products. *International Journal of Environmental Science and Technology* 22(8), 7211–7230. DOI: 10.1007/s13762-024-06202-6

Kaya, C. and Ashraf, M. (2024). Foliar fertilization: a potential strategy for improving plant salt tolerance. *Critical Reviews in Plant Sciences* 43(2), 94–115. DOI: 10.1080/07352689.2023.2270253

Kazmi, A., Niaz, I., and Jilani, G. (1993) Evaluation of some plant extracts for antifungal properties. *Pakistan Journal of Phytopathology* 5(1–2), 93–97.

Khalequzzaman, K. and Hossain, I. (2007) Effect of seed treatment with Rhizobium strains and biofertilizers on foot/root rot and yield of bushbean in *Fusarium solani* infested soil. *Journal of Agricultural Research* 45(2), 151–160. DOI: 10.58475/e6kkgd36

Khan, G., Keshavulu, K., Reddy, B.M., and Radhika, K. (2003) Effects of pre-sowing seed treatments on the establishment of sunflower. *Seed Research* 31(1), 94–97.

Khan, M.A., Khan, R., Praveen, P., Verma, A.R., and Panda, M.K. (2024) *Infrastructure Possibilities and Human-Centered Approaches with Industry 5.0*. IGI Global, Hershey, Pennsylvania.

Khilare, V. and Gangawane, L. (1997) Application of medicinal plant extracts in the management of thiophanate methyl resistant *Penicillium digitatum* causing green mold of Mosambi. *Journal of Mycology and Plant Pathology* 27(2), 134–137.

Kumar, S., Sindhu, S.S., and Kumar, R. (2022) Biofertilizers: an ecofriendly technology for nutrient recycling and environmental sustainability. *Current Research in Microbial Sciences* 3, 100094. DOI: 10.1016/j.crmicr.2021.100094

Laishram, B., Devi, O.R., Dutta, R., Senthilkumar, T., Goyal, G. *et al.* (2024) Plant–microbe interactions: PGPM as microbial inoculants/biofertilizers for sustaining crop productivity and soil fertility. *Current Research in Microbial Sciences* 8, 100333. DOI: 10.1016/j.crmicr.2024.100333

Lin, B.B. (2011) Resilience in agriculture through crop diversification: adaptive management for environmental change. *BioScience* 61(3), 183–193. DOI: 10.1525/bio.2011.61.3.4

Malusá, E. and Vassilev, N. (2014) A contribution to set a legal framework for biofertilisers. *Applied Microbiology and Biotechnology* 98(15), 6599–6607. DOI: 10.1007/s00253-014-5828-y

Mishra, V., Mudgal, N., Rawat, D., Poria, P., Mukherjee, P. *et al.* (2023) Integrating microalgae into textile wastewater treatment processes: advancements and opportunities. *Journal of Water Process Engineering* 55, 104128. DOI: 10.1016/j.jwpe.2023.104128

Ngegba, P.M., Cui, G., Khalid, M.Z., and Zhong, G. (2022) Use of botanical pesticides in agriculture as an alternative to synthetic pesticides. *Agriculture* 12(5), 600. DOI: 10.3390/agriculture12050600

Oyedoh, O.P., Compant, S., Doty, S.L., Santoyo, G., Glick, B.R. *et al.* (2024) Root colonizing microbes associated with notable abiotic stress of global food and cash crops. *Plant Stress* 15, 100714. DOI: 10.1016/j.stress.2024.100714

Pandiyan, A., Sarsan, S., Durga, G.G.S., and Ravikumar, H. (2024) Biofertilizers and biopesticides as microbial inoculants in integrated pest management for sustainable agriculture. In: Singh, R.P., Manchanda, G., Sarsan, S., Kumar, A., and Panosyan, H. (eds) *Microbial Essentialism*. Elsevier, Amsterdam, pp. 485–518.

Prabhakar, A.C. and Brar, G.P. (2024) Green revolution, agricultural performance with sustainability and bio-diversity: special reference to India. *International Journal of Economic Performance* 7(1), 281. DOI: 10.54241/2065-007-001-015

Punj, V. and Gupta, R. (1988) VA-mycorrhizal fungi and Rhizobium as biological fertilizers for *Leucaena leucocephala*. *Acta Microbiologica Polonica* 37(3/4) 327–336.

Rahim, H.U., Ali, W., Uddin, M., Ahmad, S., Khan, K. *et al.* (2025) Abiotic stresses in soils, their effects on plants, and mitigation strategies: a literature review. *Chemistry and Ecology* 41(4), 552–585. DOI: 10.1080/02757540.2024.2439830

Ray, S., Maitra, S., Sairam, M., Sravya, M., Priyadarshini, A. *et al.* (2024) An unravelled potential of foliar application of micro and beneficial nutrients in cereals for ensuring food and nutritional security. *International Journal of Experimental Research and Review* 41(Special issue), 19–42. DOI: 10.52756/ijerr.2024.v41spl.003

Rehman, A., Ullah, A., Nadeem, F., and Farooq, M. (2019) Sustainable nutrient management. In: Farooq, M. and Pisante, M. (eds) *Innovations in Sustainable Agriculture*. Springer, Cham, Switzerland, pp. 167–211.

Sarangi, P.K., Pal, P., Singh, A.K., Sahoo, U.K., and Prus, P. (2024) Food waste to food security: transition from bioresources to sustainability. *Resources* 13(12), 164. DOI: 10.3390/resources13120164

Sarker, A., Ahmmed, R., Ahsan, S.M., Rana, J., Ghosh, M.K. *et al.* (2024) A comprehensive review of food waste valorization for the sustainable management of global food waste. *Sustainable Food Technology* 2(1), 48–69. DOI: 10.1039/D3FB00156C

Selvaraju, R., Gommes, R., and Bernardi, M. (2011) Climate science in support of sustainable agriculture and food security. *Climate Research* 47(1), 95–110. DOI: 10.3354/cr00954

Shahrajabian, M.H., Sun, W., and Cheng, Q. (2022) Foliar application of nutrients on medicinal and aromatic plants, the sustainable approaches for higher and better production. *Beni-Suef University Journal of Basic and Applied Sciences* 11(1), 26. DOI: 10.1186/s43088-022-00210-6

Sharma, B., Tiwari, S., Kumawat, K.C., and Cardinale, M. (2023) *The Science of the Total Environment* 860, 160476. DOI: 10.1016/j.scitotenv.2022.160476

Shelar, A., Nile, S.H., Singh, A.V., Rothenstein, D., Bill, J. *et al.* (2023) Recent advances in nano-enabled seed treatment strategies for sustainable agriculture: challenges, risk assessment, and future perspectives. *Nano-Micro Letters* 15(1), 54. DOI: 10.1007/s40820-023-01025-5

Shivpuri, A., Sharma, O., and Jhamaria, S. (1997) Fungitoxic properties of plant extracts against pathogenic fungi. *Journal of Mycology and Plant Pathology* 27, 9–31.

Singh, J., Singh, M., Jain, A., Bhardwaj, S., Singh, A. *et al.* (2013) An introduction of plant nutrients and foliar fertilization: a review. In: Ram, T., Lohan, S.K., Singh, R., and Singh, P. (eds) *Precision Farming: A New Approach*. Daya Publishing Company, New Delhi, pp. 252–320.

Singh, M., Dotaniya, M., Mishra, A., Dotaniya, C., Regar, K. *et al.* (2016) Role of biofertilizers in conservation agriculture. In: Bisht, J.K., Meena, V.S., Mishra, P.K., and Pattanayak, A. (eds) *Conservation Agriculture: An Approach to Combat Climate Change in Indian Himalaya*. Springer, Singapore, pp. 113–134.

Souza, R. de, Ambrosini, A., and Passaglia, L.M.P. (2015) Plant growth-promoting bacteria as inoculants in agricultural soils. *Genetics and Molecular Biology* 38(4), 401–419. DOI: 10.1590/S1415-475738420150053

Sreethu, S., Chhabra, V., Kaur, G., and Ali, B. (2024) Biofertilizers as a greener alternative for increasing soil fertility and improving food security under climate change condition. *Communications in Soil Science and Plant Analysis* 55(2), 261–285. DOI: 10.1080/00103624.2023.2265945

Stafford, J.V. (2000) Implementing precision agriculture in the 21st century. *Journal of Agricultural Engineering Research* 76(3), 267–275. DOI: 10.1006/jaer.2000.0577

Stewart, Z.P., Paparozzi, E.T., Wortmann, C.S., Jha, P.K., and Shapiro, C.A. (2021) Effect of foliar micronutrients. *Plants* 10(3), 528. DOI: 10.3390/plants10030528

Tiwari, P. and Park, K.-I. (2024) Seed biotechnologies in practicing sustainable agriculture: insights and achievements in the decade 2014–2024. *Applied Sciences* 14(24), 11620. DOI: 10.3390/app142411620

Tóthné Bogdányi, F., Boziné Pullai, K., Doshi, P., Erdős, E., Gilián, L.D. *et al.* (2021) Composted municipal green waste infused with biocontrol agents to control plant parasitic nematodes – a review. *Microorganisms* 9(10), 2130. DOI: 10.3390/microorganisms9102130

Voss, M., Valle, C., Calcio Gaudino, E., Tabasso, S., Forte, C. *et al.* (2024) Unlocking the potential of agrifood waste for sustainable innovation in agriculture. *Recycling* 9(2), 25. DOI: 10.3390/recycling9020025

Vurukonda, S.S.K.P., Fotopoulos, V., and Saeid, A. (2024) Production of a rich fertilizer base for plants from waste organic residues by microbial formulation technology. *Microorganisms* 12(3), 541. DOI: 10.3390/microorganisms12030541

Vyas, P., Sharma, S., and Gupta, J. (2022) Vermicomposting with microbial amendment: implications for bioremediation of industrial and agricultural waste. *Biotechnologia* 103(2), 203–215. DOI: 10.5114/bta.2022.116213

Wakindiki, I.I., Malobane, M.E., and Nciizah, A.D. (2018) Integrating biofertilizers with conservation agriculture can enhance its capacity to mitigate climate change. In: Leal Filho, W. and Leal-Arcas, R. (eds) *University Initiatives in Climate Change Mitigation and Adaptation*. Springer, Cham, Switzerland, pp. 277–289.

Yasmin, N., Jamuda, M., Panda, A.K., Samal, K., and Nayak, J.K. (2022) Emission of greenhouse gases (GHGs) during composting and vermicomposting: measurement, mitigation, and perspectives. *Energy Nexus* 7, 100092. DOI: 10.1016/j.nexus.2022.100092

Zafar, S., Bilal, M., Ali, M.F., Mahmood, A., Kijsomporn, J. *et al.* (2024) Nano-biofertilizer an eco-friendly and sustainable approach for the improvement of crops under abiotic stresses. *Environmental and Sustainability Indicators* 24, 100470. DOI: 10.1016/j.indic.2024.100470

Zaini, N.S.M., Khudair, A.J.D., Mohsin, A.Z., Jitming Lim, E., Minato, W. *et al.* (2023) Biotransformation of food waste into biofertilisers through composting and anaerobic digestion: a review. *Plant, Soil and Environment* 69(9), 409–420. DOI: 10.17221/101/2023-PSE

Zanzotti, R., Bertoldi, D., Baldantoni, D., and Morelli, R. (2023) Soil fertility and agronomic performance of green manure in vineyard. In: *IV Convegno AISSA#under40, Fisciano (SA), Italy, 12–13 July*, p. 142.

Zhao, G., Zhu, X., Zheng, G., Meng, G., Dong, Z. *et al.* (2024a) Development of biofertilizers for sustainable agriculture over four decades (1980–2022). *Geography and Sustainability* 5(1), 19–28. DOI: 10.1016/j.geosus.2023.09.006

Zhao, L., Zhou, X., Kang, Z., Peralta-Videa, J.R., and Zhu, Y.-G. (2024b) Nano-enabled seed treatment: a new and sustainable approach to engineering climate-resilient crops. *Science of The Total Environment* 910, 168640. DOI: 10.1016/j.scitotenv.2023.168640

Zhou, W., Arcot, Y., Medina, R.F., Bernal, J., Cisneros-Zevallos, L. *et al.* (2024) Integrated pest management: an update on the sustainability approach to crop protection. *ACS Omega* 9(40), 41130–41147. DOI: 10.1021/acsomega.4c06628

10 Economics of Biofertilizers

Abha Kumari*
Centre for Biotechnology and Biochemical Engineering, Amity Institute of Biotechnology, Amity University, Noida, India

Abstract
The global biofertilizer sector has recorded intense development in the last ten years due to rising interest in sustainability and the capacity of the environment, land, and the damaging effects of the chemical fertilizers. This chapter gives an in-depth analysis of the biofertilizer industry, describing the most important developments in North and South America, the European Union, the Asia–Pacific region, and Africa. It delves into the effects of favorable government policies, such as the organic farming initiatives implemented by India and the Green Deal of the European Union, on the growth of this market. The chapter also discusses the financial benefits of biofertilizers, such as their decreased reliance on chemical fertilizers that must be imported and their comparatively low cost of production and application. Biofertilizers also benefit the environment by improving soil health and lowering greenhouse gas emissions. A more specific taxonomy of biofertilizers by source of microbe and type of production model is offered, and some discussion of technological breakthroughs and formulation typologies is given. In spite of its promise, biofertilizer continues to face challenges associated with product viability, variable performance in the field, and cost of production. The chapter ends with prospects for the future informed by innovations in the field of biotechnology, regulatory provisions, and the changing demands of the market. The present analysis is an attempt to bring to the forefront the use of biofertilizers, which may serve as one of the solutions for sustainable agriculture.

10.1 Introduction

The use of biofertilizers in agricultural systems is becoming a necessity that serves a great environmental and economic opportunity. As innovations consisting of living microorganisms, including nitrogen-fixing bacteria, phosphate-solubilizing bacteria, and mycorrhizal fungi, biofertilizers stimulate the growth of plants by increasing nutrient uptake through biological mechanisms (Vessey, 2003; Malus and Vassilev, 2014). Such inputs offer an environmentally friendly substitute to chemical inorganic fertilizers that have played a significant role in deteriorating the quality of soil, water, and the ecosystem (Tilman *et al.*, 2002; Savci, 2012). Against the background of emerging environmental issues and rising prices of fertilizer inputs, biofertilizers are becoming popular as an alternative, affordable, and environment friendly segment of an integrated nutrient management plan.

Economically, a farmer is presented with a number of cost-saving and value-creating opportunities with biofertilizers. They can also decrease

*Corresponding author: akumari@amity.edu

DOI: 10.1079/9781836991021.0010

the consumption rate of costly chemical fertilizers, thereby reducing the cost of inputs (Adesemoye *et al.*, 2009). Specifically, nitrogen-fixing bacteria, for example *Rhizobium* and *Azospirillum*, may greatly decrease reliance on the use of nitrogenous fertilizers that are one of the most expensive investments in agriculture. In addition to that, biofertilizers enhance the structure and fertility of the soil over time leading to long-term economic gains and sustained yields (Bhattacharyya and Jha, 2012).

Positive cost-benefit results are confirmed by a number of empirical studies related to the use of biofertilizers. An example here would be the study by Iwuagwu *et al.* (2013), who showed that the inoculation of maize with plant growth-promoting rhizobacteria not only increases the yield but has a positive benefit-cost ratio. Likewise, Kumari *et al.* (2017) revealed that the use of biofertilizers during rice farming based on an integrated nutrient management system resulted in increased net returns and enhanced the health of the soils compared to conventional fertilizer technology. Such results are especially relevant to smallholder or resource-poor farmers, for whom cutting input costs may have a significant effect on welfare.

Besides the microeconomic effects at farm level, macroeconomic effects of adoption of biofertilizer are also important. Biofertilizer markets have been favorable because of the national policies directed toward decreasing fertilizer subsidies and encouraging organic or low-input systems of farming (Singh *et al.*, 2011). The promotion of biofertilizers has been incorporated into more comprehensive policies of sustainable agriculture and rural development at country level in various countries, including India, Brazil, and Kenya, both through government and non-government initiatives. As an illustration of the support given to organic farming, in India, Paramparagat Krishi Vikas Yojana (PKVY) offers financial and technical incentives to farmers, including some provisions concerning the utilization of biofertilizers (Mukherjee *et al.*, 2022).

Nevertheless, the biofertilizer industry has a number of economic challenges despite such advantages. One of the greatest limitations is the unpredictability of field performance due to the influence of soil pH, microbial compatibility, and climatic conditions that discourage adoption (Lucy *et al.*, 2004). Moreover, farmer awareness, the supply chain, and extension services are limited and thus biofertilizer technologies are not scalable, particularly in low- and middle-income states (Mäder *et al.*, 2002). Another economic issue is quality control, because having low-grade products in the market limits the level of confidence that farmers have toward biofertilizers, thereby limiting their adoption (Herridge *et al.*, 2008).

Biofertilizers can be economically feasible once strong markets and regulatory systems are developed. Biofertilizer products should be standardized, certified, and labeled as this is essential to maintain product quality and help develop trust with respect to their use (Mahanty *et al.*, 2017). There must be substantial investment, both public and otherwise, in research and development (R&D) to enhance the effectiveness of strains, lengthen their shelf life, and create tailor-made formulations that are specific to varied agroecological climates. In addition, economic analysis and evaluations can guide policy makers and other investors to come up with economic scales involving biofertilizer promotion and devise-specific subsidy or credit schemes that are suitable (Giri *et al.*, 2019).

In terms of the economics of sustainability, the externality generated in the context of using biofertilizer is mostly on the positive side. The main difference between biofertilizers and chemical fertilizers is that the former either do not or are less prone to cause greenhouse gas emissions, eutrophication, or acidification of soil (Pretty *et al.*, 2006). Therefore, they can be used in accordance with the principles of a circular economy and green growth, providing the way to the sustainable intensification of agriculture (Godfray *et al.*, 2010). The evidence demonstrated by life cycle assessment (LCA) studies also shows that biofertilizers are associated with reduced embedded energy and emissions than their synthetic analogs and further reinforces the economic viability of using biofertilizers on a greater scale (Tuomisto *et al.*, 2012).

Finally, the economics behind biofertilizers is complex and includes farm profitability, policy incentives at a national scale, market factors, and environmental externality. Even though the existing empirical evidence indicates a significant economic benefit, large-scale adoption is likely to rely on overcoming technical constraints, making products more efficient, strengthening extension systems, and the provision of supporting regulatory and market infrastructure. In the context of the dual agricultural challenge of having both more productivity and greater sustainability, biofertilizers are likely to become a major player in the agricultural economics of the future.

10.2 Global Trends in the Last Decade

The biofertilizer market has undergone significant growth worldwide over the last decade. According to a thorough meta-analysis, the annual growth level of the market in terms of percentages (compound annual growth rate, CAGR) will be around 10.9% between 2022 and 2028 (Markets and Markets, 2023). Researchers cite the rise of environmental consciousness, regulation of the use of chemical fertilizers, and the upward trend in the price of conventional means of input as reasons for this (Markets and Markets, 2023). Simultaneously, microbial consortia, seed-coated, and automated production processes have diversified the biofertilizer portfolio beyond the classical nitrogen-fixing biofertilizer, to contain phosphate-solubilizing and potassium mobilizing organisms (Santos *et al.*, 2024). The diversification is associated with the general trend of relying on bio-based economy models that are more focused on minimized greenhouse gas emissions due to increased use of fertilizers and the maximization of the soil nutrient cycle (Santos *et al.*, 2024).

One of the factors that stimulated the growth in demand is the development of organic farming that greatly depended on biofertilizers to manage nutrient status and restore soil health. Measurement of organic farmland in North America has shown an increment of 10.7–13.5% in the period 2022–2023 alone, which is evidently increasing the demand of bio-based inputs (Garcha, 2023; Santos *et al.*, 2024). This long-term growth on the intercontinental scale indicates the growing effort of farmers and policy makers to decrease the use of synthetic fertilizers. Simultaneously, advancement in research and development areas has begun to remedy past issues, including temperate survival of microbes in extreme environments specific to fields, along with production of formidable multi-species inoculants (Santos *et al.*, 2024).

10.2.1 Regional market developments

10.2.1.1 North America

The world market is headed by North America, which is expected to reach US$1.08–1.30 billion in 2025 and US$2.10 billion by 2030, with a CAGR close to 10% (Garcha, 2023; Santos *et al.*, 2024). This expansion is based on the existence of a bustling organic farming industry that had surpassed 3.6 million hectares in the USA and Canada by 2023, plus high consumer demand paying premium prices for organic produce (Santos *et al.*, 2024).

Policy measures have enhanced faster adoption. The USDA Organic Transition Initiative (US$300 million fund) facilitates the transition of farmers who want to abandon the use of chemicals, and the Canadian Food Inspection Agency (CFIA) has eased the registration process for bio-based fertilizers in Canada (Garcha, 2023; North America Biological Organic Fertilizer Market, 2025). This positive policy environment plus technological developments in formulation and delivery (e.g. spore encapsulation, seed-coating) has facilitated the accessibility of biofertilizers in the organic as well as conventional agricultural operations (Santos *et al.*, 2024). According to empirical data, about 29.4% of marketed products are in the form of nitrogen-fixing microbial inoculants in the USA alone, where phosphate-solubilizers are also emerging (Santos *et al.*, 2024).

10.2.1.2 South America

In spite of the fact that South America has been less widely studied, substantial progress has been recorded. The high rate of *Bradyrhizobium* inoculants in soybean production in Brazil has led to a decrease in the use of synthetic nitrogen by up to 79% (saving approximately US$15.2 billion; Santos *et al.*, 2024). The existence of institutional support by way of the Brazilian Agricultural Research Corporation (Embrapa) and government approved strain registries has been contributing favorably toward adoption of such biofertilizers. The markets in neighboring countries, including Argentina, are also expanding via sustainable intensification initiatives, but regional market information gathered in peer-reviewed reports remains scarce.

10.2.1.3 European Union

The transformation of agricultural activities in Europe is mainly facilitated by the European Green Deal and the Farm to Fork strategy that require at least a 20% decrease in the use of chemical fertilizers by 2030 (Wesseler, 2022). Although certain facts on the adoption of biofertilizers are uncertain, key research studies have revealed that large investments are made in microbe soil PFAs (poly- and perfluoroalkyl substances) and organic nutrient sources (Santos *et al.*, 2024). Commercialization is being enhanced by legislative prohibition of some phosphorus fertilizers as well as greater investment in startups that fall into the category of sustainable inputs. A study has emphasized the effects of diversified microbial inoculant packages to enhance nutrient uptake, reduce the leaching process, and contribute to EU decarbonation targets (Santos *et al.*, 2024).

10.2.1.4 Asia–Pacific

The Asia–Pacific region is characterized by market vibrancy, led by India and China. The Indian biofertilzer market was estimated at US$152.5 million in 2025 and is projected to increase to US$233.5 million by the year 2030 (Garcha, 2023). Research and development led by the Indian Council of Agricultural Research (ICAR) has brought about the introduction of region-specific strains, which can increase yield by 10–25% and add 20–25% to conventional fertilizers (TechSci Research, 2024). There have been regulatory reforms, such as the amendment of the Fertiliser Control Order (1985) and the introduction of the National Mission for Sustainable Agriculture (NMSA), which have supported the penetration of biofertilizers in sustainable agriculture schemes (TechSci Research, 2024). Further propelling domestic demand are organic area conversions set to reach 4.48 million hectares in 2024 (GII Research, 2022; TechSci Research, 2024).

The Chinese and ASEAN markets are on the same track. Even though there is limited formal market data, past and continued research collaborations have revealed the adoption of multi-strain inoculants that enhance nutrient use efficiency in large rice and maize corridors. There are also programs and subsidies piloted by institutions, conducted in accordance with the activities of regenerative agriculture.

10.2.1.5 Africa

Africa is a new frontier in the use of biofertilizers, including by smallholder farmers. In two pilot schemes in Zambia and the Republic of Congo, production has increased by 48% using a targeted application of *Rhizobium* inoculation of legumes (Santos *et al.*, 2024). These achievements, facilitated by organizations such as the Alliance for a Green Revolution in Africa (AGRA), rely on considerable contributions from international donor funds. Scaling, however, involves establishing manufacturing capabilities, regulatory systems, and distribution channels locally, structural challenges which remain barriers to large-scale adoption in commerce.

10.3 Government Policies and Regulatory Impacts

10.3.1 National and regional policies

10.3.1.1 India: Paramparagat Krishi Vikas Yojana (PKVY) & FCO guidelines

The PKVY, launched in 2015 under India's National Mission for Sustainable Agriculture, promotes cluster-based organic farming using biofertilizers and organic inputs. A 2023 impact study in Tamil Nadu found that PKVY significantly improved farmer livelihoods, reduced reliance on chemical inputs, and encouraged eco-friendly farming techniques (Shalini *et al.*, 2023). The study documented transition toward sustainable agriculture and positive economic outcomes among smallholder farmers.

In Himachal Pradesh, Singh and Thakur (2022) analyzed the scheme's effectiveness in organic vegetable cultivation. They documented high benefit–cost ratios: 1.59 for peas and 1.39 for potatoes, signaling notable profitability in PKVY-supported clusters. This empirical evidence illustrates how PKVY contributes to both farmer incomes and local adoption of biofertilizers.

Additionally, Reddy (2020) conducted a national-level economic assessment using a difference-in-differences design. Findings revealed that while organic yields under PKVY were 12–18% lower than conventional agriculture, input costs were 14–19% lower, resulting in a net profitability increase of around 10–15%. The study also flagged implementation challenges, such as inconsistent funding and varying adoption levels across regions.

Regarding the Fertiliser Control Order (FCO), stricter regulations were introduced to ensure quality production and labeling of biofertilizers. While direct evaluated studies are still emerging, widespread farmer concerns regarding unreliable products suggest that clearer FCO enforcement could further enhance trust and adoption.

10.3.2 Impact of policies on industry growth

10.3.2.1 Stimulus through subsidy, research, certification

Public support under PKVY has reduced input costs and stimulated broader biofertilizer use. By providing ₹31,000 per hectare over 3 years specifically for biofertilizers and organic inputs (PIB, n.d.), the program eased financial barriers for small-scale farmers. Economic analyses by Reddy (2020) and Singh and Thakur (2022) confirm that economic incentives directly contributed to positive cost–benefit outcomes, even in the face of lower yields.

Improved certification through the FCO—mandating label quality and microbial strain viability—bolsters product reliability. While enforcement data are still emerging, the combined effect of tariffs, standards, and subsidy-backed adoption is strengthening the biofertilizer supply chain.

10.3.2.2 Constraints due to regulatory bottlenecks

Despite gains, challenges remain. Reddy (2020) cites delays in funding transfers and uneven implementation across states. In Tamil Nadu, Shalini *et al.* (2023) observed inconsistent cluster support and variable training quality. Singh and Thakur (2022) also noted that technical guidance on biofertilizer usage was less effective than expected, limiting nutrient management improvements for some farmers.

Quality remains a concern. FCO regulations exist to standardize products, but enforcement gaps persist. Singh and Thakur (2022) report instances of ineffective biofertilizers, which can undermine farmer trust. These issues underscore the need for tighter quality oversight and enhanced extension services.

10.4 Environmental Impact of Biofertilizers

Living microorganisms, which have been shown to enhance plant growth by improving the availability of nutrients, are a compelling reason to use biofertilizers as an alternative to chemical fertilizers, especially in sustainable agriculture. Their economic and environmental effects have been properly outlined in the recent literature. Economically, biofertilizers provide significant savings in input costs as farmers do not have to rely so much on synthetic fertilizers that are not only costly but also energy-intensive to manufacture (Kour *et al.*, 2020). The introduction of biofertilizers in the agricultural process may result in a 25–30% decline in chemical fertilizers consumption without resulting in a loss in crop production (Rajvir and Kumar, 2013). In addition, there is increased soil health when it is applied in the long term, meaning that fewer soil amendments and inputs are needed thereby making farming more cost-effective. Also, *Rhizobium*, *Azospirillum*, and phosphate-solubilizing bacteria have been shown to have the potential to improve the productivity of crops, especially legumes and cereals, which form a substantial part of staple meals consumed globally (Mahanty *et al.*, 2017). Although there could be some costs associated with training and adaptation in the very first use of biofertilizers, the long-term economic gains, with the help of subsidies and government incentive schemes promoting the use of organic inputs, makes using biofertilizers more and more appealing to small and marginal farmers (Liu *et al.*, 2022).

Biofertilizers are useful environmentally, limiting the harmful outcomes caused by chemical fertilizers: acidification of the soil and water eutrophication together with emission of greenhouse gases. It has been suggested that besides assisting in nitrogen fixation and phosphorus solubilization, biofertilizers also facilitate maintaining microbial diversity and organic matter content that are very important to the long-term fertility of soil (Adesemoye and Kloepper, 2009). Biofertilizers do not trigger profound ecological imbalances compared to synthetic fertilizers, which in most cases lead to nitrate leaching and groundwater pollution (Sahu and Brahmaprakash, 2016). Furthermore, they also produce fewer carbon emissions as they are biologically based and so less energy are needed to create and transport it (Ajmal *et al.*, 2018). A life cycle study done by Singh *et al.* (2022) indicated that greenhouse emissions produced during the production and utilization of biofertilizers are a maximum of 90% lower compared to traditional nitrogen-based fertilizers. Reliance on fossil fuel-based agricultural inputs is also decreased by use of biofertilizers, a condition that is aligned with globally agreed-upon climate mitigation strategies, such as those in the Paris Agreement of the United Nations.

Biofertilizers also contribute toward the resilience of agroecology through improved soil structure and water retention, which are important aspects in the light of the changing climate. An example of enhancement is the use of mycorrhizal fungi, which enhance the drought tolerance of plants by facilitating uptake of water and improving root architecture (Alori *et al.*, 2017). This kind of functional character not only results in the increased production of agriculture under abiotic stress but also helps maintain farm outputs, which enhances food security. This applies in particular to areas where there are extreme weather conditions or water scarcity. Biofertilizers also have environmental benefits for the preservation of biodiversity. Through decreasing the burden of toxicity on land and local ecosystems, biofertilizers are helpful in balancing the ecosystem and protecting positive soil microflora, for example earthworms and other microorganisms (Vessey, 2003). In addition, soil carbon sequestration is associated with the use of biofertilizers, increasing the accumulation of humus and soil organic electrons that are vital in tackling land degradation and desertification (Azimzadeh and Azimzadeh, 2012).

Despite these benefits, there are difficulties in the popularization of biofertilizers. Their scalability in the market has been constrained by short shelf life, strain specificity, and sensitivity to environmental conditions (Bhardwaj *et al.*, 2014). There is also an absence of uniform rules and quality assurance processes, especially in the developing world, where performance tends to vary. Kumar *et al.* (2020a) also argue that the effectiveness of biofertilizers has a strong strain-specific character and depends on earth type, soil pH level, soil moisture, and crop type. Therefore, region-specific formulations and agronomic advice are required. However, research and innovation in formulation technologies (e.g. encapsulation, co-inoculation strategies, and carrier material improvements) have been gradually resolving these constraints (Malusá and Vassilev, 2014). In terms of policy, to eliminate the barriers to adoption, it is important to consider economic and other stimuli that will motivate people to use biofertilizers, such as subsidies, awareness events, and including them in the sustainable agricultural schemes. Case studies of pilot programs in India and Brazil suggest that when adequately supported by extension services and farming education, the adoption rates of biofertilizers can considerably increase (Mazid and Khan, 2014).

To conclude, the economic and environmental effects of biofertilizers are highly positive as they offer a sustainable means of ensuring that the ecological footprint of agricultural activities is decreased or becomes neutral without affecting productivity. Biofertilizers are a vital ingredient of sustainable and economically viable agricultural systems as they reduce the need for chemical fertilizers, enhance the health of soils, and help farmers by bringing down costs. Nevertheless, this will not be achieved unless efforts are coordinated to conduct research, policy, and outreach to farmers on how to overcome current issues that revolve around product quality, efficacy in the field, and large-scale operations. Sustainable agriculture has a close connection with biologically based solutions, and biofertilizers will kick-start this change. Table 10.1 outlines the potential economic prospects and opportunities for biofertilizers.

Table 10.1. Economic prospects and opportunities for biofertilizers. Table author's own.

Prospect	Economic benefit	Reference
Reduced input costs	Decreases chemical fertilizer dependency by 20–30%, improving cost efficiency.	(Singh *et al.*, 2011)
Access to premium organic markets	Biofertilizer users can sell at higher prices in certified organic markets.	(Habanyati *et al.*, 2024)
Policy incentives and subsidies	Government support reduces economic risk and adoption barriers.	(Kumar *et al.*, 2020a)
Technological innovations	Advances in formulation (e.g. liquid biofertilizers) increase efficacy and reduce application cost.	(Mahanty *et al.*, 2017)
Long-term soil productivity	Enhanced soil health leads to sustainable yield over time, benefiting farm economics.	(Adesemoye and Kloepper, 2009)
Climate resilience and sustainability goals	Biofertilizers support low-carbon agriculture, opening access to carbon credit markets.	(Singh *et al.*, 2011)

10.5 Types and Production Models of Biofertilizers and Their Economic Significance

The use of biofertilizers involves an eco-friendly application of inoculants of microbes, improving the supply of primary nutrients to the host plant through natural mechanisms of nitrogen and phosphate fixation and phosphate solubility, and improving growth by synthesizing growth-stimulating compounds. In broad dimensions, biofertilizers can be classified as nitrogen-fixing, phosphate-solubilizing, potassium-solubilizing, and plant growth-promoting rhizobacteria (PGPR) on the basis of functionality (Vessey, 2003; Bhardwaj *et al.*, 2014). *Rhizobium*, *Azotobacter*, *Azospirillum*, and cyanobacteria like *Anabaena* and *Nostoc* are well-known nitrogen-fixing biofertilizers used on leguminous and non-leguminous crops because they convert atmospheric nitrogen into elements that can be used by plants (Bidyarani *et al.*, 2016). Insoluble phosphorus is converted into plant-available forms by secretions of enzymes and organic acids by phosphate-solubilizing microorganisms (PSMs) such as *Bacillus*, *Pseudomonas*, and *Aspergillus* spp., leading to improved phosphorus uptake of plants (Rodríguez and Fraga, 1999). Less dominant studies analyze the presence of potassium-solubilizing bacteria, which plays a crucial role in the mobilization of potassium through silicate minerals and feldspar, which have positive effects on the wellbeing and toughness of the plant (Parmar and Sindhu, 2013). Furthermore, the multifunctional properties of *Pseudomonas fluorescens* and *Bacillus subtilis*—as PGPR—which solubilize nutrients, inhibit pathogens, and stimulate hormones, has attracted attention to the holistic effects of using these particular biofertilizers (Glick, 2012).

Biofertilizer production models fit into the following categories: laboratory -scale production, pilot-scale production, and commercial-scale models. Production entails cultivation, which includes selection of effective strains, growing in large quantities with the help of fermenters, preparation of carriers, and formulation of inoculants (Malusá and Vassilev, 2014). The liquid models of biofertilizer are becoming quite popular in place of the traditional carrier-source systems because they have a longer shelf life, an increased number of microbes, and are easy to apply (Sahu and Brahmaprakash, 2016). The liquid formulation has additives including humic acid, glycerol, and polymers, which increase microbial viability and stress resistance.

Submerged fermentation is the basis of industrial-scale production, in which a chosen microbial strain is grown in large fermenters under specific conditions, centrifuged out, and then combined with suitable carriers or stabilizers (Bhattacharyya and Jha, 2012). Others use consortium-based biofertilizers creating blends of favorable microorganisms with synergistic effects and suitable for a broader agroecological range. There is also an attempt to use integrated production systems to produce more economically sustainable and cost-effective biofertilizer using agricultural waste such as molasses, whey, and fruit pulp as the substrates (Malusá *et al.*, 2012).

Economically, biofertilizers promise to substitute usage of chemical fertilizers, saving on input costs, increasing the fertility of the soil, and sustaining yields. As reported by Singh *et al.* (2011), the application of biofertilizers resulted in the reduction of synthetic fibres by 15% and increased crop yield by 10%, which resulted in a significant increase in farmers' income. In addition, the manufacturing cost of biofertilizer is comparatively lower than that of the chemical fertilizer and therefore more economically feasible for small-scale farmers in the developing world. As an example, the production cost of 1 ha of the crop grown with biofertilizers in India is approximately 20–30% lower than one grown with chemical fertilizers (Bhardwaj *et al.*, 2014). Moreover, biofertilizers offer long-term economic gains through the promotion of soil microbial biodiversity, a decrease in the occurrence of pests and diseases, and minimization of environmental degradation brought about by overuse of chemicals (Kumar *et al.*, 2020b). Biofertilizer is also gaining popularity in the global market with an expected compound annual growth rate (CAGR) of more than 12% between 2021 and 2027 due to improved awareness of sustainable farming and government subsidies for biofertilizer (MarketsandMarkets, 2021).

Linked to agribusiness, the production of biofertilizers commercially offers prospects for entrepreneurship and farm job creations in rural areas. It is low capital-intensive to conduct local units of production, and with essential training, farmers as well as young rural-based citizens can be employed in production and distribution (Bhattacharyya and Jha, 2012). The economic relevance of biofertilizers has also been enhanced by government projects like the Indian National Project on Organic Farming and even subsidies given to biofertilizer units. However, shortcomings such as unstable field performance, absence of a quality index, and minimal shelf life in classic formulation continue to hamper mass usage (Malusá and Vassilev, 2014).

10.6 Challenges and Future Prospects

The use of biofertilizers, which refers to formulations containing living microorganisms that increase nutrient availability to plants, is being embraced as a sustainable alternative to chemical fertilizers for replenishing the nutrient needs of crops. They face multifaceted challenges and opportunities, however, in regards to their economic implications, involving adoption, scalability, and long-term viability. This discussion critically examines these dimensions with the help of present and dependable studies.

Profitability is one of the major economic barriers. Although biofertilizers may cut the cost of inputs in the long term, they tend to produce smaller yields in the short term, especially in poor soil and climatic environments (Sindhu *et al.*, 2022). The level of returns on investment of biofertilizers may not be appealing in the early phases of entrepreneurship for farmers who are used to high yields and quick returns using synthetic inputs. This restricts the spread of biofertilizer use on a large scale, especially with regard to smallholder farmers, who have a low level of risk tolerance (Uzinger *et al.*, 2020). Table 10.2 outlines the potential challenges for biofertilizers.

There is also no standardization and quality control in biofertilizer production/distribution. Weak regulatory structures have resulted in the market being flooded with substandard products, thus creating poor performance and financial losses for farmers (Bhattacharyya and Jha, 2012). The resulting hesitancy concerning its effectiveness attributes to market uncertainty and low consumer confidence, which adversely influence the financial feasibility of the biofertilizer business.

There is also an economic challenge faced by their storage, transportation infrastructure, and application. Biofertilizers are sensitive to temperature and humidity, and cold chains and adequate treatment are required, which adds to

Table 10.2. Key economic challenges associated with biofertilizers. Table author's own.

Challenge	Description	Reference
Low initial profitability	Biofertilizers show lower short-term yield compared to chemical fertilizers.	(Sindhu *et al.*, 2022)
Lack of product standardization	Poor regulation results in inconsistent quality and unreliable performance.	(Bhattacharyya and Jha, 2012)
Storage and distribution costs	Biofertilizers require specific storage conditions, increasing logistical expenses.	(Kumar *et al.*, 2020b)
Limited farmer awareness	Insufficient extension services and farmer training limit effective adoption.	(Pathak *et al.*, 2022)
Risk aversion of smallholders	Farmers are hesitant to switch from chemical inputs due to economic uncertainty.	(Uzinger *et al.*, 2020)
Market penetration difficulties	Weak demand due to inconsistent performance affects commercial expansion.	(Mahanty *et al.*, 2017)

operational costs, particularly in developing economies (Kumar *et al.*, 2020a). The costs may cancel out the anticipated savings to be made from reduced use of synthetic fertilizer, thus making a cost–benefit analysis complicated.

Poor awareness and extension is also an obstacle to economic adoption. Most farmers do not know about the economic gains of the use of biofertilizers, such as better soil health and the end of long-term dependence on chemicals. Poor access to training and demonstrations also inhibit uptake by farmers (Pathak *et al.*, 2022).

In spite of these problems, there are good long-term economic prospects for biofertilizers. They are able to greatly reduce input costs after some time by minimizing the use of expensive chemical fertilizers (Adesemoye and Kloepper, 2009). It has been estimated that the application of biofertilizers can reduce fertilizer costs by 20–30%, with comparable yields after some seasons of cropping (Singh *et al.*, 2011).

Also, the increased popularity of organic and sustainable farming has created a market pull for biofertilizer-based production systems. Products can effectively be marketed at premium prices as the demand for organic food by consumers has increased, thus increasing profitability (Habanyati *et al.*, 2024).

International organizations and governments are also intervening to offer policy incentives and subsidies in the light of the potential of biofertilizers to improve the environment and the economy. Programs like the India's PKVY provide financial assistance and training to engender organic inputs including biofertilizers (Thomas and Singh, 2019). Such programmes increase penetration and eliminate the economic risks for early adopters.

Moreover, advances in technology are increasing the shelf life, efficacy, and versatility of biofertilizers for diverse agroecological climates. The enhanced versions, like liquid biofertilizers and consortia inoculants, are relatively stronger and more effective, with a reduced cost–benefit (Mahanty *et al.*, 2017).

10.7 Conclusion

The economics of biofertilizers is a challenging combination of short-run constraints and long-run prospects. Although they provide a sustainable solution to chemical fertilizers by improving soil fertility, lowering input costs, and assisting in environmental preservation, their short-run financial rewards are less attractive to most farmers, especially smallholders. Challenges including low initial yield responses, non-standardization of the product, poor regulatory enforcement, and deficient production and distribution infrastructure continue to impede extensive economic viability. In addition, low farmer awareness and technical support further limit adoption, leaving a gap between maximal and actual benefit.

Yet, possibilities for biofertilizer adoption are promising. Rising world demand for organic and environmentally friendly agricultural products offers significant economic incentives for adoption. Government subsidies and policies that encourage sustainable practice are also reducing economic risk for early adopters. Technical developments in microbial technologies are enhancing the effectiveness, shelf life, and flexibility of biofertilizer products, which are becoming easier to use and more cost-effective. In the long term, the collective economic gains—such as lowered dependence on costly chemical inputs, enhanced soil fertility, and access to high-paying organic markets—can outbalance the initial constraints.

For the full realization of the economic value of biofertilizers, coordinated action by policy makers, researchers, industry, and extension services is needed. Investment in quality control, farmer training, and public–private partnership will be essential in establishing trust and increasing market access. Finally, the integration of biofertilizers into conventional farming practices can result in an economically more resilient and environmentally more sustainable agricultural system, especially in areas that are experiencing ecological degradation and increasing input prices. The shift might involve strategic interventions and a long-term perspective, but the economic argument for biofertilizers is gaining traction slowly.

References

Adesemoye, A.O. and Kloepper, J.W. (2009) Plant–microbes interactions in enhanced fertilizer-use efficiency. *Applied Microbiology and Biotechnology* 85(1), 1–12. DOI: 10.1007/s00253-009-2196-0

Adesemoye, A.O., Torbert, H.A., and Kloepper, J.W. (2009) Enhanced plant nutrient use efficiency with PGPR and AMF in an integrated nutrient management system. *Canadian Journal of Microbiology* 55(8), 875–883. DOI: 10.1139/w08-081

Ajmal, M., Ali, H.I., Saeed, R., Akhtar, A., Tahir, M., Mehbob, M.Z., and Ayub, A. (2018) Biofertilizer as an alternative for chemical fertilizers. *Research & Reviews: Journal of Agriculture and Allied Sciences*. Available at: https://www.rroij.com/open-access/biofertilizer-as-an-alternative-for-chemical-fertilizers.pdf (accessed September 9, 2025).

Alori, E.T., Glick, B.R., and Babalola, O.O. (2017) Microbial phosphorus solubilization and its potential for use in sustainable agriculture. *Frontiers in Microbiology* 8, 971. DOI: 10.3389/fmicb.2017.00971

Azimzadeh, S.M. and Azimzadeh, S.J. (2012) Study on replacement probability of biofertilizer with chemical fertilizer in bread wheat (*Triticum astivum* L.). *Advances in Environmental Biology* 6(10), 2602–2610.

Bhardwaj, D., Ansari, M.W., Sahoo, R.K., and Tuteja, N. (2014) Biofertilizers function as key player in sustainable agriculture by improving soil fertility, plant tolerance and crop productivity. *Microbial Cell Factories* 13(1), 66. DOI: 10.1186/1475-2859-13-66

Bhattacharyya, P.N. and Jha, D.K. (2012) Plant growth-promoting rhizobacteria (PGPR): emergence in agriculture. *World Journal of Microbiology and Biotechnology* 28(4), 1327–1350. DOI: 10.1007/s11274-011-0979-9

Bidyarani, N., Prasanna, R., Babu, S., Hossain, F., and Saxena, A.K. (2016) Enhancement of plant growth and yields in chickpea (*Cicer arietinum* L.) through novel cyanobacterial and biofilmed inoculants. *Microbiological Research* 188–189, 97–105. DOI: 10.1016/j.micres.2016.04.005

Chandrasekaran, M., Ambrose, G., and Jayabalan, N. (2014) Influence of biofertilizers on the growth and yield of maize (*Zea mays* L.). *Plant Archives* 14(1), 311–316.

Garcha, S. (2023) Present scenario: status of the biofertilizer industry in India. In: Kaur, S., Dwibedi, V., Sahu, P.K., and Kocher, G.S. (eds) *Metabolomics, Proteomes and Gene Editing Approaches in Biofertilizer Industry*. Springer, Singapore, pp. 21–36.

GII Research (2022) *India Biofertilizer Market Share Analysis, Industry Trends & Statistics, Growth Forecasts*. Global Information, Inc., Kawasaki, Japan.

Giri, B., Prasad, R., Wu, Q.-S., and Varma, A. (eds) (2019) *Biofertilizers for Sustainable Agriculture and Environment*. Springer, Cham, Switzerland.

Glick, B.R. (2012) Plant growth-promoting bacteria: mechanisms and applications. *Scientifica* 2012, 963401. DOI: 10.6064/2012/963401

Godfray, H.C.J., Beddington, J.R., Crute, I.R., Haddad, L., Lawrence, D. *et al.* (2010). *Science* 327(5967), 812–818. DOI: 10.1126/science.1185383

Habanyati, E.J., Paramasivam, S., Seethapathy, P., and Manalil, S. (2024) Assessing organic farming adoption in selected districts of Tamil Nadu: challenges, practices, and pathways for growth. *Agronomy* 14, 2537. DOI: 10.3390/agronomy14112537

Herridge, D.F., Peoples, M.B., and Boddey, R.M. (2008) Global inputs of biological nitrogen fixation in agricultural systems. *Plant and Soil* 311(1–2), 1–18. DOI: 10.1007/s11104-008-9668-3

Iwuagwu, M., Ks, C., Uka, U.N., and Amandianeze, M. (2013) Effects of Biofertilizers on the Growth of *Zea mays* L. *Asian Journal of Microbiology Biotechnology and Environmental Sciences* 15(2), 235–240.

Kour, D., Rana, K.L., Yadav, A.N., Yadav, N., Kumar, M. *et al.* (2020) Microbial biofertilizers: bioresources and eco-friendly technologies for agricultural and environmental sustainability. *Biocatalysis and Agricultural Biotechnology* 23, 101487. DOI: 10.1016/j.bcab.2019.101487

Kumar, V., Behl, R.K., and Narula, N. (2020a) Establishment of phosphate-solubilizing strains of *Azotobacter chroococcum* in the rhizosphere and their effect on wheat cultivars under greenhouse conditions. *Microbiological Research* 155(4), 311–316. DOI: 10.1078/0944-5013-00081

Kumar, A., Singh, R., and Yadav, A.N. (2020b) Microbial inoculants in sustainable agricultural productivity: recent development and future prospects. In: Yadav, A.K., Kumar, A., Singh, S., and Upadhyay, D.K. (eds) *Plant Microbiomes for Sustainable Agriculture*. Springer, Cham, Switzerland, pp. 57–76.

Kumari, R., Kumar, S., Kumar, R., Das, A., Kumari, R., Choudhary, C.D., and Sharma, R.P. (2017) Effect of long-term integrated nutrient management on crop yield, nutrition and soil fertility under rice-wheat system. *Journal of Applied and Natural Science* 9(3), 1801–1807. DOI: 10.31018/jans.v9i3.1442

Liu, Y., Wang, J., and Liang, L. (2022) Evaluation of economic viability and farmer perceptions of biofertilizers in developing countries: a meta-analysis. *Agricultural Systems* 196, 103313. DOI: 10.1016/j.agsy.2021.103313

Lucy, M., Reed, E., and Glick, B.R. (2004) Applications of free living plant growth-promoting rhizobacteria. *Antonie van Leeuwenhoek* 86(1), 1–25. DOI: 10.1023/B:ANTO.0000024903.10757.6e

Mäder, P., Fliessbach, A., Dubois, D., Gunst, L., Fried, P. *et al.* (2002) Soil fertility and biodiversity in organic farming. *Science* 296(5573), 1694–1697. DOI: 10.1126/science.1071148

Mahanty, T., Bhattacharjee, S., Goswami, M., Bhattacharyya, P., Das, B. *et al.* (2017) Biofertilizers: a potential approach for sustainable agriculture development. *Environmental Science and Pollution Research International* 24(4), 3315–3335. DOI: 10.1007/s11356-016-8104-0

Malusá, E. and Vassilev, N. (2014) A contribution to set a legal framework for biofertilisers. *Applied Microbiology and Biotechnology* 98(15), 6599–6607. DOI: 10.1007/s00253-014-5828-y

Malusá, E., Sas-Paszt, L., and Ciesielska, J. (2012) Technologies for beneficial microorganisms inocula used as biofertilizers. *The Scientific World Journal* 2012, 491206. DOI: 10.1100/2012/491206

MarketsandMarkets (2021) Biofertilizers Market by Type, Crop Type, Microorganism, Application Mode, Form, and Region – Global Forecast to 2028. MarketsandMarkets. Available at: https://www.marketsandmarkets.com/Market-Reports/compound-biofertilizers-customized-fertilizers-market-856.html (accessed August 26, 2025).

MarketsandMarkets (2023) Biofertilizers Market Size, Share and Future Growth Expectations. MarketsandMarkets. Available at: https://www.marketsandmarkets.com/ResearchInsight/biofertilizers-market-size-and-share.asp (accessed August 26, 2025).

Mazid, M. and Khan, T.A. (2014) Future of bio-fertilizers in Indian agriculture: an overview. *International Journal of Agricultural and Food Research* 3(3).

Mukherjee, K., Konar, A., and Ghosh, P. (2022) Organic farming in India: a brief review. *International Journal of Research in Agronomy* 5(2B). DOI: 10.33545/2618060X.2022.v5.i2b.120

Parmar, P. and Sindhu, S.S. (2013) Potassium solubilization by rhizosphere bacteria: influence of nutritional and environmental conditions. *Journal of Microbiology Research* 3(1), 25–31. DOI: 10.5923/j.microbiology.20130301.04

Pathak, H., Rao, D.L.N., and Reddy, S.R. (2022) Adoption and economic potential of biofertilizers in India. *Current Science* 123(7), 831–838.

PIB (n.d.) Promoting organic farming under Paramparagat Krishi Vikas Yojana (PKVY). Press Information Bureau, Government of India. Available at: https://indiapressrelease.com/pib-press-releases/promoting-organic-farming-under-paramparagat-krishi-vikas-yojana-pkvy/ (accessed September 6, 2025).

Pretty, J.N., Noble, A.D., Bossio, D., Dixon, J., Hine, R.E. *et al.* (2006) Resource-conserving agriculture increases yields in developing countries. *Environmental Science & Technology* 40(4), 1114–1119. DOI: 10.1021/es051670d

Rajvir, S. and Kumar, S. (2013). Role of bio-fertilizer in organic agriculture: a review. *Research Journal of Recent Sciences* 2, 39.

Reddy, A.A. (2020) *Impact Study of Paramparagath Krishi VikasYojana (Organic Agriculture)* (MANAGE Report). National Institute of Agricultural Extension Management, Rajendranagar, Hyderabad, India. DOI: 10.31220/osf.io/64t5j

Rodríguez, H. and Fraga, R. (1999) Phosphate solubilizing bacteria and their role in plant growth promotion. *Biotechnology Advances* 17(4–5), 319–339. DOI: 10.1016/s0734-9750(99)00014-2

Sahu, P.K. and Brahmaprakash, G.P. (2016) Formulations of biofertilizers – approaches and advances. In: Singh, D.P., Singh, H.B., and Prabha, R. (eds) *Microbial Inoculants in Sustainable Agricultural Productivity*, Vol 2. Springer, New Delhi, pp. 179–198. DOI: 10.1007/978-81-322-2644-4_12

Santos, F., Melkani, S., Oliveira-Paiva, C., Bini, D., Pavuluri, K. *et al.* (2024) Biofertilizer use in the United States: definition, regulation, and prospects. *Applied Microbiology and Biotechnology* 108(1), 511. DOI: 10.1007/s00253-024-13347-4

Savci, S. (2012) An agricultural pollutant: chemical fertilizer. *International Journal of Environmental Science and Development* 3(1), 73–80. DOI: 10.7763/IJESD.2012.V3.191

Shalini, J.R., Selvarani, G., Dhamodaran, T., Senthilnathan, S., and Prabakaran, K. (2023) Impact of Paramparagath Krishi Vikas Yojana scheme on the livelihood of beneficiaries. *Asian Journal of Agricultural Extension, Economics & Sociology* 41(10), 26–31. DOI: 10.9734/ajaees/2023/v41i102136

Sindhu, S.S., Sehrawat, A., Phour, M., and Kumar, R. (2022) Nutrient acquisition and soil fertility: contribution of rhizosphere microbiomes in sustainable agriculture. In: Arora, N.K. and Bouizgarne, B. (eds) *Microbial BioTechnology for Sustainable Agriculture Volume 1*. Springer, Singapore, pp. 1–41. Available at: https://ijhfonline.org/index.php/ijhf/article/view/32 (accessed September 6, 2025).

Singh, A. and Thakur, R.K. (2022). Status of organic farming and economics of organic vegetable cultivation in Himachal Pradesh under Paramparagat Krishi Vikas Yojana (PKVY). *Indian Journal of Hill Farming* 35(1), 123–130.

Singh, J.S., Pandey, V.C., and Singh, D.P. (2011) Efficient soil microorganisms: a new dimension for sustainable agriculture and environmental development. *Agriculture, Ecosystems & Environment* 140(3–4), 339–353. DOI: 10.1016/j.agee.2011.01.017

Singh, R. Singh, A., Sheoran, P. Fagodiya, R.K., Rai, A.K. *et al.* (2022) Energy efficiency and carbon footprints of rice-wheat system under long-term tillage and residue management practices in western Indo-Gangetic Plains in India. *Energy* 244(Part A), 122655. DOI: 10.1016/j.energy.2021.122655

TechSci Research (2024) In: *India Biofertilizers Market by Size, Share & Forecast to 2030F*. TechSci Research. Available at: https://www.techsciresearch.com/report/india-biofertilizers-market/7818.html (accessed September 9, 2025).

Thomas, L. and Singh, I. (2019) Microbial biofertilizers: types and applications. In: Giri, B., Prasad, R., Wu, Q.-S., and Varma, A. (eds) *Biofertilizers for Sustainable Agriculture and Environment*. Springer, Cham, Switzerland, pp. 1–19.

Tilman, D., Cassman, K.G., Matson, P.A., Naylor, R., and Polasky, S. (2002) Agricultural sustainability and intensive production practices. *Nature* 418(6898), 671–677. DOI: 10.1038/nature01014

Tuomisto, H.L., Hodge, I.D., Riordan, P., and Macdonald, D.W. (2012) Does organic farming reduce environmental impacts? A meta-analysis of European research. *Journal of Environmental Management* 112, 309–320. DOI: 10.1016/j.jenvman.2012.08.018

Uzinger, N., Takács, T., Szili-Kovács, T., Radimszky, L., Füzy, A. *et al.* (2020). Fertility impact of separate and combined treatments with biochar, sewage sludge compost and bacterial inocula on acidic sandy soil. *Agronomy* 10(10), 1612. DOI: 10.3390/agronomy10101612.

Vessey, J.K. (2003) Plant growth promoting rhizobacteria as biofertilizers. *Plant and Soil* 255(2), 571–586. DOI: 10.1023/A:1026037216893

Wesseler, J. (2022) The EU's farm-to-fork strategy: an assessment from the perspective of agricultural economics. *Applied Economic Perspective and Policy* 44(4), 1826–1843. DOI: 10.1002/aepp.13239

11 Scope of Biofertilizers in India

Abha Kumari*
Centre for Biotechnology and Biochemical Engineering, Amity Institute of Biotechnology, Amity University, Noida, India

Abstract
Millions of people in India depend on agriculture for their living, and it makes a substantial contribution to the country's GDP. However, overuse of chemical fertilizers has resulted in environmental issues, decreased microbial activity, and soil deterioration. As a result, biofertilizers have become a viable and environmentally responsible way to improve crop yield and soil fertility. Biofertilizers containing living microorganisms increase the availability of vital nutrients, which in turn stimulates plant development. Among these are mycorrhizal fungus, phosphate-solubilizing microorganisms, and nitrogen-fixing bacteria. The use of biofertilizers is increasing in India as a result of the country's increased emphasis on sustainable agriculture, organic farming, and environmental preservation. The present state, advantages, difficulties, and potential of biofertilizers in India are examined in this chapter. It draws attention to market developments, government activities, and the function of biofertilizers, in total summing up the scope.

11.1 Introduction

Agriculture remains the backbone of India's economy, supporting nearly half of the population's livelihood. However, excessive reliance on chemical fertilizers has led to soil degradation, environmental pollution, and diminishing crop yields over time. In this context, biofertilizers have emerged as a sustainable alternative, offering eco-friendly and cost-effective solutions to enhance soil fertility and crop productivity. Biofertilizers, composed of beneficial microorganisms such as nitrogen-fixing bacteria, phosphate-solubilizing bacteria, and mycorrhizal fungi, play a crucial role in improving nutrient availability and soil health. There are various biofertilizers found in different ways and each has its own history. The commercial history of biofertilizers began with the launch of "Nitragin," a laboratory culture of rhizobia, in 1895 by Nobbe and Hiltner. This was followed by the discovery of *Azotobacter* and blue green algae (BGA) and a host of other microorganisms. *Azospirillum* and vesicular-arbuscular mycorrhizae (VAM) are fairly recent discoveries. In India, the first study on legume–*Rhizobium* symbiosis was conducted by N.V. Joshi, and commercial production began in 1956 (Ghosh, 2004). Rhizospheric microorganisms have recently become a viable and eco-friendly way to mitigate the adverse impacts of overuse of artificial fertilizers. With a variety of useful characteristics, these helpful microorganisms are being

*Corresponding author: akumari@amity.edu

DOI: 10.1079/9781836991021.0011

used more and more as biofertilizers to maintain soil and environmental health while boosting agroecosystem productivity in a sustainable manner (Kaur *et al.*, 2023).

The scope of biofertilizers in India is significant, given the country's vast agricultural landscape, government initiatives promoting organic farming, and increasing awareness among farmers regarding sustainable practices. With a growing emphasis on reducing chemical inputs and improving long-term agricultural sustainability, biofertilizers present a viable option to enhance food security while maintaining environmental balance. This chapter explores the potential of biofertilizers in India, highlighting their benefits, challenges, and prospects in the context of modern agricultural practices.

11.2 Types of Biofertilizers

Biofertilizers consist of living microorganisms that, when applied to seeds, roots, or soil, enhance plant growth by increasing nutrient availability or providing phytohormones (Singh *et al.*, 2022; Alzate Zuluaga *et al.*, 2024). The main categories include nitrogen-fixers, solubilizers of phosphate, potassium, and zinc, vesicular-arbuscular mycorrhizal (VAM) fungi, and plant growth-promoting rhizobacteria (PGPR) (Table 11.1).

Table 11.1. Types of biofertilisers, their mechanisms of action, and representative microorganisms. Table author's own.

Biofertilizer type	Mechanism of action	Example organisms	Reference
Nitrogen-fixing bacteria	Convert atmospheric nitrogen (N_2) to ammonia via nitrogenase enzyme; enhance root architecture via IAA production.	*Rhizobium*, *Azotobacter*, *Azospirillum*	(Glick, 2020)
Phosphate-solubilizing microbes	Secrete organic acids and phosphatases to mobilize insoluble phosphate into plant-available forms.	*Pseudomonas fluorescens*, *Bacillus megaterium*	(Sánchez-Castro *et al.*, 2023)
Potassium-solubilizing bacteria	Release organic acids that weather K-bearing minerals, liberating potassium ions.	*Frateuria aurantia*, *Bacillus mucilaginosus*	(Qiaoyi *et al.*, 2022)
Zinc-mobilizing microbes	Produce acids and chelators to solubilize zinc, facilitating uptake in zinc-deficient soils.	*Bacillus subtilis*, *Pseudomonas spp.*	(Sánchez-Castro *et al.*, 2023)
Arbuscular mycorrhizal fungi (AMF)	Extend root absorption via hyphal networks; improve nutrient uptake (especially P, Zn) and drought tolerance.	*Rhizophagus irregularis*, *Glomus mosseae*	(Savastano and Bais, 2024)
PGPR with ACC deaminase	Reduce plant ethylene levels under stress by degrading ACC, a precursor to ethylene; enhance root elongation under abiotic stress.	*Pseudomonas putida*, *Bacillus spp.*	(Etesami and Beattie, 2023)
PGPR with ISR-inducing ability	Produce elicitors that activate plant defense pathways (e.g. jasmonate, ethylene), leading to resistance against pathogens.	*Pseudomonas fluorescens*, *Bacillus subtilis*	(Savastano and Bais, 2024)
Dual inoculants (PGPR + AMF)	Act synergistically to enhance nutrient acquisition, modulate hormone levels, and improve resilience under biotic/abiotic stress.	*Bacillus + Glomus spp.* co-inoculation	(Wang *et al.*, 2025)

11.2.1 Nitrogen-fixing biofertilizers

Nitrogen-fixing microorganisms convert atmospheric nitrogen into ammonia, a usable form for plants. Symbiotic diazotrophs, such as *Rhizobium*, form nodules on legumes. These nodules create a microaerobic environment ideal for nitrogenase activity, contributing 50–80% of the total biologically fixed nitrogen worldwide—approximately 70 million tons annually (Alzate Zuluaga *et al.*, 2024)

Recent insights into *Rhizobium leguminosarum* highlight genetic and regulatory factors underpinning nodule formation and nitrogen fixation efficiency, offering the potential to improve biofertilizer strains (Zhang *et al.*, 2023). Free-living diazotrophs include *Azotobacter*, *Azospirillum*, and cyanobacteria. *Azotobacter chroococcum* and related species fix up to 60 kg N/ha annually, often alongside soil amendments (Kumar *et al.*, 2021). They also produce phytohormones like indole-3-acetic acid (IAA) and gibberellin, vitamins (B-complex), and siderophores, improving root development, nutrient uptake, and disease resistance (Rafay and Usman, 2023; Singh *et al.*, 2022). Field trials using *Azotobacter chroococcum* show yield increases of up to 72 % when combined with organic amendments, while individual applications boost yields by 10–40 % across cereals (Kumar *et al.*, 2021; Rafay and Usman, 2023).

Azospirillum, another non-symbiotic nitrogen fixer, particularly benefits C4 plants such as maize, sugarcane, sorghum, and millet. It produces IAA, gibberellins, and cytokinins, enhancing root growth, nutrient uptake, and yields (Jalal *et al.*, 2022; Rafay and Usman, 2023). In maize, *Azospirillum brasilense* has been shown to alter root architecture and boost phytohormone levels, improving nutrient absorption and crop output (Jalal *et al.*, 2022).

Cyanobacteria, such as *Anabaena*, *Nostoc*, and *Aulosira*, and the fern–cyanobacterium symbiosis in *Azolla*, fix 20–30 kg N/ha in flooded rice systems and contribute vitamins and growth regulators (Kumar *et al.*, 2021; Nosheen *et al.*, 2021). These cyanobacteria also improve soil organic matter, water retention, and phosphorous availability, and suppress weeds and salinity (Kumar *et al.*, 2021; Castleberry, 2023). In summary, nitrogen-fixing biofertilizers—whether symbiotic or free-living—offer substantial nitrogen inputs (20–80 kg N/ha/year), plus growth-enhancing compounds, reduced reliance on chemical fertilizers, and improved soil health (Kumar *et al.*, 2021; Alzate Zuluaga *et al.*, 2024)

11.2.2 Phosphate-solubilizing biofertilizers

Phosphorus often remains unavailable due to soil binding. Microbes like *Pseudomonas*, *Bacillus*, and *Aspergillus* solubilize phosphate through organic acids and phosphatases. While specific quantitative data were outside our recent search range, organic farming literature consistently reports 20–40 % yield gains in cereals and legumes with these inoculants (Singh *et al.*, 2022). *Pseudomonas fluorescens* and *Bacillus megaterium* enhance phosphorus availability and crop growth via robust organic acid production and root colonization (Alzate Zuluaga *et al.*, 2024).

11.2.3 Potassium- and zinc-solubilizing biofertilizers

Potassium and zinc are crucial yet often locked-in soil minerals. *Frateuria aurantia* and various *Bacillus* species can solubilize these nutrients: *F. aurantia* releases organic acids that mobilize potassium from mica and K-feldspar; *Bacillus mucilaginosus* and *B. edaphicus* solubilize zinc compounds, improving micronutrient uptake and enhancing plant vigor (reviewed in Singh *et al.*, 2022). Although large-scale field data are still emerging, greenhouse studies show significant yield improvements, particularly in micronutrient-deficient soils.

11.2.4 Vesicular-arbuscular mycorrhizal (VAM) fungi

Vesicular-arbuscular mycorrhizal (VAM) fungi, including *Glomus* and *Rhizophagus* species, form symbiotic associations with the majority of terrestrial plants. Their fine hyphal networks expand the root absorption zone, boosting phosphorus, zinc, and water uptake. Recent meta-analyses (2023–2024) reveal mycorrhization can enhance nutrient uptake by 20–50%, drought resilience, and yield by up to 30%, especially in challenging soils (Alzate Zuluaga *et al.*, 2024). VAM fungi are particularly valuable in low-input agriculture, adding sustainability and resilience.

11.2.5 Plant growth-promoting rhizobacteria (PGPR)

PGPR encompass both nitrogen-fixers and solubilizers, but also include strains like *Pseudomonas* and *Bacillus* that produce phytohormones, induce systemic resistance, and secrete enzymes that suppress pathogens. The review by Alzate Zuluaga *et al.* (2024) highlights their multifaceted benefits: hormone production, nutrient mobilization, pathogen antagonism, and stress tolerance enhancement.

Inoculation trials demonstrate that combined PGPR consortia outperform single-strain inoculants, yielding 15–40% increases in crops like maize, rice, and wheat under stress or nutrient-limited conditions (Singh *et al.*, 2022; Rafay and Usman, 2023).

11.3 Mechanism of Action of Biofertilizers

Biofertilizers encompass a diverse array of beneficial microorganisms—such as nitrogen-fixers, phosphate- and potassium-solubilizers, zinc-mobilizing bacteria, mycorrhizal fungi, and PGPR—which enhance plant growth through direct and indirect mechanisms. Their multifaceted actions include improving nutrient acquisition, modulating phytohormones, alleviating abiotic stress, and inducing systemic resistance, thereby offering a sustainable alternative to chemical fertilizers (Glick, 2020; Sánchez-Castro *et al.*, 2023).

At the core of many biofertilizers are PGPR, free-living rhizobacteria that colonize the root surface and promote plant growth via nutrient mobilization, hormone production, and disease suppression. These bacteria engage in nitrogen fixation, converting atmospheric N_2 to ammonia. They also solubilize phosphate and potassium through secretion of organic acids, which lower soil pH and release bound nutrients. Moreover, PGPR synthesize siderophores that chelate iron, facilitating its uptake by plants. Collectively, such activities lead to improved nutrient absorption and enhanced plant growth (Glick, 2020; Qiao *et al.*, 2022; Sánchez-Castro *et al.*, 2023).

Nitrogen fixation by biofertilizers occurs through both symbiotic and non-symbiotic pathways. Symbiotic diazotrophs, such as *Rhizobium* in legumes, form root nodules in which nitrogenase enzymes fix atmospheric N_2. Meanwhile, free-living bacteria like *Azotobacter* and *Azospirillum* fix nitrogen independently in the rhizosphere. These bacteria not only provide fixed nitrogen but also produce phytohormones such as IAA, which stimulates root growth and thereby enhances nutrient uptake. Consequently, plant biomass and yield increase under nitrogen-limiting conditions (Glick, 2020; Qiao *et al.*, 2022).

Phosphate-solubilizing microorganisms (PSM), including species of *Pseudomonas*, *Bacillus*, and *Aspergillus*, release organic acids—such as gluconic, citric, or oxalic acid—into the rhizosphere. These acids chelate calcium ions that bind phosphate, releasing it in a plant-available form. Many bacteria also secrete phosphatase enzymes, which hydrolyze organic phosphate compounds. This dual strategy significantly enhances plant phosphorus uptake, especially in soils with high pH or high calcium content that immobilize inorganic phosphate (Glick, 2020; Sánchez-Castro *et al.*, 2023).

Similarly, potassium-solubilizing bacteria (KSB), such as *Bacillus mucilaginosus* and *Frateuria aurantia*, secrete acids that weather K-bearing minerals like mica or feldspar, liberating potassium ions. Zinc-mobilizing bacteria also produce acids and chelating agents that solubilize zinc, supplying essential micronutrients vital for enzyme function and protein synthesis. These micronutrient-mobilizing microbes complement major nutrient solubilizers, enabling more balanced nutrition (Glick, 2020; Sánchez-Castro *et al.*, 2023).

A key group of biofertilizers are mycorrhizal fungi, particularly AMF such as *Glomus*, *Rhizophagus*, and *Funneliformis*. AMF create a symbiotic relationship with most terrestrial plants, extending extraradical hyphae far beyond the root depletion zone. These hyphae are physically finer than root hairs and penetrate small soil pores, vastly expanding nutrient-absorbing surface area. Chemically, AMF release organic acids and enzymes that enhance solubilization of phosphorus, zinc, and other minerals. In return, plants supply the fungus with carbohydrates. The result is improved plant nutrient uptake, water relations, and stress tolerance—especially under drought or nutrient-poor conditions (Qiao *et al.*, 2022; Savastano and Bais, 2024).

Importantly, AMF often act synergistically with PGPR. Many PGPR also function as mycorrhiza helper bacteria (MHB), enhancing AMF establishment and function. These helper bacteria induce root branching, stimulate hyphal growth, and mobilize nutrients that support fungal colonization. Consequently, dual inoculation with AMF and PGPR improves plant nutrient uptake, biomass, and resilience more than either organism alone (Sánchez-Castro *et al.*, 2023; Savastano and Bais, 2024).

Plant growth promotion by biofertilizers is also mediated through phytohormones. Many PGPR synthesize auxins (IAA), gibberellins, cytokinins, or abscisic acid, which influence root architecture, cell division, and stress responses. For example, IAA-producing bacteria stimulate root branching and length, increasing the root–soil interface. Gibberellin-producing strains promote shoot elongation and leaf area, while cytokinin-producing organisms delay senescence. These phytohormones act locally and systemically, enhancing plant vigor and stress tolerance (Glick, 2020; Sánchez-Castro *et al.*, 2023).

Another important enzymatic mechanism is bacterial ACC deaminase activity, which degrades 1-aminocyclopropane-1-carboxylic acid (ACC), the precursor of ethylene. Under stress (e.g. drought, salinity), ethylene levels rise and inhibit root growth. ACC deaminase-producing PGPR reduce ethylene synthesis by degrading ACC, alleviating growth inhibition. This mechanism is critical in supporting root development under adverse conditions (Glick, 2020; Sánchez-Castro *et al.*, 2023).

Biofertilizers also contribute to induced systemic resistance (ISR). Certain PGPR, notably *Pseudomonas* and *Bacillus* species, produce elicitors—such as lipopolysaccharides, flagellin, or volatile organic compounds—that prime plant defenses. ISR activation enhances jasmonic acid and ethylene signaling, increasing expression of pathogenesis-related (PR) genes. As a result, plants exhibit stronger resistance to pathogens and pests without direct antimicrobial action by bacteria (Sánchez-Castro et al., 2023; Savastano and Bais, 2024).

In addition, siderophore production by PGPR not only improves iron uptake for plants, but also restricts availability of iron to soil pathogens—an indirect biocontrol mechanism. Similarly, some PGPR produce hydrogen cyanide (HCN), antibiotics, or hydrolytic enzymes (e.g. chitinases, proteases) that suppress soil-borne fungi and bacterial pathogens. These biocontrol functions reduce disease incidence and support healthier root systems (Sánchez-Castro *et al.*, 2023; Savastano and Bais, 2024).

Biofertilizers also mitigate abiotic stress. PGPR and AMF enhance drought tolerance through improved water uptake—via expanded root systems and fungal hyphae—as well as by upregulating antioxidant enzymes (e.g. superoxide dismutase, catalase) that neutralize reactive oxygen species (Etesami *et al.*, 2023; Savastano and Bais, 2024). These organisms may also modulate osmolyte accumulation (e.g. proline), improving cell turgor and stress resilience (Etesami *et al.*, 2023).

At the molecular level, biofertilizers influence gene expression in plants. PGPR colonization can modulate expression of genes related to nutrient transporters, stress response proteins, phytohormone pathways, and defense-related metabolite biosynthesis. Similarly, communication between plant roots, AMF, and microbial MHB involves signaling compounds—such as Myc and nodulation factors—that activate the Common Symbiosis Signaling Pathway (CSSP). This genetic circuitry governs colonization and nutrient exchange, optimizing symbiosis (Qiao *et al.*, 2022; Savastano and Bais, 2024).

The cumulative effect of these mechanisms is improved nutrient uptake, enhanced vegetative growth, increased yield, and stronger stress resilience. Field trials show that combined application of nitrogen-fixers, PSM, KSB, zinc-mobilizers, PGPR, and AMF can significantly improve plant performance over single inoculations (Qiao *et al.*, 2022; Wang *et al.*, 2025). Yet, responses vary with soil type, plant species, microbial strains, and environmental conditions—highlighting the need for tailored biofertilizer consortia and site-specific application techniques.

Despite their promise, biofertilizer efficacy can be affected by formulation challenges (e.g. carrier quality, shelf life), inoculum compatibility, and field application methods. Genetic manipulation and strain selection have led to next-generation biofertilizers with enhanced nitrogen fixation, phosphate solubilizing, phytohormone, and ACC deaminase traits (Glick, 2020; Qiao *et al.*, 2022).

11.4 Schemes of Government of India

Agrochemicals are still used in crop management techniques to supply nutrients and control pests and diseases in order to meet the global population's ever-increasing demand for food. One of the biggest fertilizer consumers is India, and maintaining agricultural productivity without compromising soil fertility is a major challenge. Despite the fact that high-input chemical fertilizers boost output, their careless application degrades soil, water, and air quality and reduces nutrient-use efficiency (Swarnalakshmi *et al.*, 2016). Biofertilizers have multiple beneficial impacts on the soil and can be relatively cheap and convenient to use. The government is seeking to encourage their use in agriculture and also to promote private initiative and commercial viability of production (Ghosh, 2004).

In an effort to improve sustainable farming methods and lessen dependency on chemical fertilizers, the Indian government has put in place a number of programs to encourage the use of biofertilizers. Important projects include the following.

11.4.1 Paramparagat Krishi Vikas Yojana (PKVY)

This program, which was introduced in 2015–2016, promotes organic farming by offering ₹15,000 per acre (approximately ₹37,065 per hectare) over three years as financial aid. This assistance, which is provided via Direct Benefit Transfer (DBT), includes the purchase of organic inputs including vermicompost, biofertilizers, and biopesticides. Additionally, the program provides farmers with practical instruction on how to make and use organic fertilizers.

11.4.2 The Mission Organic Value Chain Development for North Eastern Region (MOVCDNER)

MOVCDNER, which was also started in 2015–2016, is dedicated to encouraging organic farming in the states of the North-east. Over a 3-year period, farmers receive ₹25,000 per hectare in financial help for organic inputs, including organic manure and biofertilizers, that are used both on and off the farm. The scheme provides end-to-end support, from production to marketing of organic produce.

11.4.3 Capital Investment Subsidy Scheme (CISS)

CISS provides the following to increase the production of biofertilizer: state government agencies receive 100% assistance up to ₹160 lakh per unit to set up cutting-edge biofertilizer production facilities with a 200-t annual capacity; 25% of the price, up to ₹40 lakh per unit, is paid to private organizations or individuals via the National Bank for Agriculture and Rural Development (NABARD). Additionally, up to ₹85 lakh for new laboratories and up to ₹45 lakh for existing ones are offered as support for the establishment or improvement of Biofertilizer Testing Quality Control Laboratories (BOQCL).

11.4.4 PM Programme for Restoration, Awareness, Nourishment and Amelioration of Mother Earth (PM-PRANAM)

This scheme aims to promote the balanced use of fertilizers, encouraging the adoption of biofertilizers and organic fertilizers. States that reduce chemical fertilizer consumption compared to the previous three years' average receive grants amounting to 50% of the subsidy savings.

11.4.5 Market Development Assistance (MDA) Scheme

Approved in June 2023, the MDA Scheme incentivizes the production and use of organic fertilizers, including those produced under the Galvanizing Organic Bio-Agro Resources Dhan (GOBARdhan) initiative. With a total outlay of ₹1,451.84 crore for the financial year 2023/24–2025/26, it provides an incentive of ₹1,500 per metric tonne of biofertilizers.

11.4.6 Namo Drone Didi Scheme

Designed to empower women's self-help groups (SHGs) and give them access to contemporary agricultural technology, this program makes it easier for SHGs to receive drones, which are used to spray water-soluble and nano fertilizers, increasing the effectiveness and reach of applying biofertilizer. Fig. 11.1 depicts the general outline of schemes of the government of India.

11.5 Biofertilizer Promotion Policies in South-east Asia

11.5.1 Memorandum Order No. 32 (Philippines)—Balanced Fertilization Program (2023)

This order was issued by the Department of Agriculture, Philippines. It integrates the use of biofertilizers with organic and inorganic inputs. The objective is to enhance yield in a sustainable manner and reduce dependency on imported synthetic fertilizers. Locally produced biofertilizers are preferred so as to build domestic supply and reduce imports.

11.5.2 Legislative advocacy: House Bill 9751 (Philippines)

This was filed on December 12, 2023. The key objectives of this bill were to transition from synthetic to organic and biofertilizer usage for staple crops like rice, maize, and sugar and to establish a National Organic and Biofertilizers Support Program under the Department of Agriculture. This bill targets farmers who own up to 5 ha of land and is intended to reach smallholder rice, maize, and sugar producers across the country. This bill was designed bearing in mind high fertilizer import dependence, soil degradation from chemical overuse, and unstable global fertilizer prices.

11.5.3 Subsidies for organic fertilizers (including biofertilizers)

Indonesia has recently reintroduced subsidies for organic fertilizers, including biofertilizers, as part of its shift toward sustainable agriculture. Following President Joko Widodo's directive, the Ministry of Agriculture amended regulations in 2024 to include these eco-friendly inputs alongside traditional urea and NPK. Targeting smallholder farmers with up to 2 ha of land, the program ensures access through registered farmer groups and the national e-RDKK (Electronic Definitive Group Needs Plan) system. This move reflects Indonesia's growing commitment to soil health, local production, and reducing reliance on imported chemical fertilizers.

11.5.4 12th Malaysia Plan (2021–2025)

Under the 12th Malaysia Plan (2021–2025), the Malaysian government—through the Ministry of Agriculture and Food Security (MAFS)—has made sustainable farming a top priority. A key part of this effort is the SMART Farming (PINTAR) initiative, which encourages modern, environmentally friendly agricultural practices. The plan supports farmers in adopting Good Agricultural Practices (GAP), and integrating bio-based solutions like biofertilizers and biopesticides into their routine. These steps not only improve soil health and crop resilience but also

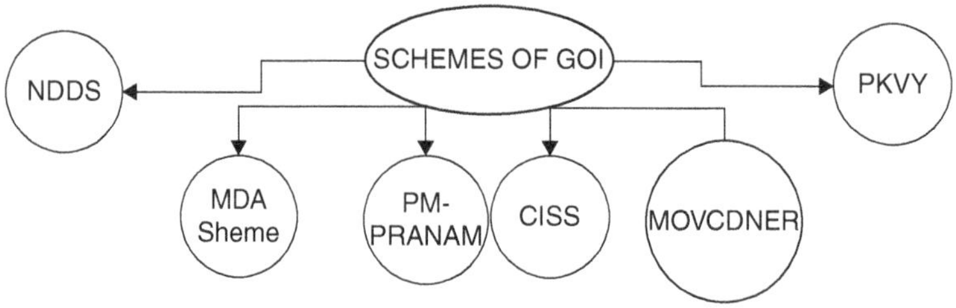

Fig. 11.1. General outline of schemes of the government of India. Figure author's own.

reduce dependence on chemical inputs. The government is also aligning its procurement policies to favor farms and producers who follow these sustainable practices. By doing this, MAFS hopes to create an eco-friendlier food production system that benefits both farmers and the environment—making agriculture smarter, safer, and more sustainable for future generations. This plan is a big step toward long-term food security and climate-resilient farming in Malaysia.

11.6 Biofertilizer Promotion Policies in the West

11.6.1 Sustainable Agriculture Research and Education (SARE) program

The United States Department of Agriculture (USDA)'s Sustainable Agriculture Research and Education (SARE) program plays a key role in promoting environmentally responsible farming practices across the United States. By funding research, education, and outreach initiatives, SARE supports innovations that improve farm profitability while enhancing soil health and environmental sustainability. A significant focus of the program is on advancing the use of organic and biological soil amendments, including compost, microbial inoculants, and biofertilizers. These inputs offer viable alternatives to synthetic fertilizers, helping to reduce chemical dependency and improve long-term soil fertility. Through its farmer-led trials and academic partnerships, SARE encourages adoption of low-input, regenerative farming practices that are tailored to local conditions. The program also helps disseminate knowledge to producers and extension agents, ensuring widespread awareness and practical application of these sustainable methods. Overall, SARE fosters a shift toward bio-based agriculture that aligns with broader goals of climate resilience and long-term food security.

11.6.2 Organic certification program

This program, administered by the National Organic Program (NOP), is a critical policy tool that promotes environmentally sustainable farming across the USA. This program sets strict standards for organic farming, prohibiting the use of synthetic fertilizers, pesticides, and genetically modified organisms (GMOs). Instead, it encourages the use of natural and bio-based inputs that improve soil health and ecological balance. Among these are biofertilizers such as compost teas, microbial inoculants, and seaweed extracts, which enhance nutrient availability and support beneficial soil microorganisms. By promoting these alternatives, the program reduces environmental pollution, builds long-term soil fertility, and supports biodiversity. Farmers who meet these standards can label their products as "USDA Organic," giving them access to a premium market and a growing consumer base that values sustainable practices. Ultimately, the program not only benefits organic producers but also contributes to broader goals of climate resilience, food safety, and environmental stewardship.

11.6.3 Horizon Europe (EU R&D funding)

This is the European Union's flagship research and innovation funding program, which plays a vital role in advancing sustainable agriculture through its strong focus on climate action and the bioeconomy. With a budget exceeding €95 billion, the program supports cutting-edge research into eco-friendly farming practices and the development of innovative bio-based solutions. Within this framework, numerous projects are dedicated to the creation and scaling of microbial fertilizers, PGPR, and precision biofertilizer application techniques. These initiatives aim to reduce reliance on synthetic agrochemicals, improve soil health, and enhance crop resilience in the face of climate change. Horizon Europe also fosters cross-border collaboration between research institutions, agri-tech companies, and farmers, ensuring that scientific advances translate into practical, scalable solutions. By funding such initiatives, the program not only accelerates the adoption of sustainable inputs like biofertilizers but also supports the EU's broader Green Deal and Farm to Fork Strategy objectives of building a climate-resilient and sustainable food system.

11.6.4 Bulhnova project

This project, funded under the European Union's small and medium-sized enterprise (SME) innovation initiative in Spain, exemplifies applied research in the development of sustainable agricultural inputs. The project focuses on formulating a liquid biofertilizer utilizing nitrogen-fixing bacteria—*Azospirillum brasilense* and *Pantoea dispersa*—to enhance soil fertility and crop productivity. These microbial strains are known for their ability to convert atmospheric nitrogen into plant-available forms, thereby reducing dependence on synthetic nitrogen fertilizers. In addition to nutrient provision, the biofertilizer contributes to improved soil structure, microbial biodiversity, and overall plant health. Bulhnova's approach aligns with the EU's Green Deal and Farm to Fork Strategy, which prioritize the reduction of chemical inputs and environmental impact in food production systems. By offering a scalable and field-ready bio-based solution, the project supports both climate-smart agriculture and the broader bioeconomy transition. Bulhnova thus represents a significant advancement in the commercialization of microbial biofertilizers within the European Union's regulatory and sustainability frameworks.

11.7 Need for Inoculum Quality Check for Biofertilizer

For biofertilizers to effectively encourage plant development and enhance soil fertility, the quality of the microbial inoculants employed in them must be guaranteed. Inadequate inoculants can lead to contamination, decreased productivity, and uneven negative agricultural results.

1. With inoculum quality verification, high microbial viability is guaranteed. The survival and activity of helpful microorganisms determines how effective biofertilizers are. The viability and potency of the microbial population are maintained by routine tests.
2. Quality checks prevent contamination by contaminants such as dangerous bacteria, fungi, or undesired microbes, which can diminish efficacy and damage crops.
3. Verification also ensures nutrient availability and that the microorganisms can efficiently fix nitrogen, solubilize phosphorus, or unleash potassium.
4. Verification also enhances stability and shelf life, confirming that the inoculant is still viable after being stored and used.
5. Adherence to regulatory guidelines involving periodic testing is necessary to meet the stringent quality criteria for biofertilizers in many nations.

11.8 Limitations of Setting Up a Biofertilizer Industry in India

The Indian biofertilizer business faces a number of obstacles that prevent large-scale production and commercialization, despite the rising demand for environmentally acceptable agricultural goods.

1. Insufficient knowledge and acceptance and low farmer awareness: For immediate results, many farmers choose chemical fertilizers and are ignorant of the advantages of biofertilizers. Many also have a traditional mindset and are opposed to switching from conventional methods to organic ones.
2. Inadequate standardization and quality control and inconsistent product quality: Ineffective items on the market are the result of lax quality standards.
3. Limited infrastructure for quality testing: Not enough testing facilities are available to guarantee the purity and viability of microorganisms.
4. Lack of advanced technology and infrastructure and limited fermentation facilities: Small-scale producers lack advanced fermentation and production units.
5. Poor carrier material quality: Inefficient carrier materials reduce microbial survival and shelf life.
6. Storage and transportation challenges and the short shelf life of biofertilizers: Microbial viability decreases over time due to improper storage.
7. Inadequate cold storage facilities: There is a need for temperature-controlled transport systems.
8. Regulatory and certification barriers and lack of uniform policies: Different states have varying regulatory frameworks.

9. Time-consuming approval process: There are delays in obtaining licenses and quality certification from regulatory bodies.
10. Investment and financial restraints and high initial capital investment: Establishing packaging facilities, quality control labs, and fermentation units comes at a high cost.
11. Limited government assistance and subsidies: Small-scale biofertilizer manufacturers do not receive enough financial incentives.
12. Chemical fertilizer competition and strong market domination by the chemical fertilizer industry: Biofertilizers are less competitive due to the availability of highly subsidized chemical fertilizers. Low profit margins are caused by low sales and high production expenses.

11.9 Conclusion and Perspectives

Biofertilizers have a lot of potential for the future of Indian agriculture, given the growing need for sustainable farming methods and the move away from chemical fertilizers. These environmentally acceptable substitutes, which are frequently made from organic waste and helpful microbes, provide a workable way to satisfy crop nutritional requirements and fill in any possible nutrient deficiencies. The contribution of biofertilizer to soil health and sustainable farming is becoming more widely acknowledged. Long-term soil fertility management requires the use of a variety of microbial inoculants, especially those that increase the availability of nitrogen and phosphorus. It is anticipated that government programs supporting organic farming and growing farmer knowledge of the agronomic and environmental advantages of biofertilizers will propel their widespread use. Biofertilizers support environmental preservation as well as agricultural sustainability by lowering chemical runoff, minimizing soil deterioration, and increasing crop output. But in order to realize their full potential, issues including poor distribution infrastructure, uneven product quality, and low farmer knowledge, must be successfully resolved.

References

Alzate Zuluaga, M.Y., Fattorini, R., Cesco, S., and Pii, Y. (2024) Plant–microbe interactions in the rhizosphere for smarter and more sustainable crop fertilization: the case of PGPR-based biofertilizers. *Frontiers in Microbiology* 15, 1440978. DOI: 10.3389/fmicb.2024.1440978

Castleberry, C. (2023) How salinity influences soil organisms: earthworms, archaea, bacteria and fungi. Master's thesis, North Dakota State University, Fargo, North Dakota.

Etesami, H. and Beattie, G.A. (2023) Plant growth-promoting rhizobacteria improve drought tolerance of plants through physiological and molecular mechanisms: a review. *Plant Soil and Environment* 69(1), 13–29.

Ghosh, N. (2004) Promoting biofertilisers in Indian agriculture. *Economic and Political Weekly* 39(52), 5617–5625. DOI: 10.2307/4415978

Glick, B.R. (2020). Introduction to plant growth-promoting bacteria. In: *Beneficial Plant-Bacterial Interactions*, 2nd edn. Springer International Publishing, Cham, Switzerland, pp. 1–37.

Jalal, A., Filho, M.C.M.T., da Silva, E.C., da Silva Oliveira, C.E., Freitas, L.A., and do Nascimento, V. (2022). Plant growth-promoting bacteria and nitrogen fixing bacteria: sustainability of non-legume crops. In: Maheshwari, D.K., Dobhal, R., and Dheeman, S. (eds) *Nitrogen Fixing Bacteria: Sustainable Growth of Non-Legumes*. Springer Nature, Singapore, pp. 233–275.

Kaur, J., Sharma, B., and Taman (2023) Constraints in biofertilizer industry and future scope. In: Kaur, S., Dwibedi, V., Sahu, P.K., and Kocher, G.S. (eds) *Metabolomics, Proteomes and Gene Editing Approaches in Biofertilizer Industry*. Springer Nature, Singapore, pp. 1–19. DOI: 10.1007/978-981-99-3561-1_1

Kumar, S., Sindhu, S.S., and Kumar, R. (2021) Biofertilizers: an ecofriendly technology for nutrient recycling and environmental sustainability. *Current Research in Microbial Sciences* 3, 100094. DOI: 10.1016/j.crmicr.2021.100094

Nosheen, S., Ajmal, I., and Song, Y. (2021) Microbes as biofertilizers, a potential approach for sustainable crop production. *Sustainability* 13(4), 1868. DOI: 10.3390/su13041868

Qiaoyi, Z., Liyang, G., Jieting, G., Shumei, L, Yuchao, X., Jiamin, Z., and Caijin, L. (2022). Effects of amino acid bacterial fertilizer on tea plant growth and soil nutrients. *Soils and Crops* 11(1), 81–87.

Rafay, M. and Usman, M. (2023). Soil salinity hinders plant growth and development and its remediation – a review. *Journal of Agricultural Research* 61(3), 189–200.

Sánchez-Castro, I., Molina, L., Prieto-Fernández, M.Á., and Segura, A. (2023). Past, present and future trends in the remediation of heavy-metal contaminated soil – remediation techniques applied in real soil-contamination events. *Heliyon*, 9(6): e16692.

Savastano, N. and Bais, H. (2024) Synergism or antagonism: do arbuscular mycorrhizal fungi and plant growth-promoting rhizobacteria work together to benefit plants? *International Journal of Plant Biology* 15(4), 944–958. DOI: 10.3390/ijpb15040067

Singh, R., Kumar, A., and Sharma, P. (2022) Microbes as biofertilizers: potential and challenges. *Environmental Agriculture Reviews* 11(1), 30–45.

Swarnalakshmi, K., Vandana, Y., Senthilkumar, M., and Dolly, W.D. (2016) Bio fertilizers for higher pulse production in India: scope, accessibility and challenges. *Indian Journal of Agronomy* 61(Special issue), 173–181.

Wang, L., Xiang, D., Li, X., Gao, Q., Chen, Y. *et al.* (2025) Effects of combined inoculation of arbuscular mycorrhizal fungi and plant growth-promoting rhizobacteria on maize under drought stress. *Frontiers in Microbiology* 15, 1432637. DOI: 10.3389/fmicb.2024.1475485

Zhang, X., Li, J., Wang, Y., and Chen, Q. (2023) Determinants of nitrogen-fixation efficiency in *Rhizobium leguminosarum*: regulatory insights. *Plant and Soil* 489(1), 137–150.

Index

www.ingramcontent.com/pod-product-compliance
Lightning Source LLC
Chambersburg PA
CBHW040137200326
41458CB00025B/6290